Nonlinear Optics - From Solitons to Similaritons

Edited by İlkay Bakırtaş and Nalan Antar

Published in London, United Kingdom

IntechOpen

Supporting open minds since 2005

Nonlinear Optics - From Solitons to Similaritons
http://dx.doi.org/10.5772/intechopen.87907
Edited by İlkay Bakırtaş and Nalan Antar

Contributors
Zhao Xingdong, Kejian Yang, Yaling Yang, Shengzhi Zhao, Jingliang He, Carmen-Georgeta Ristoscu, Mihai Oane, Ion N. Mihailescu, Antônio Carlos Amaro De Faria, Alexander Volyar, Mikhail Bretsko, Yana Akimova, Yuriy Egorov, Mahadevappa Kariduraganavar, Radha Doddamani, Balachandar Waddar, Saidi Reddy Parne, Youssoufa Mati, Ousmanou Dafounansou, Alidou Mohamadou, Guo-Quan Zhou, Soumo Tchio Michel-Rostand, Abdoulkary Saidou, Theodoros Horikis, Dimitrios J. Frantzeskakis, Georgios Koutsokostas, İlkay Bakırtaş, Nalan Antar

First published in London, United Kingdom, 2021 by IntechOpen
IntechOpen is the global imprint of INTECHOPEN LIMITED, registered in England and Wales, registration number: 11086078, 5 Princes Gate Court, London, SW7 2QJ, United Kingdom
Printed in Croatia

British Library Cataloguing-in-Publication Data
A catalogue record for this book is available from the British Library

Additional hard and PDF copies can be obtained from orders@intechopen.com

Nonlinear Optics - From Solitons to Similaritons
Edited by İlkay Bakırtaş and Nalan Antar
p. cm.
Print ISBN 978-1-83962-651-7
Online ISBN 978-1-83962-652-4
eBook (PDF) ISBN 978-1-83962-653-1

For EU product safety concerns:
IN TECH d.o.o., Prolaz Marije Krucifikse Kozulić 3, 51000 Rijeka, Croatia, info@intechopen.com or visit our website at intechopen.com.

Meet the editors

Dr. İlkay Bakırtaş is working as a Professor of applied mathematics at the Istanbul Technical University (ITU), Department of Mathematics. She received her Ph.D. degree in Mechanics from ITU in 2003. She completed her postdoctoral studies at the University of Colorado at Boulder, USA. She has published 18 research papers in peer-reviewed journals in SCI, 4 book chapters, 21 conference proceedings in the fields of perturbation methods, nonlinear wave propagation in arteries, optical solitons and wave collapse in optics and water waves problems. She is the editor of the book "Perturbation Methods with Applications in Science and Engineering" by IntechOpen. Dr. Bakırtaş is a member of the Scientific Committee of TUMTMK (Turkish National Committee of Theoretical and Applied Mechanics) and she was awarded "Dr. Serhat Ozyar, Young Scientist Of The Year Award" in 2004 and "Best Ph.D. Dissertation Award" by TUMTMK in 2003.

Dr. Nalan Antar is working as a Professor of applied mathematics for the Istanbul Technical University, Department of Mathematical Engineering. She received her Ph.D. degree in mechanics from the Istanbul Technical University in 1999. She has completed her post-doctoral studies at the University of Alberta, Department of Mathematics and Statistics, Canada and later she participated in academic research projects at the University of Colorado at Boulder, USA. She has published 32 research papers in peer-reviewed journals in SCI, 13 conference proceedings in the fields of nonlinear wave propagation in arteries, optical solitons in nonlinear optics and water waves problems, in particular gravity currents. She has been the supervisor of many graduate students in applied mathematics. Dr. Antar is a member of the Scientific Committee of TUMTMK (Turkish National Committee of Theoretical and Applied Mechanics).

Contents

Preface

With this book, we aim to bring the recent advances in nonlinear optics to the attention of the scientific community and present the reader with the bright spectrum of optics at the frontiers of research.

The eleven selected chapters in the book are authored by twenty-eight authors from ten different countries: Greece, Turkey, China, Romania, Brazil, Russia, India, Cameroon, Germany, and Italy.

Each chapter in this book is independent and self-contained, providing a contemporary overview of the methodology and approach used in theoretical and applied sciences. The reference list at the end of each chapter provides the reader with a selected list of journal papers, books, and conference proceedings.

The chapters can be summarized as follows:

In the first chapter, the two-dimensional nonlocal nonlinear Schrödinger model is studied and the connection between water waves and nonlinear optics is discussed.

In the second and third chapters, the derivative nonlinear Schrödinger equation (DNLS) with vanishing boundary conditions is studied using a novel approach in inverse scattering transform (IST). Based upon this new inverse scattering transform method, different kinds of soliton solutions such as the light/dark soliton, the pure soliton, and the mixed breather type are found for the derivative nonlinear Schrödinger equation with non-vanishing boundary conditions.

In the fourth chapter, the existence of magnetic solitons in atomic spinor Bose-Einstein condensates confined in an optical lattice are demonstrated.

The fifth chapter reviews the development of the Q-switched solid-state lasers, which are obtained using flash and diode pumping schemes based on active and passive Q-switchers.

In the sixth chapter, the laser cladding process is reported by proposing a semi-analytical model. The cladding process for different input powers and various laser beam velocities is studied.

In the seventh chapter, the main foundations for the conception, design, and the projection of optical sensors that explore the effects of nonlinear and quantum optics are presented.

In the eighth chapter, the digital sorting technique of vortex spectra is constructed. A new approach for measuring vortex spectra based on digital analyzing high-order intensity moments of complex beams is introduced.

In the ninth chapter, the main foundations of optical switches are reviewed and the design and fabrication of nonlinear optical switches are discussed from different aspects of the current technology.

In the tenth chapter, the evolution of light beams in a cubic-quintic-septic-nonical medium is investigated. As the model equation, an extended form of the well-known nonlinear Schrödinger equation is taken into account. By using a special ansatz, exact analytical solutions describing bright/dark and kink solitons are constructed.

In the last chapter, the authors investigated the modulational instability (MI) in Kerr media, induced by cross-phase modulation of two optical beams in nonlinear fiber including the effect of higher-order dispersion.

This book is intended to reach researchers, scientists, and postgraduate students in academia, as well as in industry, to give impressions and suggestions on a variety of topics and is published as an open access book in order to increase the impact of the contained information.

Dr. İlkay Bakırtaş and Dr. Nalan Antar
Department of Mathematics,
Istanbul Technical University,
Istanbul, Turkey

Chapter 1

Water Waves and Light: Two Unlikely Partners

Georgios N. Koutsokostas, Theodoros P. Horikis, Dimitrios J. Frantzeskakis, Nalan Antar and İlkay Bakırtaş

Abstract

We study a generic model governing optical beam propagation in media featuring a nonlocal nonlinear response, namely a two-dimensional defocusing nonlocal nonlinear Schrödinger (NLS) model. Using a framework of multiscale expansions, the NLS model is reduced first to a bidirectional model, namely a Boussinesq or a Benney-Luke-type equation, and then to the unidirectional Kadomtsev-Petviashvili (KP) equation – both in Cartesian and cylindrical geometry. All the above models arise in the description of shallow water waves, and their solutions are used for the construction of relevant soliton solutions of the nonlocal NLS. Thus, the connection between water wave and nonlinear optics models suggests that patterns of water may indeed exist in light. We show that the NLS model supports intricate patterns that emerge from interactions between soliton stripes, as well as lump and ring solitons, similarly to the situation occurring in shallow water.

Keywords: Kadomtsev-Petviashvili equation, nonlocal nonlinear Schrödinger equation, line solitons, soliton interactions, lump solitons, patterns of light

1. Introduction

For many decades, solitons, namely robust localized waveforms that propagate undistorted in nonlinear dispersive media, have been a topic of particular interest in physics [1, 2] and applied mathematics [3, 4]. These waveforms have a variety of fascinating properties that should be mentioned: due to their particle-like nature, solitons collide elastically, thus preserving their shape after the collision process. In some cases, as e.g. in shallow water, line soliton collisions give rise to the emergence of various wave patterns, including X-, H-, or Y-shaped waves, as well as other, even more complicated, waveforms – see, e.g., Ref. [5] for a set of remarkable examples and observations. Such patterns can be described by multidimensional line soliton solutions of the Kadomtsev-Petviashvili II (KPII) equation, a variant of the KP equation [6] with negative dispersion – as is the case of water waves with small surface tension. In particular, the KP equation can be expressed as:

$$\frac{\partial}{\partial x}\left(\frac{\partial u}{\partial t}+6u\frac{\partial u}{\partial x}+\frac{\partial^3 u}{\partial x^3}\right)+3\sigma\frac{\partial^2 u}{\partial y^2}=0, \tag{1}$$

where $\sigma=\pm1$: for $\sigma=+1$ (negative dispersion/small surface tension) Eq. (1) is a KPII, while for $\sigma=-1$ (positive dispersion/large surface tension) Eq. (1) is a KPI.

We should mention that both versions of the KP belong to the – rather limited – class of completely integrable equations in $(2+1)$-dimensions [3].

Important to the collision-induced emergence of different types of wave patterns in shallow water is the fact that quasi (line) solitons of the KPII are stable. Furthermore, of paramount importance is the effect of soliton resonances [7–10], whereby two (or more) colliding solitons resonate under certain conditions, thus creating novel stable structures. It is, thus, not surprising that the collision dynamics of many robust solitons, can give rise to a wealth of complex wave patterns [11–16] (see also [1]).

The above discussion motivates the following question: can solitons, similar to those governed by the KPII equation and feature rich collision dynamics (as in shallow water waves), be also predicted – and, possibly, observed – in nonlinear optics? In this context, key model is the nonlinear Schrödinger (NLS) equation, which is known to support soliton solutions (see, e.g., Ref. [17] and references therein). As was shown some time ago [18], the $(2+1)$-dimensional defocusing NLS can be asymptotically reduced to the KPI equation, which allows one to approximate dark soliton stripes of the NLS as line solitons of KPI; then, the fact that line KPI solitons are unstable [3], can be used to investigate the transverse instability of dark solitons [18–20]. On the other hand, as was recently shown [21], the presence of nonlocality introduces an effective *surface tension* in the system. Then, in the case where the surface tension is small – corresponding to the case of *strong nonlocality* – it can be shown that the nonlocal NLS model can be asymptotically reduced to a *KPII equation*. This result allows for the derivation of approximate *stable* line soliton solutions of the nonlocal NLS, which behave similarly to the stable line solitons of KPII, i.e., they undergo elastic collisions and form patterns similar to those observed in shallow water.

The scope of this work is to review the above ideas and present a variety of soliton solutions and their dynamics in the framework of a defocusing nonlocal NLS model. These solutions stem from the soliton solutions of KP models, which are derived from the nonlocal NLS via a two-step multiscale expansion method. We show that, similarly to the water wave case, the nonlocal NLS can also support complicated wave patterns, arising from "complex interactions" between two or more antidark soliton stripes that are supported in the strong nonlocality regime. In addition, we present other soliton solutions, namely dark lumps (pertinent to the KPI equation), as well as ring dark and antidark solitons (pertinent to the cylindrical KP (cKP), *alias* Johnson's Equation [22, 23]). In particular, a summary of our main results, as well as the organization of this work, are as follows.

First, in Section 2, we introduce the nonlocal NLS model, present its continuous-wave (cw) solution, and study its stability. Then, we perform a two-step multiscale analysis. First, at an intermediate stage, we derive a Boussinesq – or a Benney-Luke (BL) [24] – type equation, which is a long-wave, multi-dimensional and bi-directional, shallow water wave model. Then, from the Boussinesq/BL-type equation, we derive its far field, pertinent to the long time behavior, which is the KP equation, both in Cartesian and in cylindrical geometry. The reduction to the KP model allows us to construct approximate soliton solutions of the nonlocal NLS, which are presented in Section 3. In the strong nonlocality regime, we derive antidark soliton stripes, satisfying an effective KPII equation, which are shown to form patterns similar to those observed in shallow water; these include X-, Y, and H-shaped waveforms, as well as other complicated patterns arising from resonant interactions between more than two solitons. We also present dark lump solitons, as well as ring dark and antidark soliton solutions of the NLS, which are numerically found to propagate undistorted in the framework of the NLS model. Finally, in Section 4, we briefly discuss the conclusions of this work.

2. Nonlocal NLS and multiscale analysis

2.1 Introduction of the model and linear regime

The model under consideration is a two-dimensional (2D) defocusing NLS, with a nonlocal nonlinearity, which is expressed in the following dimensionless form:

$$\mathrm{i}u_t + \frac{1}{2}\Delta u - nu = 0, \tag{2}$$

$$d\Delta n - n + |u|^2 = 0. \tag{3}$$

where subscripts denote partial derivatives, $u \in \mathbb{C}$, $n \in \mathbb{R}$,

$$\Delta \equiv \partial_x^2 + \partial_y^2, \quad \text{or} \quad \Delta \equiv \frac{1}{r}\partial_r(r\partial_r) + \frac{1}{r^2}\partial_\theta^2, \tag{4}$$

is the Laplacian in Cartesian or polar coordinates respectively, while the parameter $d \in \mathbb{R}_+$ measures the degree of nonlocality (see below). The above system finds a number of physical applications. For instance, in the context of optics, this model describes the evolution of the complex electric field envelope u, via a nonlinear wave equation in the paraxial regime, coupled with a diffusion-type equation for the nonlinear correction to the refractive index n depending on the intensity $I = |u|^2$ (in this case, t represents the propagation coordinate) [25, 26]. In this context, the system (2)–(3) has been used to model experiments on liquid solutions exhibiting thermal nonlinearities [27, 28]. In addition, it has also been used in studies of plasmas (in this case, n represents the relative electron temperature perturbation) [29, 30], as well as in nematic liquid crystals (in this case, n is the optically induced angle perturbation) [31, 32].

As mentioned above, parameter d represents the degree of nonlocality: for $d = 0$, one recovers the local limit, whereby the system (2)–(3) reduces to the defocusing 2D NLS equation with a local cubic (Kerr-type) nonlinearity:

$$\mathrm{i}u_t + \frac{1}{2}\Delta u - |u|^2 u = 0, \tag{5}$$

while for $d \neq 0$, Eqs. (2) and (3) feature a spatially nonlocal nonlinearity. Notice that in our perturbative analytical approach below, d will be treated as a free parameter, of order $O(1)$, thus allowing the model to acquire an arbitrary degree of nonlocality.

Our analytical approach relies on the analysis of the hydrodynamic form of the model. This can be derived upon using the Madelung transformation:

$$u = u_0\sqrt{\rho}\exp(\mathrm{i}\phi), \tag{6}$$

where u_0 is a complex constant, while the unknown real functions ρ and ϕ denote the amplitude and phase of the field u, respectively. Substituting into Eqs. (2) and (3), we obtain the following hydrodynamical system of equations:

$$\phi_t + n + \frac{1}{2}(\nabla\phi)^2 - \frac{1}{2}\rho^{-1/2}\Delta\rho^{1/2} = 0, \tag{7}$$

$$\rho_t + \nabla\cdot(\rho\nabla\phi) = 0, \tag{8}$$

$$d\Delta n - n + |u_0|^2\rho = 0, \tag{9}$$

where $\nabla \equiv (\partial_x, \partial_y)$ or $\nabla \equiv (\partial_r, \frac{1}{r}\partial_\theta)$ is the gradient operator in Cartesian or in polar coordinates respectively.

The simplest, non trivial, solution of Eqs. (2) and (3) is of the form:

$$\rho = 1, \quad \phi = -|u_0|^2 t, \quad n = |u_0|^2, \tag{10}$$

which corresponds to the continuous wave (cw) solution $u = u_0 \exp(-\mathrm{i}|u_0|^2 t)$ and the constant function $n = |u_0|^2$. This solution will serve as a "pedestal", on top of which we will seek nonlinear excitations, namely solitons. It is thus necessary to investigate whether the solution (7) is modulationally stable. This can be done upon introducing to Eqs. (7)–(9) the following perturbation ansatz:

$$\rho = 1 + \varepsilon\tilde{\rho}, \quad \phi = -|u_0|^2 t + \varepsilon\tilde{\phi}, \quad n = |u_0|^2 + \varepsilon\tilde{n}, \tag{11}$$

where $0 < \varepsilon \ll 1$ is a formal small parameter. This way, we obtain, at $O(\varepsilon)$, the following linear system:

$$\tilde{\phi}_t + \tilde{n} - \frac{1}{4}\Delta\tilde{\rho} = 0, \tag{12}$$

$$\tilde{\rho}_t + \Delta\tilde{\phi} = 0, \tag{13}$$

$$d\Delta\tilde{n} - \tilde{n} + |u_0|^2\tilde{\rho} = 0. \tag{14}$$

The above system can straightforwardly be decoupled as follows: solving Eq. (12) for $\tilde{n}$, as well as Eq. (13) for $\Delta\tilde{\phi}$, and substituting back in Eq. (13), the following linear equation for $\tilde{\rho}$ is obtained:

$$\tilde{\rho}_{tt} - |u_0|^2\Delta\tilde{\rho} - d\Delta\tilde{\rho}_{tt} + \frac{1}{4}(1 - d\Delta)\Delta^2\tilde{\rho} = 0. \tag{15}$$

Since the above solution is linear we can examine its stability in first order by examining the dispersion relation of plane wave solutions, i.e., $\propto \exp[\mathrm{i}(\boldsymbol{k}\cdot\boldsymbol{r} - \omega t)]$, of wavenumber $k = \|\boldsymbol{k}\|$ and frequency ω (here, e.g., in Cartesian geometry, $\boldsymbol{r} = (x, y)$ and $\boldsymbol{k} = (k_x, k_y)$). Then, it can readily be found that these plane waves are characterized by the following dispersion relation:

$$\omega^2 = \frac{|u_0|^2 k^2}{1 + dk^2} + \frac{1}{4}k^4. \tag{16}$$

From the above equation we can obtain the following information. First, it is observed that $\omega \in \mathbb{R}\ \forall k \in \mathbb{R}$, which indicates that the steady-state solution is modulationally stable. This result is important since, below, we will seek for soliton solutions on top of the stable cw background (7). Second, in the long-wavelength limit, such that $dk^2 \ll 1$, Eq. (16) can be reduced to the following Bogoliubov-type dispersion relation:

$$\omega^2 \approx k^2 C^2 + \frac{1}{4}\alpha k^4, \tag{17}$$

where

$$C^2 = |u_0|^2, \tag{18}$$

is the squared "speed of sound", namely the velocity of linear waves propagating on top of the cw solution. In addition, parameter α is given by:

$$\alpha = 1 - 4d|u_0|^2, \tag{19}$$

and plays the role of an effective surface tension for our original system. Indeed, as discussed in Ref. [21], this can be inferred by the fact that Eq. (17) is reminiscent of the dispersion relation for shallow water waves [4]:

$$\omega^2 \approx k^2 c_0^2 + \frac{1}{3}(3\hat{T} - 1)c_0^2 h^2 k^4. \tag{20}$$

Here, $c_0^2 = gh$ is the velocity (g is the acceleration of gravity and h the depth of water at rest), while $\hat{T} = T/(\rho g h^2)$, where ρ is the density and T the surface tension. Comparing Eqs. (17) and (20), one can identify the correspondence: $3\hat{T} \to 4d|u_0|^2$, which indeed suggests that a *surface tension* analogue in our problem is $\propto d|u_0|^2$, depending on the nonlocality parameter d and the cw intensity $|u_0|^2$.

Obviously, regarding the magnitude of the effective surface tension, one may identify the following two regimes:

- $\alpha < 0 \Rightarrow 4d|u_0|^2 > 1$: weak surface tension – strong nonlocality (for fixed $|u_0|^2$),
- $\alpha > 0 \Rightarrow 4d|u_0|^2 < 1$: strong surface tension – weak nonlocality (for fixed $|u_0|^2$).

Below it will be shown that these two regimes correspond, respectively, to a KP-II and a KP-I equation, with the former giving rise to complex interactions of line solitons (that we will study in detail).

2.2 The nonlinear regime – asymptotic analysis

2.2.1 The intermediate stage – Bousinesq/Benney-Luke equation

We now proceed by analyzing the fully nonlinear regime. We start by seeking solutions of Eqs. (7)–(9) in the form of the following asymptotic expansions:

$$\rho = 1 + \varepsilon\rho_1 + \varepsilon^2\rho_2 + \cdots, \tag{21}$$

$$\phi = -|u_0|^2 t + \varepsilon^{1/2}\Phi \tag{22}$$

$$n = |u_0|^2 + \varepsilon n_1 + \varepsilon^2 n_2 + \cdots, \tag{23}$$

where $0 < \varepsilon \ll 1$ is a formal small parameter and the unknown functions ρ_j, Φ and n_j ($j = 1, 2, \ldots$) depend on the following stretched variables:

$$X = \varepsilon^{1/2}x, \quad Y = \varepsilon^{1/2}y, \quad T = \varepsilon^{1/2}t, \tag{24}$$

$$R = \varepsilon^{1/2}r, \quad \theta = \theta, \quad T = \varepsilon^{1/2}t, \tag{25}$$

for the Cartesian and polar geometry, respectively (the angular coordinate θ in the polar geometry remains unchanged). By substituting the expansions (21)–(23) into Eqs. (7)–(9), and equating terms of the same order in ε, we obtain the following results. First, Eq. (13), at $\mathcal{O}(\varepsilon^{3/2})$ and $\mathcal{O}(\varepsilon^{5/2})$, yields:

$$\rho_{1T} + \tilde{\Delta}\Phi = 0, \quad \rho_{2T} + \tilde{\nabla} \cdot \left(\rho_1 \tilde{\nabla}\Phi\right) = 0. \tag{26}$$

Furthermore, Eq. (14) at $\mathcal{O}(\varepsilon)$ and $\mathcal{O}(\varepsilon^2)$ leads to the equations:

$$n_1 = |u_0|^2 \rho_1, \quad d\tilde{\Delta} n_1 - n_2 + |u_0|^2 \rho_2 = 0, \tag{27}$$

which connect the amplitudes $\rho_{1,2}$ and $n_{1,2}$ with the phase Φ; note that, here,

$$\tilde{\Delta} \equiv \partial_X^2 + \partial_Y^2, \quad \tilde{\nabla} \equiv (\partial_X, \partial_Y), \text{ or } \tilde{\Delta} \equiv \frac{1}{R}\partial_R(R\partial_R) + \frac{1}{R^2}\partial_\theta^2, \quad \tilde{\nabla} \equiv \left(\partial_R, \frac{1}{R}\partial_\theta\right), \tag{28}$$

in Cartesian and polar coordinates, respectively. In addition, Eq. (14) yields:

$$\Phi_T + n_1 + \varepsilon\left[\frac{1}{2}\partial_T\left(\tilde{\nabla}\Phi\right)^2 + n_2 - \frac{1}{4}\tilde{\Delta}\rho_1\right] = \mathcal{O}(\varepsilon^2). \tag{29}$$

To this end, combining Eq. (29) with Eqs. (26) and (27), yields the following biderectional nonlinear dispersive wave equation for Φ:

$$\Phi_{TT} - C^2\tilde{\Delta}\Phi + \varepsilon\left[\frac{1}{4}\alpha\tilde{\Delta}^2\Phi + \frac{1}{2}\partial_T\left(\tilde{\nabla}\Phi\right)^2 + \tilde{\nabla} \cdot \left(\Phi_T \tilde{\nabla}\Phi\right)\right] = \mathcal{O}(\varepsilon^2), \tag{30}$$

where C^2 is the squared speed of sound given by Eq. (18) and the parameter α is given by Eq. (19). Here it is worth observing the following. First, at the leading-order in ε, Eq. (30) is the standard second-order wave equation, while at order $\mathcal{O}(\varepsilon)$, the linear part of Eq. (30) corresponds to the dispersion relation (16) for the small-amplitude linear waves of Eqs. (2) and (3) propagating on top of the steady state with $|u| = |u_0|$ and $n = |u_0|^2$. The full Eq. (30), incorporates fourth-order dispersion and quadratic nonlinear terms, resembling the Boussinesq and Benney-Luke [24] equations. These models, are used to describe bidirectional shallow water waves, in the framework of small-amplitude and long-wave approximations [4].

2.2.2 Long-time behavior – KP equation

Using a multiscale expansion method, similar to the one employed in the water wave problem [4], we now derive the KP equation, which is obtained under the additional assumptions of *quasi-two-dimensionality* and *unidirectional propagation*. In particular, we introduce the asymptotic expansion:

$$\Phi = \Phi_0 + \varepsilon\,\Phi_1 + \cdots, \tag{31}$$

where the unknown functions Φ_ℓ ($\ell = 0, 1, \dots$) depend on the variables:

$$\chi = X - CT, \quad \tilde{\chi} = X + CT, \quad \mathcal{Y} = \varepsilon^{1/2}Y, \quad \mathcal{T} = \varepsilon T, \tag{32}$$

$$\varrho = R - CT, \quad \tilde{\varrho} = R + CT, \quad \Theta = \varepsilon^{-1/2}\theta, \quad \mathcal{T} = \varepsilon T, \tag{33}$$

for the Cartesian and polar geometry, respectively. Substituting the expansion (31) into Eq. (30), at the leading-order in ε, we obtain the equations:

$$\Phi_{0\chi\tilde{\chi}} = 0, \tag{34}$$

$$\Phi_{0\varrho\tilde{\varrho}} = 0, \tag{35}$$

in Cartesian and polar coordinates respectively. The above equations imply that Φ_0 can be expressed as a superposition of two waves. In the Cartesian geometry, these waves are a right-going one, $\Phi_0^{(R)}$, which depends on χ, and a left-going one, $\Phi_0^{(L)}$, depending on $\tilde{\chi}$. Similarly, in the polar case, Φ_0 can be expressed as superposition of a radially expanding wave (depending on ϱ), and a radially contracting one (depending on $\tilde{\varrho}$). Thus, the solutions of Eqs. (34) and (35) read:

$$\Phi_0 = \Phi_0^{(R)}(\chi, \mathcal{Y}, \mathcal{T}) + \Phi_0^{(L)}(\tilde{\chi}, \mathcal{Y}, \mathcal{T}), \tag{36}$$

$$\Phi_0 = \Phi_0^{(R)}(\varrho, \Theta, \mathcal{T}) + \Phi_0^{(L)}(\tilde{\varrho}, \Theta, \mathcal{T}). \tag{37}$$

In addition, at order $\mathcal{O}(\varepsilon)$, we obtain the following equations:

$$\begin{aligned} 4C^2\Phi_{1\chi\tilde{\chi}} = &-C\left(\Phi_{0\chi\chi}^{(R)}\Phi_{0\tilde{\chi}}^{(L)} - \Phi_{0\chi}^{(R)}\Phi_{0\tilde{\chi}\tilde{\chi}}^{(L)}\right) \\ &+\left[\partial_\chi\left(-2C\Phi_{0\mathcal{T}}^{(R)} + \frac{\alpha}{4}\Phi_{0\chi\chi\chi}^{(R)} - \frac{3C}{2}\Phi_{0\chi}^{(R)2}\right) - C^2\Phi_{0\mathcal{Y}\mathcal{Y}}^{(R)}\right] \\ &+\left[\partial_{\tilde{\chi}}\left(2C\Phi_{0\mathcal{T}}^{(L)} + \frac{\alpha}{4}\Phi_{0\tilde{\chi}\tilde{\chi}\tilde{\chi}}^{(L)} + \frac{3C}{2}\Phi_{0\tilde{\chi}}^{(L)2}\right) - C^2\Phi_{0\mathcal{Y}\mathcal{Y}}^{(L)}\right], \end{aligned} \tag{38}$$

for the Cartesian geometry, and

$$\begin{aligned} 4C^2\Phi_{1\varrho\tilde{\varrho}} = &-C\left(\Phi_{0\varrho\varrho}^{(R)}\Phi_{0\tilde{\varrho}}^{(L)} - \Phi_{0\varrho}^{(R)}\Phi_{0\tilde{\varrho}\tilde{\varrho}}^{(L)}\right) \\ &+\left[\partial_\rho\left(-2C\Phi_{0\mathcal{T}}^{(R)} + \frac{\alpha}{4}\Phi_{0\varrho\varrho\varrho}^{(R)} - \frac{3C}{2}\Phi_{0\varrho}^{(R)2} - \frac{C}{\mathcal{T}}\Phi_{0\tilde{\varrho}}^{(R)}\right) - \frac{1}{\mathcal{T}^2}\Phi_{0\Theta\Theta}^{(R)}\right] \\ &+\left[\partial_{\tilde{\varrho}}\left(2C\Phi_{0\mathcal{T}}^{(L)} + \frac{\alpha}{4}\Phi_{0\tilde{\varrho}\tilde{\varrho}\tilde{\varrho}}^{(L)} + \frac{3C}{2}\Phi_{0\tilde{\varrho}}^{(L)2} - \frac{C}{\mathcal{T}}\Phi_{0\tilde{\varrho}}^{(L)}\right) - \frac{1}{\mathcal{T}^2}\Phi_{0\Theta\Theta}^{(R)}\right], \end{aligned} \tag{39}$$

for the polar geometry. Here, it is important to observe the following. Once Eq. (38) is integrated in χ or $\tilde{\chi}$ [and, similarly, Eq. (39) in ϱ or $\tilde{\varrho}$], secular terms arise: the secular therms are those in the square brackets in the right-hand side, because are functions of χ or $\tilde{\chi}$ (ϱ or $\tilde{\varrho}$ in polar) alone, not both. Removal of these secular terms leads to two uncoupled nonlinear evolution equations for $\Phi_0^{(R)}$ and $\Phi_0^{(L)}$ in both geometries. Furthermore, using the equation $\Phi_{\mathcal{T}} = -n_1$ (i.e., the linear part of Eq. (29), together with Eqs. (26) and (27), it is found that the amplitude ρ_1 can also be decomposed to a left- and a right-going wave; this means that

$$\rho_1 = \rho_1^{(R)} + \rho_1^{(L)},$$

with the fields $\rho_1^{(R)}$ and $\rho_1^{(L)}$ satisfying a pair of two uncoupled KP equations. In the Cartesian geometry, the KP equations are of the form:

$$\partial_\chi\left(\rho_{1\mathcal{T}}^{(R)} - \frac{\alpha}{8C}\rho_{1\chi\chi\chi}^{(R)} + \frac{3C}{2}\rho_1^{(R)}\rho_{1\chi}^{(R)}\right) + \frac{C}{2}\rho_{1\mathcal{Y}\mathcal{Y}}^{(R)} = 0, \tag{40}$$

$$\partial_{\tilde{\chi}}\left(\rho_{1\mathcal{T}}^{(L)} + \frac{\alpha}{8C}\rho_{1\tilde{\chi}\tilde{\chi}\tilde{\chi}}^{(L)} - \frac{3C}{2}\rho_1^{(L)}\rho_{1\tilde{\chi}}^{(L)}\right) - \frac{C}{2}\rho_{1\mathcal{Y}\mathcal{Y}}^{(L)} = 0, \tag{41}$$

while in the polar geometry, the relevant equations take the form of the so-called cylindrical KP (cKP), or Johnson's equations (see Refs. [22, 23]):

$$\partial_{\varrho}\left(\rho_{1\mathcal{T}}^{(R)} - \frac{\alpha}{8C}\rho_{1\varrho\varrho\varrho}^{(R)} + \frac{3C}{2}\rho_{1}^{(R)}\rho_{1\chi}^{(R)} + \frac{1}{2\mathcal{T}}\rho_{1}^{(R)}\right) + \frac{1}{2C\mathcal{T}^{2}}\rho_{1\Theta\Theta}^{(R)} = 0, \tag{42}$$

$$\partial_{\tilde{\varrho}}\left(\rho_{1\mathcal{T}}^{(L)} + \frac{\alpha}{8C}\rho_{1\tilde{\varrho}\tilde{\varrho}\tilde{\varrho}}^{(L)} - \frac{3C}{2}\rho_{1}^{(L)}\rho_{1\tilde{\varrho}}^{(L)} - \frac{1}{2\mathcal{T}}\rho_{1}^{(L)}\right) - \frac{1}{2C\mathcal{T}^{2}}\rho_{1\Theta\Theta}^{(L)} = 0. \tag{43}$$

2.2.3 Versions of the KP equations

It is now convenient to further simplify the equations above, namely the KP and the cKP equations, in order to express them in their standard form [4]. We consider right-going waves in the Cartesian geometry, and radially expanding ones in the polar geometry, and introduce the following transformations,

$$\mathcal{Y} \mapsto \sqrt{\frac{3|\alpha|}{4C^{2}}}\mathcal{Y}, \quad \Theta \to \sqrt{\frac{3|\alpha|}{4}}\Theta, \quad \mathcal{T} \mapsto -\frac{\alpha}{8C}\mathcal{T}, \quad \rho_{1}^{(R)} = -\frac{\alpha}{2C^{2}}U, \tag{44}$$

which cast the respective KP and cKP equations into the following form:

$$\partial_{\mathcal{X}}(U_{\mathcal{T}} + 6UU_{\mathcal{X}} + U_{\mathcal{XXX}}) + 3\sigma^{2}U_{\mathcal{YY}} = 0, \tag{45}$$

$$\partial_{\varrho}\left(U_{\mathcal{T}} + 6UU_{\varrho} + U_{\varrho\varrho\varrho} + \frac{1}{2\mathcal{T}}U\right) + \frac{3\sigma^{2}}{\mathcal{T}^{2}}U_{\Theta\Theta} = 0. \tag{46}$$

Here, the parameter σ^2 is defined as:

$$\sigma^{2} = -\operatorname{sgn}\alpha = \operatorname{sgn}\left(4d|u_{0}|^{2} - 1\right). \tag{47}$$

It is now clear that, in the Cartesian case, Eq. (45) includes both versions of the KP equation, KPI and KPII [33]. Indeed, for $\sigma^2 = 1 \Rightarrow \alpha < 0$, i.e., for $|u_0|^2 > 1/(4d)$ or $d > 1/(4|u_0|)^2$, Eq. (45) is a KPII equation; this corresponds to a small effective surface tension. On the other hand, for $\sigma^2 = -1 \Rightarrow \alpha > 0$, i.e., $|u_0|^2 < 1/(4d)$ or $d < 1/(4|u_0|)^2$, Eq. (45) is a KPI equation; this corresponds to the case of large effective surface tension (see end of Section 2.1 and discussion below).

In other words, for a fixed degree of nonlocality d, a larger (smaller) background amplitude $|u_0|$, as defined by the sign of α, corresponds to KPII (KPI); similarly, for a fixed background amplitude $|u_0|$, a strong (weak) nonlocality, as defined by the above regimes of d, corresponds to KPII (KPI). Notice that in the local limit of $d = 0$, the asymptotic analysis leads only to the KPI model, in which line solitons are unstable; this fact was used to better understand self-focusing and transverse instability of plane dark solitons of the defocusing NLS Equation [19, 20]. Notably, the same parameter α can also been shown to distinguish solutions in 1D [34] and radially symmetric [35] systems; a similar case, pertinent to the polar geometry, will be considered below.

Once again, it is important to highlight the fact that the existence of these regimes resembles the situation occurring in shallow water. In this context, weak surface tension corresponds to $\sigma^2 = 1$ in Eq. (45) (i.e., KPII), while strong surface tension is pertinent to $\sigma^2 = -1$ (i.e., KPI). Thus, there exists an immediate connection between the original problem with the one of shallow water waves: relatively large (small) background amplitude or nonlocality corresponds to weak (strong) surface tension, leading to KPII (KPI).

3. Approximate soliton solutions

We now employ the results of the above analysis and construct the solution of our original problem. This can be done as follows. Once a solution of Eq. (45) [or (46)] is known, an approximate solution of Eqs. (2) and (3) has the form:

$$u \approx u_0 \left(1 - \varepsilon \frac{\alpha}{2|u_0|^2} U\right)^{1/2} \exp\left(-i|u_0|^2 t + \frac{i}{2}\alpha\varepsilon^{-1/2}\int_0^{\mathcal{T}} U \mathrm{d}\mathcal{T}'\right), \tag{48}$$

$$n \approx |u_0|^2 - \frac{1}{2}\varepsilon\alpha U. \tag{49}$$

It is important to note that Eq. (48) describes two different types of solitons, namely dark and antidark ones. Indeed, for $\alpha < 0$, i.e., for solutions satisfying the KPII equation (corresponding to the *strong nonlocality regime*) the solution (48) has the form of a *hump* on top of the cw background and is, thus, an *antidark soliton*. On the other hand, for $\alpha > 0$, i.e., for solutions satisfying the KPI equation (corresponding to the *weak nonlocality regime*) the solution (48) has the form of a *dip* on top of the cw background and is, thus, a *dark soliton*. Note that in the local limit, with $d = 0$, it turns out that $\alpha = 1$ and the solutions have always the form of dark solitons. In other words, antidark solitons are only supported by a sufficiently strong nonlocality.

Below, we will use known solutions of the KP and cKP equations and present corresponding solutions of the nonlocal NLS. In addition, we will perform direct numerical simulations to examine the evolution and – in some cases – the interaction dynamics of the approximate soliton solutions (48).

3.1 Antidark stripe solitons and their interactions

We start with the Cartesian geometry and consider, in particular, the simplest soliton solutions of the KP equation, the so-called line solitons. Specifically, the one-line soliton solution of Eq. (45), traveling at an angle to the Y-axis, reads [33]:

$$\begin{gathered} U(\mathcal{X}, \mathcal{Y}, \mathcal{T}) = 2\kappa^2 \mathrm{sech}^2(Z), \\ Z \equiv \kappa\left[\mathcal{X} + \lambda\mathcal{Y} - \left(\kappa^2 + 3\lambda^2\right)\mathcal{T} + \delta\right], \end{gathered} \tag{50}$$

where κ, λ and δ are free parameters. We are particularly interested in the case where the corresponding solution of the NLS takes the form of an antidark soliton, with U obeying the KPII equation: since, in this case, line solitons are stable, we may expect that their interactions will give rise to patterns resembling those observed in shallow water [5]. An example of such an antidark soliton is ahown in **Figure 1**, for $t = 0$. For this example, the following parameter values were used: $u_0 = 1$ and $d = 1/\sqrt{3}$ (leading to $\alpha = -1/3 < 0$), as well as $\kappa = \lambda = 1$, $\delta = 0$, and $\varepsilon = 0.2$. These values will also be used below.

In what follows, we choose line solitons with specific parameters, so that their angle during interaction will determine the type of pattern. In the simulations, we evolve the initial configuration up to $t = 600$, so that solitons have enough time to interact and generate interesting interaction patterns. In all cases, the numerical integration of Eqs. (2) and (3) is performed via a high-accuracy pseudo-spectral

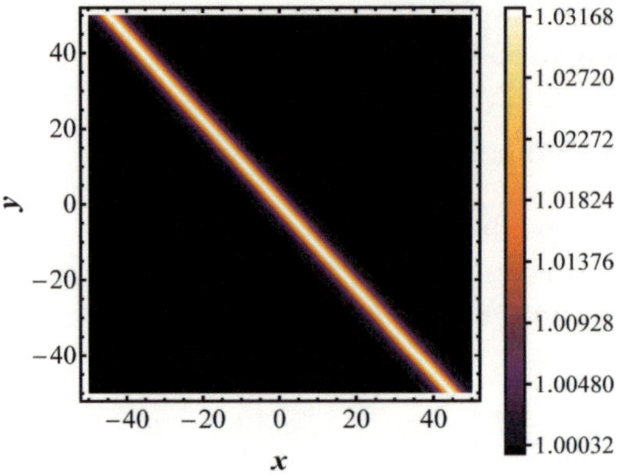

Figure 1.
A typical single antidark soliton solution of the nonlocal NLS ($u(x,y,0)$), with $\kappa = \lambda = 1$, $\delta = 0$; other parameter values are: $u_0 = C = 1$, $d = 1/\sqrt{3}$, and $\varepsilon = 0.2$.

method in space and fourth-order in time. To adjust to the periodic boundary conditions, the scheme requires we place the solutions on top of a wide super-Gaussian background that decays at infinity and makes the initial condition periodic. As it will be seen below, interacting antidark NLS solitons follow closely the KP dynamics, giving rise to patterns that are usually observed in water.

We start with the interaction of two solitons, and express the two-line soliton solution of KPII as follows:

$$\begin{aligned} U(\mathcal{X},\mathcal{Y},\mathcal{T}) &= 2\partial_{\mathcal{X}}^2 \ln F(\mathcal{X},\mathcal{Y},\mathcal{T}), \\ F &\equiv 1 + \exp(Z_1) + \exp(Z_2) + \exp(Z_1 + Z_2 + A_{12}), \\ \exp(A_{12}) &= \frac{(\kappa_1 - \kappa_2)^2 - (\lambda_1 - \lambda_2)^2}{(\kappa_1 + \kappa_2)^2 - (\lambda_1 - \lambda_2)^2}, \end{aligned} \tag{51}$$

with $Z_i \equiv \kappa_i\left[\mathcal{X} + \lambda_i\mathcal{Y} - (\kappa_i^2 + 3\lambda_i^2)\mathcal{T} + \delta_i\right]$.

We first focus on interacting line solitons that form Y- and X-shaped waveforms, and proceed with more complex ones later in the text. In **Figure 2**, we show a typical Y-shaped pattern, whereby the resonant two-soliton interaction gives rise to the emergence of one soliton, with its maximum height being four times that of the incoming solitons (this resonant Y-shaped solution was originally found by

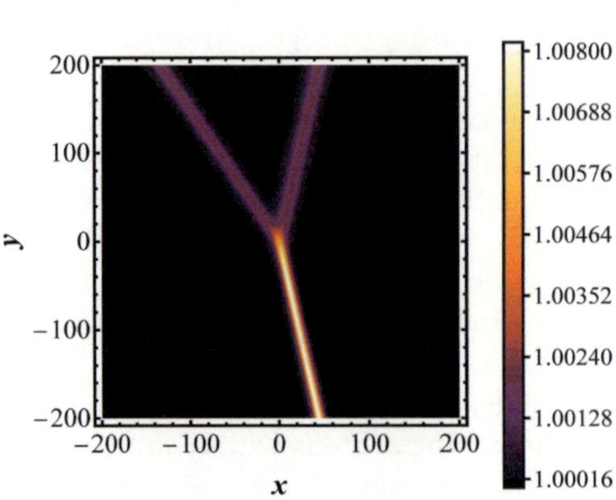

Figure 2.
A typical Y-type interaction, $u(x,y,0)$, whereby out of two incoming colliding solitons, a single line soliton emerges. Parameter values are: $\kappa_1 = \kappa_2/2 = 1/2$, $\lambda_1 = 3\lambda_2 = 3/4$.

Miles [7, 8]). This feature is clearly observed in the figure. Parameter values in this case are: $\kappa_1 = \kappa_2/2 = 1/2$, $\lambda_1 = 3\lambda_2 = 3/4$ (and for all cases, also pertaining to X-type interactions below, $\delta_1 = \delta_2 = 0$).

Next, we proceed with the X-type interactions, which can be discriminated according to the resulting "stem": an interaction with a *short stem* (**Figure 3**-left), an interaction with a long stem where the stem height is higher than the incoming line solitons (**Figure 3**-middle and right respectively).

In the same category falls an X-type interaction with a long stem, with the stem height being lower than the tallest incoming line soliton, which we refer to here as an H-interaction. A typical such interaction is given in **Figure 4**, and parameter values: $\kappa_1 = \kappa_2/2 = 1/2$, $\lambda_1 = 1/2 - 10^{-7}$, and $\lambda_2 = 0$.

An immediate generalization of the above is a three-wave interaction, where now

$$F = 1 + \sum_{1 \le i \le 3} e^{\eta_i} + \sum_{1 \le i < j \le 3} e^{\eta_i + \eta_j + A_{ij}} + e^{\eta_1 + \eta_2 + \eta_3 + A_{12} + A_{13} + A_{23}}, \tag{52}$$

and

$$\exp\left(A_{ij}\right) = \frac{\left(\kappa_i - \kappa_j\right)^2 - \left(\lambda_i - \lambda_j\right)^2}{\left(\kappa_i + \kappa_j\right)^2 - \left(\lambda_i - \lambda_j\right)^2}, \tag{53}$$

so that the resulting interaction is as in **Figure 5**, for $t = 200$. Here, $\kappa_1 = 1$, $\kappa_2 = 2$, $\kappa_3 = 3$, $\lambda_1 = \lambda_2/2 = \lambda_3/5 = -1/3$ and, as before, $\delta_1 = \delta_2 = \delta_3 = 0$.

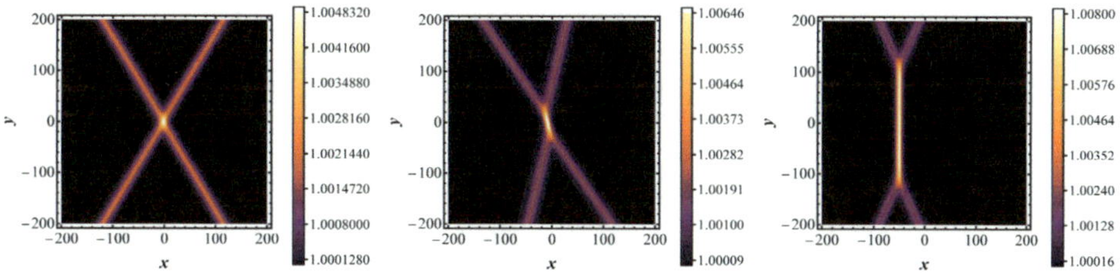

Figure 3.
X-type interactions, $u(x,y,0)$, with: $\kappa_1 = \kappa_2 = 1/2$, $\lambda_1 = -\lambda_2 = -2/3$ (left), $\kappa_1 = \kappa_2 = 1/2$, $\lambda_1 = 1/4 - 0.01$, $\lambda_2 = 3/4$ (middle), $\kappa_1 = \kappa_2 = 1/2$, $\lambda_1 = 1/2$, $\lambda_2 = -1/2 - 10^{-10}$ (right). Observe that the length of the stem is increased from left to right (see also text).

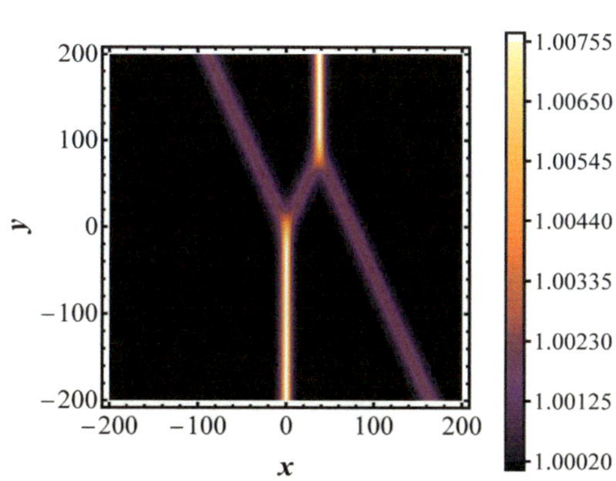

Figure 4.
An H-type interaction, $u(x,y,0)$, with $\kappa_1 = \kappa_2/2 = 1/2$, $\lambda_1 = 1/2 - 10^{-7}$, $\lambda_2 = 0$.

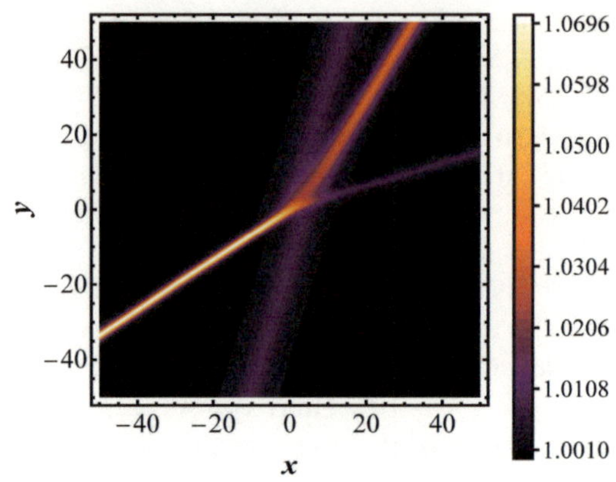

Figure 5.
A 2–3 wave interaction, $u(x, y, 0)$, with $\kappa_1 = 1$, $\kappa_2 = 2$, $\kappa_3 = 3$, $\lambda_1 = \lambda_2/2 = \lambda_3/5 = -1/3$ and $\delta_1 = \delta_2 = \delta_3 = 0$.

More intricate structures may still form whose mathematical description is more complex. Such novel weblike structures for the KP equation, termed N-in M-out, have been found using Wronskian methods [11, 13, 15] and the so-called τ-function. We illustrate the basic construction of these solitons but do not go into further detail as this is beyond the scope of this work. The most convenient way to write the line soliton solutions of the KPII system is through the Wronskian, or τ-function, which replaces the F function above. Now, the solutions of the KPII equation are written in a similar manner as above, as

$$U(\mathcal{X}, \mathcal{Y}, \mathcal{T}) = 2\partial_{\mathcal{X}}^2 \ln \tau(\mathcal{X}, \mathcal{Y}, \mathcal{T}), \tag{54}$$

where this new τ-function represents the Wronskian determinant

$$\tau(\mathcal{X}, \mathcal{Y}, \mathcal{T}) = \begin{pmatrix} f_1 & f_2 & \cdots & f_N \\ f_1' & f_2' & \cdots & f_N' \\ \vdots & \vdots & \cdots & \vdots \\ f_1^{(N-1)} & f_2^{(N-1)} & \cdots & f_N^{(N-1)} \end{pmatrix}. \tag{55}$$

Here superscripts denote differentiation with respect to X, and the set of f_n functions constitute the set of linearly independent solutions of the system:

$$f_{\mathcal{Y}} = f_{\mathcal{X}\mathcal{X}}, \quad f_{\mathcal{T}} = f_{\mathcal{X}\mathcal{X}\mathcal{X}}. \tag{56}$$

In particular, for line solitons these are defined as:

$$f_n(\mathcal{X}, \mathcal{Y}, \mathcal{T}) = \sum_{m=1}^{M} a_{nm} e^{\theta_m}, \quad n = 1, 2, \ldots, N, \tag{57}$$

where $\theta_m = k_m \mathcal{X} + k_m^2 \mathcal{Y} + k_m^3 \mathcal{T} + \theta_{0m}$ with distinct real parameters k_m 's with the property: $k_1 < k_2 < \cdots < k_M$. The parameters θ_{0m} are real constants. Importantly, the coefficients a_{nm} define an $N \times M$ matrix of rank N, due to the linear independence of the functions f_n, such that (a_{nm}=:A). Below we provide two examples, and their evolution for a 2–2 and 3–3 interaction (see **Figures 6** and 7, respectively).

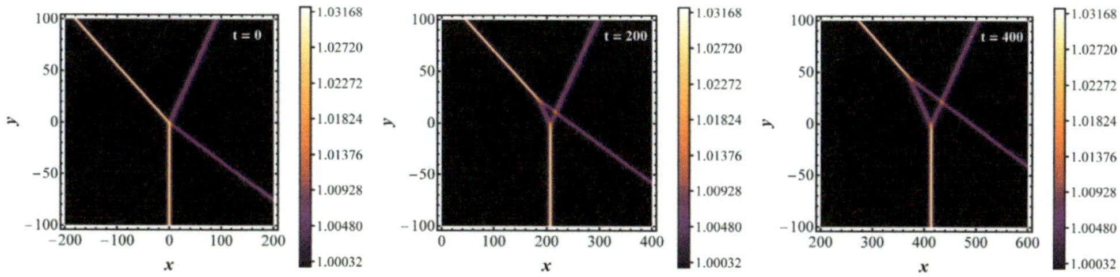

Figure 6.
A typical 2–2 wave evolution for $t = 0$ (left), $t = 200$ (middle) and $t = 400$ (right). All figures refer to the variable $u(x, y, t)$.

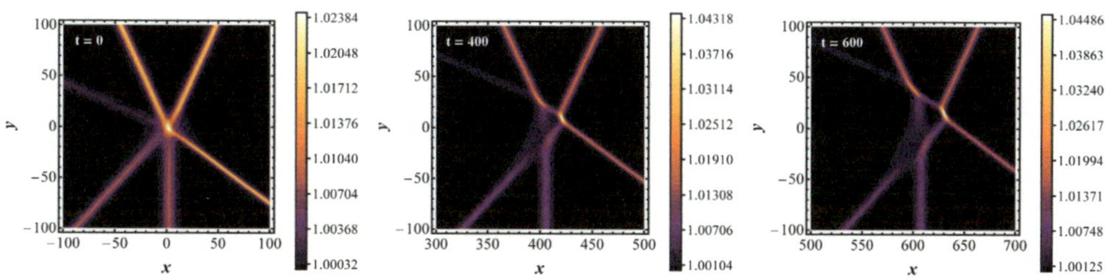

Figure 7.
A typical 3–3 wave evolution, for $t = 0$ (left), $t = 400$ (middle), and $t = 600$ (right). All figures refer to the variable $u(x, y, t)$.

We start with a 2–2 interaction, by defining the matrix A, such that

$$A = \begin{pmatrix} 1 & 0 & -1 & -1 \\ 0 & 1 & 1 & 1 \end{pmatrix}, \tag{58}$$

and the relative real parameters $k_1 = -1$, $k_2 = 0$, $k_3 = 1$ and $k_4 = 2$ corresponding to the relative $\theta_m|_{m=1}^{4}$. Then, the evolution of this interaction is depicted in **Figure 6**.

Similarly, for a 3–3 wave interaction one has:

$$A = \begin{pmatrix} 1 & 0 & -1 & -1 & 0 & 2 \\ 0 & 1 & 2 & 1 & 0 & -1 \\ 0 & 0 & 0 & 0 & 1 & 1 \end{pmatrix}, \tag{59}$$

and $k_1 = -1$, $k_2 = -1/2$, $k_3 = 0$, $k_4 = 1/2$, $k_5 = 1$ and $k_6 = 3/2$ to produce the following interaction pattern of **Figure 7**.

It is important to note here that the matrices A for the 2–2 and 3–3 solitons, are not arbitrarily chosen. The $N \times N$ maximal minors of A must be non-negative in order for the KP solution q to be regular for all X, Y and T. In the literature, such matrices are called totally non-negative matrices.

3.2 Dark lump solitons

So far, we have focused on structures obeying the effective KPII equation. These structures, namely the antidark stripe solitons, are quasi one-dimensional (1D) states, which are stable. The respective states that satisfy the effective KPI equation, namely dark stripe solitons, are expected to be transversely unstable – see, e.g., Ref. [36] for a stability analysis of KP line solitons.

Nevertheless, the KPI equation, that corresponds to the weakly nonlocal regime, supports stable, purely 2D, soliton states known as *lumps*. These solitons are weakly localized, namely they decay algebraically as $|x|, |y| \to \infty$. The respective solitons of the nonlocal NLS, which are of the dark type, i.e., they have the form of dark lumps, can be effectively described as solutions of Eq. (45) for the field U:

$$U(\mathcal{X}, \mathcal{Y}, \mathcal{T}) = 16 \frac{16q^2\eta^2 - 4(\xi - 2k\eta)^2 + 1/q^2}{\left[16q^2\eta^2 + 4(\xi - 2k\eta)^2 + 1/q^2\right]^2}, \qquad (60)$$
$$\xi = \mathcal{X} - 12(k^2 + q^2)\mathcal{T}, \quad \eta = \mathcal{Y} - 12k\mathcal{T},$$

where q and k are free real parameters. Notice that solutions of this type have not yet been observed in water due to the fact that the surface tension is small.

Similarly to the case of antidark stripe solitons, once Eq. (60) is substituted into Eq. (48) gives rise to the approximate dark lump soliton of the nonlocal NLS. Using the relevant analytical expression, it is straightforward to prepare the corresponding initial condition by setting $t = 0$, and numerically integrate Eqs. (2) and (3) to examine the evolution of dark lumps. In **Figure 8**, shown is the result of such a simulation for the dark lump; for this simulation, we have used the parameter values $u_0 = 1$, $d = 1/5$ (so that $\alpha = 1/5 > 0$, corresponding to dark soliton states), as well as $k = 0$ and $q = 1$. It can readily be observed that the dark lump solitons propagate undistorted up to $t = 20$, which is the "numerical horizon" for this particular simulation. This result indicates that, similarly to antidark stripe solitons, dark lumps can also be supported by the nonlocal NLS system.

3.3 Ring dark and antidark solitons

We now turn our attention to the cKP Eq. (46), and focus on radially symmetric solutions that do not depend on Θ. In such a case, Eq. (46) reduces to a quasi-1D equation, namely the cylindrical Korteweg-de Vries (cKdV), which is also a

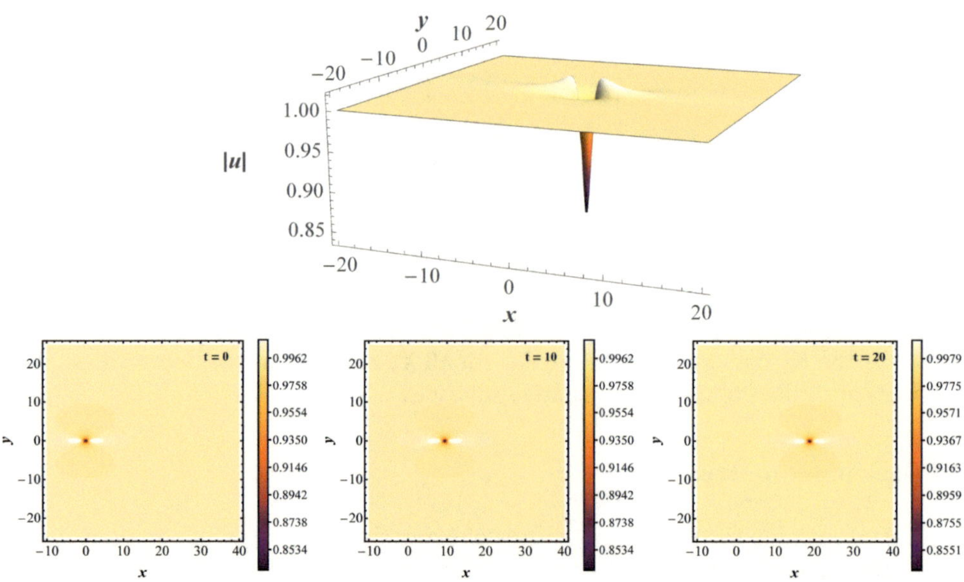

Figure 8.
A typical dark lump solution, at $t = 0$, with $k = 0$, $q = 1$ (top panel) and its evolution for $t = 0$, 10 and 20. The initial condition shown in the top panel is given by Eq. (60).

completely integrable model. The cKdV is a generic equation that can been used to describe cylindrical solitons in shallow water, plasmas, etc. [1]. As was shown in Ref. [37], the cKdV equation admits an exact soliton solution which, in the framework of Eq. (46), can be expressed as follows:

$$U(\varrho, \mathcal{T}) = \frac{\varrho}{12\mathcal{T}} + \frac{2\kappa^2}{\mathcal{T}} \operatorname{sech}^2\left(\frac{8\kappa^3}{\sqrt{\mathcal{T}}} + \frac{\kappa\varrho}{\sqrt{\mathcal{T}}} + \varrho_0\right) \tag{61}$$

where η and ϱ_0 are free parameters of the soliton. It is clear that Eq. (61) has the form of a sech^2-pulse on top of a rational background. A clearer picture regarding the structure of the cKdV equation, can be obtained by means of an asymptotic asymptotic analysis [38, 39]. Indeed, as shown in these works, to leading-order, in the regime $|\mathcal{T}| \gg |\varrho|$, the primary wave $U(\varrho, \mathcal{T})$, decays to zero at both upstream and downstream infinity. The primary wave has a form similar to that of Eq. (50), but with the following change of variables: $\mathcal{X} \mapsto \varrho$ and $\delta \mapsto \varrho_0$. However, there is a very important difference: κ now becomes a slowly-varying function of $\mathcal{T}$, due to the presence of the term $U/(2\mathcal{T})$ in the cKdV. Following the analysis of [38, 39], and using the original coordinates, it can be concluded that:

$$\kappa^2 = \kappa_0^2 \left(\frac{t_0}{t}\right)^{2/3}, \tag{62}$$

where κ_0^2 is a constant setting the solitary wave amplitude at $t = t_0$. We can know express an approximate solution of Eqs. (2) and (3), for the polar chase, and for the primary solitary wave. This is of the form of Eq. (48), with the soliton amplitude and velocity varying as $t^{-2/3}$, and the width varying as $t^{1/3}$, as follows from Eqs. (50) and (62).

Notice that this approximate solution is a ring-shaped solitary wave, on top of cw background, which can be either dark (for $\alpha > 0$) or anti-dark (for $\alpha < 0$). Notice, also that ring dark solitons were predicted to occur in optical media exhibiting either Kerr [40] or non-Kerr [41] nonlinearities, and were later observed in experiments [42]. On the other hand, ring anti-dark solitons were only predicted to occur in non-Kerr – e.g., saturable media [41, 43]. This picture is complemented by this analysis, according to which a relatively strong [i.e., $d > (1/4|u_0|)^2$] nonlocal nonlinearity can also support ring anti-dark solitary waves.

As in the Cartesian case, we have performed direct numerical simulations to examine the evolution of ring-shaped solitons in the nonlocal NLS model. First, in **Figure 9**, shown are 3D plots depicting the profiles of the ring dark solitons and ring antidark solitons. Here, we use $d = 1/5$ for the dark soliton ($\alpha > 0$) and $d = 1/3$ for

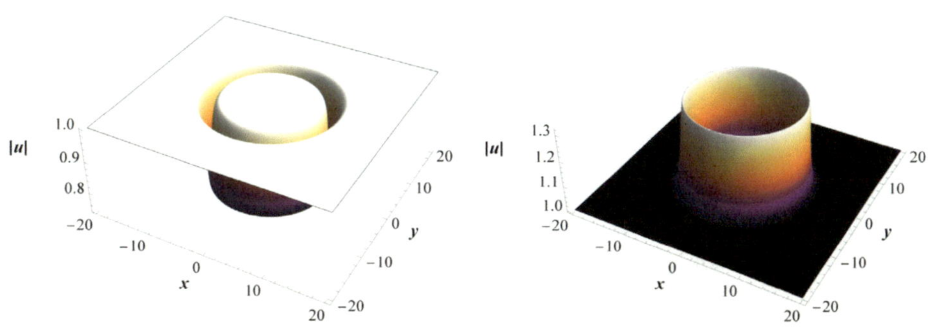

Figure 9.
(Color online) typical ring dark (left) and anti-dark (right) solitons.

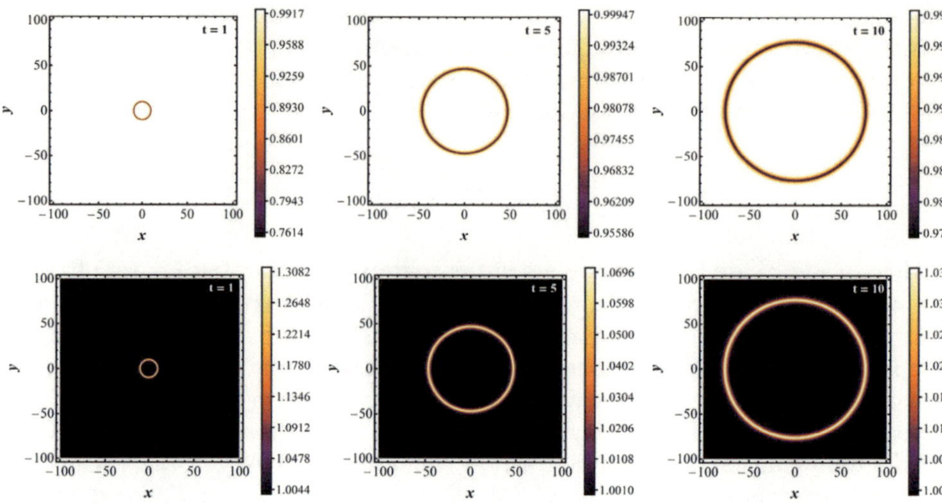

Figure 10.
(Color online) the evolution of the dark (top) and anti-dark (bottom) ring solitons of ***Figure 9****.*

the antidark ($\alpha < 0$), while the rest of the parameters are set equal to unity except $\varepsilon = 0.2$ as before. Furthermore, both solitons have an initial radius of $r(0) = 10$.

In addition, in **Figure 10**, we show the evolution of the ring solitons depicted in **Figure 9**. Once again, it is observed, that both types of ring solitons propagate undistorted up to end of the simulation ($t = 10$), which indicates that these states are also supported by the nonlocal NLS.

4. Conclusions and discussion

In conclusion, we have studied a 2D defocusing NLS model with a spatially nonlocal nonlinearity. The considered model is relevant to optical beam propagation in a variety of physical settings, including thermal media, plasmas, and nematic liquid crystals. We have analytically treated this nonlocal NLS by means of a two-stage multiscale analysis. This led to a Boussinesq/Benney-Luke-type equation, and a KP equation (in both Cartesian and polar geometries), which are models arising in the context of shallow water waves. This analysis revealed that the nonlocal NLS is characterized by an *effective surface tension*, which is related to the degree of nonlocality: small (large) effective surface tension corresponds to the strong (weak) nonlocality regime, whereby the asymptotic reduction leads to a KPII (KPI) equation.

Apart from the type of KP (KPII or KPI), the effective surface tension also controls the type of soliton: antidark (dark) solitons on top (off) the cw pedestal can only be formed in the strong (weak) nonlocality regime. We have thus found a number of different soliton states that can be supported by the nonlocal NLS. These include antidark stripe solitons, dark lump solitons, as well as ring dark and antidark solitons (in the polar case). We also used numerical simulations to examine the existence, evolution, as well as interaction dynamics. For the latter, we focused on the case of antidark solitons, which were shown to form a plethora of patterns coming out of their collisions. These include Y-, X-, H-shaped waveforms, as well as more complicated patterns involving more than two solitons; all these, have been observed in shallow water [5], a fact that highlights the deeper connection between water waves and light! In addition, we examined, by means of direct simulations, the existence and propagation of other soliton states, namely dark lump solitons, as well as ring dark and antidark solitons. We found that all these states do exist in the framework of the nonlocal NLS, and can propagate undistorted up to times set by the simulations.

At this point, it is also relevant to comment on the possibility of observing the predicted patterns in real experiments. In that regard, first we consider the case of antidark solitons. The observation of these states took place some time ago, as it was first reported in Ref. [44]; this fact indicates that the necessary experimental setting is already available. As concerns the experimental set up needed for the observation of the Y-, X- and H-waves and other related patterns, it may be similar to that of Ref. [44]. In particular, one may employ at first a cw laser beam, which is split into two parts via a beam-splitter. One branch goes through a cavity system to form a pulse (as happens in typical pulsed lasers); this pulse branch undergoes phase-engineering, i.e., passes through a phase mask so that the characteristic phase jump of the antidark soliton is inscribed. Then, the cw and the phase-engineered pulse are incoherently coupled inside the nonlocal medium, e.g., a nematic liquid crystal. This process forms one antidark soliton, as in Ref. [44]. To observe the Y-, X- or H-patterns predicted above, two such antidark solitons have to be combined inside the crystal. The angle between the two incident beams, which should be appropriately chosen so that a specific pattern be formed, can be controlled by a rotating mirror in one of the branches of the beam-splitter.

As concerns the possibility of the experimental observation of the other soliton states that were predicted by our analysis, we note the following. First, regarding the weakly localized dark lump solitons, they can be experimentally observed as follows. In a real experiment, it is straightforward to create a dark soliton by using, e.g., a proper phase mask (see, e.g., the review [45] and references therein); this can also be done even in the case of *ring dark solitons* [46]. Such structures are prone to the transverse modulational instability in 2D, and are known to decay into vortices [45]. However, as was predicted in Ref. [19], sufficiently weak gray solitons can decay to *dark lumps* satisfying an effective KPI equation. This result suggests that, to observe dark lumps, one only need to embed a shallow dark soliton (like the ones we considered herein) into a 2D space, e.g., in a planar waveguide; then, the transverse instability will set in, and dark lump solitons will emerge. Finally, regarding the ring solitons, as mentioned above, they have already been observed in a variety of experiments (see, e.g., Ref. [46] and the review [45]) and, thus they can readily be created, e.g., in thermal media or in nematic liquid crystals. Thus, a variety of patterns that can be observed in water may also emerge in optics.

The obtained results relied on the formal reduction of the nonlocal NLS model to the KP (cKP) equation. It would be interesting to investigate if other types of nonlocality, arising, e.g., in the context of chromium Bose-Einstein condensates [47], or in defocusing colloidal media [48] can support relevant soliton states. Such studies are in progress and pertinent results will be presented elsewhere.

Author details

Georgios N. Koutsokostas[1]*, Theodoros P. Horikis[2], Dimitrios J. Frantzeskakis[1], Nalan Antar[3] and İlkay Bakırtaş[3]

1 Department of Physics, University of Athens, Panepistimiopolis, Zografos, Athens 15784, Greece

2 Department of Mathematics, University of Ioannina, Ioannina 45110, Greece

3 Department of Mathematics, Istanbul Technical University, Maslak 34469, Istanbul, Turkey

*Address all correspondence to: koutsokwstasgewrgios@gmail.com

References

[1] Infeld E, Rowlands G. Nonlinear Waves, Solitons, and Chaos. Cambridge University Press; Cambridge, 1990.

[2] Dauxois T, Peyrard M. Physics of Solitons. Cambridge University Press; Cambridge, 2006.

[3] Ablowitz MJ, Clarkson PA. Solitons, nonlinear evolution equations and inverse scattering. Cambridge University Press; Cambridge, 1991.

[4] Ablowitz MJ. Nonlinear dispersive waves: Asymptotic analysis and solitons. Cambridge University Press; Cambridge, 2011.

[5] Ablowitz MJ, Baldwin DE. Nonlinear shallow ocean-wave soliton interactions on flat beaches. Phys Rev E. 2012;86: 036305.

[6] Kadomtsev BB, Petviashvili VI. On the stability of solitary waves in weakly dispersing media. Sov Phys Dokl. 1970; 15:539–541.

[7] Miles JW. Obliquely interacting solitary waves. J Fluid Mech. 1977;79: 157–169.

[8] Miles JW. Resonantly interacting solitary waves. J Fluid Mech. 1977;79: 171–179.

[9] Newell AC, Redekopp LG. Breakdown of Zakharov-Shabat Theory and Soliton Creation. Phys Rev Lett. 1977;38:377–380.

[10] Ohkuma K, Wadati M. The Kadomtsev-Petviashvili Equation: the Trace Method and the Soliton Resonances. J Phys Soc Jpn. 1983;52: 749–760.

[11] Biondini G, Kodama Y. On a family of solutions of the Kadomtsev-Petviashvili equation which also satisfy the Toda lattice hierarchy. J Phys A: Math Gen. 2003;36:10519.

[12] Biondini G, Chakravarty S. Soliton solutions of the Kadomtsev-Petviashvili II equation. J Math Phys. 2006;47: 033514.

[13] Chakravarty S, Kodama Y. Classification of the line-soliton solutions of KPII. J Phys A: Math Theor. 2008;41:275209.

[14] Chakravarty S, Kodama Y. Soliton Solutions of the KP Equation and Application to Shallow Water Waves. Stud Appl Math. 2009;123:83–151.

[15] Chakravarty S, Lewkow T, Maruno KI. On the construction of the KP line-solitons and their interactions. Applicable Analysis. 2010;89:529–545.

[16] McDowell T, Osborne M, Chakravarty S, Kodama Y. On a class of initial value problems and solitons for the KP equation: A numerical study. Wave Motion. 2017;72:201–227.

[17] Kivshar YS, Agrawal GP. Optical Solitons: From Fibers to Photonic Crystals. Academic Press; New York, 2003.

[18] Kuznetsov EA, Turitsyn SK. Instability and collapse of solitons in media with a defocusing nonlinearity. JETP. 1988;67:1583–1588.

[19] Pelinovsky DE, Stepanyants YA, Kivshar YS. Self-focusing of plane dark solitons in nonlinear defocusing media. Phys Rev E. 1995;51:5016–5026.

[20] Kivshar YS, Pelinovsky DE. Self-focusing and transverse instabilities of solitary waves. Phys Rep. 2000;331: 117–195.

[21] Horikis TP, Frantzeskakis DJ. Light Meets Water in Nonlocal Media: Surface Tension Analogue in Optics. Phys Rev Lett. 2017;118:243903.

[22] Johnson RS. Water waves and Korteweg-de Vries equations. J Fluid Mech. 1980;97:701–719.

[23] Johnson RS. A modern introduction to the mathematical theory of water waves. Cambridge University Press; 1997.

[24] Benney DJ, Luke JC. On the interactions of permanent waves of finite amplitude. J Math and Phys. 1964; 43:309–313.

[25] Rotschild C, Cohen O, Manela O, Segev M, Carmon T. Solitons in nonlinear media with an infinite range of nonlocality: first observation of coherent elliptic solitons and of vortex-ring solitons. Phys Rev Lett. 2005;95: 213904.

[26] Krolikowski W, Bang O, Nikolov NI, Neshev D, Wyller J, Rasmussen JJ, et al. Modulational instability, solitons and beam propagation in spatially nonlocal nonlinear media. J Opt B: Quantum Semiclass Opt. 2004;6:S288–S294.

[27] Ghofraniha N, Conti C, Ruocco G, Trillo S. Shocks in nonlocal media. Phys Rev Lett. 2007;99:043903.

[28] Conti C, Fratalocchi A, Peccianti M, Ruocco G, Trillo S. Observation of a gradient catastrophe generating solitons. Phys Rev Lett. 2009;102: 083902.

[29] Litvak AG, Mironov VA, Fraiman GM, Yunakovskii AD. Thermal self-effect of wave beams in a plasma with a nonlocal nonlinearity. Sov J Plasma Phys. 1975;1:60–71.

[30] Yakimenko AI, Zaliznyak YA, Kivshar YS. Stable vortex solitons in nonlocal self-focusing nonlinear media. Phys Rev E. 2005;71:065603(R).

[31] Conti C, Peccianti M, Assanto G. Route to nonlocality and observation of accessible solitons. Phys Rev Lett. 2003; 91:073901.

[32] Assanto G. Nematicons: Spatial Optical Solitons in Nematic Liquid Crystals. New Jersey: Wiley-Blackwell; 2012.

[33] Ablowitz MJ, Clarkson PA. Solitons, nonlinear evolution equations and inverse scattering. Cambridge University Press; Cambridge, 1991.

[34] Horikis TP. Small-amplitude defocusing nematicons. J Phys A: Math Theor. 2015;48:02FT01.

[35] Horikis TP, Frantzeskakis DJ. Ring dark and antidark solitons in nonlocal media. Opt Lett. 2016;41:583–586.

[36] Ablowitz MJ, Segur H. Solitons and the inverse scattering transform. SIAM; 1981.

[37] Hirota R. Exact solutions to the equation describing “cylindrical solitons”. Phys Lett A. 1979;71:393–394.

[38] Johnson RS. A note on an asymptotic solution of the cylindrical Korteweg-de Vries equation. Wave Motion. 1999;30:1–16.

[39] Ko K, Kuehl HH. Cylindrical and spherical Korteweg-deVries solitary waves. Phys Fluids. 1979;22:1343–1348.

[40] Kivshar YS, Yang X. Ring dark solitons. Phys Rev E. 1994;50:R40–R43.

[41] Frantzeskakis DJ, Malomed BA. Multiscale expansions for a generalized cylindrical nonlinear Schrödinger equation. Phys Lett A. 1999;264: 179–185.

[42] Dreischuh A, Neshev D, Paulus GG, Grasbon F, Walther H. Ring dark solitary waves: Experiment versus theory. Phys Rev E. 2002;66:066611.

[43] Nistazakis HE, Frantzeskakis DJ, Malomed BA, Kevrekidis PG. Head-on

collisions of ring dark solitons. Phys Lett A. 2001;285:157–164.

[44] Coskun TH, Christodoulides DN, Kim YR, Chen Z, Soljacic M, M S. Bright Spatial Solitons on a Partially Incoherent Background. Phys Rev Lett. 2000;84: 2374–2377.

[45] Kivshar YS, Luther-Davies B. Dark optical solitons: physics and applications. Phys Rep. 1998;298:81–197.

[46] Dreischuh A, Neshev D, Paulus GG, Grasbon F, Walther H. Ring dark solitary waves: Experiment versus theory. Phys Rev E. 2002;66:066611.

[47] Lahaye T, Menotti C, Santos L, Lewenstein M, Pfau T. The physics of dipolar bosonic quantum gases. Rep Prog Phys. 2009;72:126401.

[48] An X, Marchant TR, Smyth NF. Optical dispersive shock waves in defocusing colloidal media. Physica D. 2017;342:45–56.

Chapter 2

Soliton and Rogue-Wave Solutions of Derivative Nonlinear Schrödinger Equation - Part 1

Zhou Guo-Quan

Abstract

Based upon different methods such as a newly revised version of inverse scattering transform, Marchenko formalism, and Hirota's bilinear derivative transform, this chapter aims to study and solve the derivative nonlinear Schrödinger (DNLS for brevity) equation under vanishing boundary condition (VBC for brevity). The explicit one-soliton and multi-soliton solutions had been derived by some algebra techniques for the VBC case. Meanwhile, the asymptotic behaviors of those multi-soliton solutions had been analyzed and discussed in detail.

Keywords: soliton, nonlinear equation, derivative nonlinear Schrödinger equation, inverse scattering transform, Zakharov-Shabat equation, Marchenko formalism, Hirota's bilinear derivative transform, rogue wave

1. Introduction

Derivative nonlinear Schrödinger (DNLS for brevity) equation is one of the several rare kinds of integrable nonlinear models. Research of DNLS equation has not only mathematic interest and significance, but also important physical application background. It was first found that the Alfven waves in space plasma [1–3] can be modeled with DNLS equation. The modified nonlinear Schrödinger (MNLS for brevity) equation, which is used to describe the sub-picosecond pulses in single mode optical fibers [4–6], is actually a transformed version of DNLS equation. The weak nonlinear electromagnetic waves in ferromagnetic, anti-ferromagnetic, or dielectric systems [5–9] under external magnetic fields can also be modeled by DNLS equation.

Although DNLS equation is similar to NLS equation in form, it does not belong to the famous AKNS hierarchy at all. As is well known, a nonlinear integrable equation can be transformed to a pair of Lax equation satisfied by its Jost functions, the original nonlinear equation is only the compatibility condition of the Lax pair, that is, the so-called zero-curvature condition. Another fact had been found by some scholars that those nonlinear integrable equations which have the same first operator of the Lax pair belong to the same hierarchy and can deal with the same inverse scattering transform (IST for brevity). As a matter of fact, the DNLS equation has a squared spectral parameter of λ^2 in the first operator of its Lax pair, while the famous NLS equation, one typical example in AKNS hierarchy, has a spectral parameter of λ. Thus, the IST of the DNLS equation is greatly different from that of the NLS equation which is familiar to us. In a word, it deserves us to demonstrate several different approaches of solving it as a typical integrable nonlinear equation.

In this chapter, we will solve the DNLS equation under two kinds of boundary condition, that is, the vanishing boundary condition (VBC for brevity) and the non-vanishing boundary condition (NVBC for brevity), by means of three different methods – the revised IST method, the Marchenko formalism, and the Hirota's bilinear derivative method. Meanwhile, we will search for different types of special soliton solution to the DNLS equation, such as the light/dark solitons, the pure solitons, the breather-type solitons, and the rogue wave solution, in one- or multi-soliton form.

2. An N-soliton solution to the DNLS equation based on a revised inverse scattering transform

For the VBC case of DNLS equation, which is just the concerned theme of the section, some attempts and progress have been made to solve the DNLS equation. Since Kaup and Newell proposed an IST with a revision in their pioneer works [10, 11], one-soliton solution was firstly attained and several versions of raw or explicit multi-soliton solutions were also obtained by means of different approaches [12–20]. Huang and Chen have got a $N-$ soliton solution by means of Darboux transformation [15]. Steudel has derived a formula for $N-$ soliton solution in terms of Vandermonde-like determinants by means of Bäcklund transformation [13]; but just as Chen points out in Ref. [16], Steudel's multi-soliton solution is difficult to demonstrate collisions among solitons and still has a too complicate form to be used in the soliton perturbation theory of DNLS equation, although it can easily generate compute pictures. Since the integral kernel in Zakharov-Shabat (Z-S for brevity) equation does not tend to zero in the limit of spectral parameter λ with $|\lambda| \to \infty$, the contribution of the path integral along the big circle (the out contour) is also nonvanishing, the usual procedure to perform inverse scattering transform encounters difficulty and is invalid. Kaup thus proposed a revised IST by multiplying an additional weighing factor before the Jost solution $E(x, \ \lambda)$, so that it tends to zero as $|\lambda| \to \infty$, thus the modified *Z-S* kernel should lead to vanishing contribution of the integral along the big circle of Cauchy contour. Though the one-soliton solution has been found by the obtained Z-S equation of their IST, it is very difficult to derive directly its multi-soliton solution by their IST due to the existence of a complicated phase factor which is related to the solution itself [11]. We thus consider proposing a new revised IST to avoid the excessive complexity. Our N-soliton solution obviously has a standard multi-soliton form. It can be easily used to discuss its asymptotic behaviors and then develop its direct perturbation theory. On the other hand, in solving Z-S equation for DNLS with VBC, unavoidably we will encounter a problem of calculating determinant $\det(I + Q_1 Q_2)$, for two $N \times N$ matrices Q_1 and Q_2, where I is a $N \times N$ identity matrix. Our work also shows Binet-Cauchy formula and some other linear algebra techniques, (**Appendices** A.1–4 in Part 2), play important roles in the whole process, and actually also effective for some other nonlinear integrable models [21].

2.1 The revised inverse scattering transform and the Zakharov-Shabat equation for DNLS equation with VBC

2.1.1 The fundamental concepts for the IST theory of DNLS equation

DNLS equation for the one-dimension wave function $u(x,t)$ is usually expressed as

$$iu_t + u_{xx} + i\left(|u|^2 u\right)_x = 0 \tag{1}$$

with VBC, where the subscripts stand for partial derivative. Eq. (1) is also called Kaup-Newell (KN for brevity) equation. Its Lax pair is given by

$$L = -i\lambda^2\sigma_3 + \lambda U, U = \begin{pmatrix} 0 & u \\ -\overline{u} & 0 \end{pmatrix} \tag{2}$$

and

$$M = -i2\lambda^4\sigma_3 + 2\lambda^3 U - i\lambda^2 U^2\sigma_3 - \lambda(-U^3 + iU_x\sigma_3) \tag{3}$$

where λ is a spectral parameter, and σ_3 is the third one of Pauli matrices σ_1, σ_2, σ_3, and a bar over a letter, (e.g., $\overline{u}$ in (2)), represents complex conjugate. The first Lax equation is

$$\partial_x f(x, \ \lambda) = L(x, \ \lambda) f(x, \ \lambda) \tag{4}$$

In the limit of $|x| \to \infty$, $u \to 0$, and

$$L \to L_0 = -i\lambda^2\sigma_3; \ M \to M_0 = -i2\lambda^4\sigma_3 \tag{5}$$

The free Jost solution is a 2×2 matrix.

$$E(x, \ \lambda) = e^{-i\lambda^2 x\sigma_3}; E_{\bullet 1}(x,\lambda) = \begin{pmatrix} 1 \\ 0 \end{pmatrix} e^{-i\lambda^2 x}, E_{\bullet 2}(x,\lambda) = \begin{pmatrix} 0 \\ 1 \end{pmatrix} e^{i\lambda^2 x} \tag{6}$$

The Jost solutions of (4) are defined by their asymptotic behaviors as $x \to \pm\infty$.

$$\Psi(x,\lambda) = (\tilde{\psi}(x,\lambda), \psi(x,\lambda)) \to E(x,\lambda), \text{as } x \to \infty \tag{7}$$

$$\Phi(x,\lambda) = \left(\phi(x,\lambda), \tilde{\phi}(x,\lambda)\right) \to E(x,\lambda), \text{as } x \to -\infty \tag{8}$$

where $\psi(x,\lambda) = (\psi_1(x,\lambda), \psi_2(x,\lambda))^{\mathrm{T}}$, $\tilde{\psi}(x,\lambda) = (\tilde{\psi}_1(x,\lambda), \tilde{\psi}_2(x,\lambda))^{\mathrm{T}}$, etc., and superscript "T" represents transposing of a matrix here and afterwards.

Since the first Lax equation of DNLS is similar to that of NLS, there are some similar properties of the Jost solutions. The monodromy matrix $\mathrm{T}(\lambda)$ is defined as

$$\Phi(x,\lambda) = \Psi(x,\lambda) \ \mathrm{T}(\lambda), \tag{9}$$

where

$$\mathrm{T}(\lambda) = \begin{pmatrix} a(t,\lambda) & -\tilde{b}(t,\lambda) \\ b(t,\lambda) & \tilde{a}(t,\lambda) \end{pmatrix} \tag{10}$$

It is easy to find from (2) and (9) that

$$\sigma_2\ \overline{L(\overline{\lambda})}\ \sigma_2 = L(\lambda),\ \ \sigma_2\ \overline{T(\overline{\lambda})}\ \sigma_2 = T(\lambda) \tag{11}$$

$$\sigma_2\ \overline{\Psi(x,\overline{\lambda})}\ \sigma_2 = \Psi(x,\lambda),\ \ \sigma_2\ \overline{\Phi(x,\overline{\lambda})}\ \sigma_2 = \Phi(x,\lambda) \tag{12}$$

and

$$\sigma_3\Psi(x,\lambda)\sigma_3 = \Psi(x,-\lambda),\ \ \sigma_3\Phi(x,\lambda)\sigma_3 = \Phi(x,-\lambda) \tag{13}$$

$$\sigma_3 L(\lambda)\sigma_3 = L(-\lambda),\ \ \sigma_3 T(\lambda)\sigma_3 = T(-\lambda) \tag{14}$$

Then we can get the following reduction relation and symmetry properties

$$i\sigma_2\overline{\psi(x,\overline{\lambda})} = \tilde{\psi}(x,\lambda) \tag{15}$$

$$-i\sigma_2\overline{\varphi}(x,\overline{\lambda}) = \tilde{\varphi}(x,\lambda) \tag{16}$$

$$\overline{\tilde{a}}\left(\overline{\lambda}\right) = a(\lambda);\ \ \overline{\tilde{b}}\left(\overline{\lambda}\right) = b(\lambda) \tag{17}$$

and

$$\psi(x,-\lambda) = -\sigma_3\psi(x,\lambda) \tag{18}$$

$$\tilde{\psi}(x,-\lambda) = \sigma_3\tilde{\psi}(x,\lambda) \tag{19}$$

$$\begin{aligned} a(-\lambda) = a(\lambda);\ b(-\lambda) = -b(\lambda) \\ \tilde{a}(-\lambda) = \tilde{a}(\lambda);\ \tilde{b}(-\lambda) = -\tilde{b}(\lambda) \end{aligned} \tag{20}$$

2.1.2 *Relation between Jost functions and the solutions to the DNLS equation*

The asymptotic behaviors of the Jost solutions in the limit of $|\lambda| \to \infty$ can be obtained by simple derivation. Let $v = (v_1, v_2)^T \equiv \tilde{\psi}(x,\lambda)$; Eq. (4) can be rewritten as

$$v_{1x} + i\lambda^2 v_1 = \lambda u v_2,\ v_{2x} - i\lambda^2 v_2 = -\lambda\overline{u} v_1 \tag{21}$$

Then we have

$$v_{1xx} - u_x\left(v_{1x} + i\lambda^2 v_1\right)/u + \lambda^4 v_1 + \lambda^2|u|^2 v_1 = 0 \tag{22}$$

In the limit $|\lambda| \to \infty$, we assume $\tilde{\psi}_1(x,\lambda) = e^{-i\lambda^2 x + g}$, substituting it into Eq. (22), then we have

$$\left(-i\lambda^2 + g_x\right)^2 + g_{xx} - u_x g_x/u + \lambda^4 + \lambda^2|u|^2 = 0 \tag{23}$$

In the limit $|\lambda| \to \infty, g_x$ can be expanded as series of $\left(\lambda^{-2}\right)^j, j = 1, 2, \cdots$.

$$ig_x \equiv \mu = \mu_0 + \mu_2\left(2\lambda^2\right)^{-1} + \cdots \tag{24}$$

and

$$\mu_0 = |u|^2/2, \mu_2 = -i\overline{u}_x u/2 - |u|^4/4, \cdots \tag{25}$$

Eq. (21) leads to $g_x v_1 = \lambda u v_2$. Considering (25), in the limit of $|\lambda| \to \infty$, we find a useful formula

$$\bar{u} = i2 \lim_{|\lambda| \to \infty} \lambda \tilde{\psi}_2(x, \lambda) / \tilde{\psi}_1(x, \lambda) \tag{26}$$

which expresses the conjugate of solution u in terms of the Jost solutions as $|\lambda| \to \infty$.

On the other hand, the zeros of $a(\lambda)$ appear in pairs and can be designed by λ_n, $n = 1, 2, \cdots, N$ in the I quadrant, and $\lambda_{n+N} = -\lambda_n$ in the III quadrant. The discrete part of $a(\lambda)$ is [21–23].

$$a(\lambda) = \prod_{n=1}^{N} \frac{\lambda^2 - \lambda_n^2}{\lambda^2 - \overline{\lambda_n^2}} \bullet \frac{\overline{\lambda_n^2}}{\lambda_n^2} \tag{27}$$

where $a(0) = 1$. It comes from our consideration of the fact that, from the sum of two Cauchy integrals

$$\frac{\ln a(\lambda)}{\lambda} + 0 = \frac{1}{2\pi i} \int_\Gamma d\lambda' \frac{\ln a(\lambda') \tilde{a}(\lambda')}{\lambda'(\lambda' - \lambda)}, \Gamma = (0, \infty) \cup (i\infty, i0) \cup (0, -\infty) \cup (-i\infty, i0),$$

in order to maintain that $\ln a(\lambda) \to 0$, as $\lambda \to 0$, and $\ln a(\lambda)$ is finite as $|\lambda| \to \infty$, we then have to introduce a factor $\overline{\lambda}_n^2 / \lambda_n^2$ in (27). At the zeros of $a(\lambda)$, we have

$$\phi(x, \lambda_n) = b_n \psi(x, \lambda_n), \dot{a}(-\lambda_n) = -\dot{a}(\lambda_n), b_{n+N} = -b_n \tag{28}$$

Due to $\mu_0 \neq 0$ in (24) and (25), the Jost solutions do not tend to free Jost solutions $E(x, \lambda)$ in the limit of $|\lambda| \to \infty$. This is their most typical property which means that the usual procedure of constructing the equation of IST by a Cauchy contour integral must be invalid and abortive, thus a newly revised procedure to derive a suitable IST and the corresponding Z-S equation is proposed in our group.

2.1.3 The revised IST and Zakharov-Shabat equation for DNLS equation with VBC

The 2×1 column function $\Theta(x, \lambda)$ can be introduced as usual

$$\Theta(x, \lambda) = \begin{cases} \phi(x, \lambda)/a(\lambda), & \text{as } \lambda \text{ in I, III quadrants.} \\ \tilde{\psi}(x, \lambda), & \text{as } \lambda \text{ in II, IV quadrants.} \end{cases} \tag{29}$$

An alternative form of IST equation is proposed as

$$\frac{1}{\lambda^2} \{\Theta_1(x, \lambda) - E_{11}(x, \lambda)\} e^{i\lambda^2 x} = \frac{1}{2\pi i} \int_\Gamma d\lambda' \frac{1}{\lambda' - \lambda} \frac{1}{\lambda'^2} \{\Theta_1(x, \lambda') - E_{11}(x, \lambda')\} e^{i\lambda'^2 x} \tag{30}$$

Because in the limit of $|\lambda| \to \infty$, $\lim_{|\lambda| \to \infty} e^{i\lambda^2 x} = 0$,

as $\begin{cases} x > 0, & \text{Im } \lambda^2 > 0, & (\lambda \text{ in the I, III quadrants}), \\ x < 0, & \text{Im } \lambda^2 < 0, & (\lambda \text{ in the II, IV quadrants}), \end{cases}$ then the integral path Γ should be chosen as shown in **Figure 1**, where the radius of big circle tends to infinite, while the radius of small circle tends to zero. And the factor λ^{-2} is introduced to ensure the contribution of the integral along the big arc is vanishing. Meanwhile,

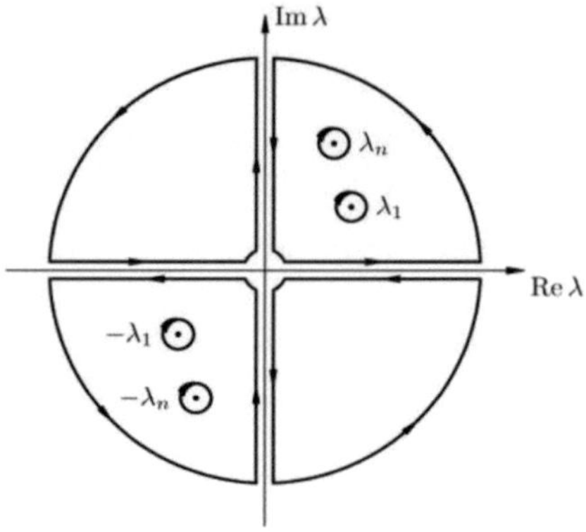

Figure 1.
The integral path for IST of the DNLS.

our modification produces no new poles since Lax operator $L \to 0$, as $\lambda \to 0$. In the reflectionless case, the revised IST equation gives

$$\tilde{\psi}_1(x,\lambda) = e^{-i\lambda^2 x} + \sum_{n=1}^{2N} \frac{1}{\lambda_n^2} \frac{\lambda^2}{\lambda - \lambda_n} \frac{b_n}{\dot{a}(\lambda_n)} \psi_1(x,\lambda_n) e^{i\lambda_n^2 x} e^{-i\lambda^2 x} \tag{31}$$

where $\dot{a}(\lambda_n) = \mathrm{d}a(\lambda)/\mathrm{d}\lambda|_{\lambda=\lambda_n}$. Similarly, an alternative form of IST equation is proposed as follows:

$$\frac{1}{\lambda}\{\Theta_2(x,\lambda)\}e^{i\lambda^2 x} = \frac{1}{2\pi i}\int_\Gamma d\lambda' \frac{1}{\lambda' - \lambda}\frac{1}{\lambda'}\{\Theta_2(x,\lambda')\}e^{i\lambda'^2 x} \tag{32}$$

where a factor λ^{-1} is introduced for the same reason as λ^{-2} in Eq. (30). Then in the reflectionless case, we can attain

$$\tilde{\psi}_2(x,\lambda) = \sum_{n=1}^{2N} \frac{1}{\lambda_n} \frac{\lambda}{\lambda - \lambda_n} \frac{b_n}{\dot{a}(\lambda_n)} \psi_2(x,\lambda_n) e^{i\lambda_n^2 x} e^{-i\lambda^2 x} \tag{33}$$

Taking the symmetry and reduction relation (18) and (28) into consideration, from (31) and (33), we can obtain the revised Zakharov-Shabat equation for DNLS equation with VBC, that is,

$$\tilde{\psi}_1(x,\lambda) = e^{-i\lambda^2 x} + \sum_{n=1}^{N} \frac{2\lambda^2}{\lambda_n(\lambda^2 - \lambda_n^2)} \frac{b_n}{\dot{a}(\lambda_n)} \psi_1(x_1,\lambda_n) e^{i\lambda_n^2 x} e^{-i\lambda^2 x} \tag{34}$$

$$\tilde{\psi}_2(x,\lambda) = \sum_{n=1}^{N} \frac{2\lambda}{\lambda^2 - \lambda_n^2} \frac{b_n}{\dot{a}(\lambda_n)} \psi_2(x_s,\lambda_n) e^{i\lambda_n^2 x} e^{-i\lambda^2 x} \tag{35}$$

2.2 The raw expression of N-soliton solution

Substituting Eqs. (34) and (35) into formula (26), we thus attain the N-soliton solution

$$\overline{u}_N = -i2U_N/V_N \tag{36}$$

where

$$U_N = \sum_{n=1}^{N} \frac{2b_n}{\dot{a}(\lambda_n)} \psi_2(x, \lambda_n) e^{i\lambda_n^2 x} \tag{37}$$

$$V_N = 1 + \sum_{n=1}^{N} \frac{2b_n}{\lambda_n \dot{a}(\lambda_n)} \psi_1(x, \lambda_n) e^{i\lambda_n^2 x} \tag{38}$$

Let $\lambda = \overline{\lambda}_m$, $m = 1, 2, \dots, N$, respectively, in Eqs. (34) and (35), and make use of the symmetry and reduction relation (15), we can attain

$$\overline{\psi_2}(x, \lambda_m) = e^{-i\overline{\lambda_m^2} x} + \sum_{n=1}^{N} \frac{2\overline{\lambda_m^2}}{\lambda_n \left(\overline{\lambda_m^2} - \lambda_n^2\right)} c_n \psi_1(x, \lambda_n) e^{i\lambda_n^2 x} e^{-i\overline{\lambda_m^2} x} \tag{39}$$

$$\overline{\psi}_1(x, \lambda_m) = -\sum_{n=1}^{N} \frac{2\overline{\lambda_m^2}}{\overline{\lambda_m^2} - \lambda_n^2} c_n \psi_2(x, \lambda_n) e^{i\lambda_n^2 x} e^{-i\overline{\lambda_m^2} x}; \; m = 1, 2, \dots, N. \tag{40}$$

where $c_n = b_n / \dot{a}(\lambda_n)$. We also define

$$f_n = \sqrt{2c_n} e^{i\lambda_n^2 x}, \quad (w_j)_n = \sqrt{2c_n} \psi_j(\lambda_n) \; j = 1, 2; \text{and } n = 1, 2, \dots N. \tag{41}$$

$$(B_1)_{mn} = \overline{f}_m \frac{\overline{\lambda_m^2}}{\left(\overline{\lambda_m^2} - \lambda_n^2\right)\lambda_n} f_n, (B_2)_{mn} = \overline{f}_m \frac{\overline{\lambda_m}}{\overline{\lambda_m^2} - \lambda_n^2} f_n; m, n = 1, 2, \dots, N \tag{42}$$

$$\begin{aligned} W_1 &= \left((w_1)_1, (w_1)_2, \cdots, (w_1)_N\right)^{\mathrm{T}}, W_2 = \left((w_2)_1, (w_2)_2, \cdots, (w_2)_N\right)^{\mathrm{T}}, \\ F &= \left(f_1, f_2, \cdots, f_N\right)^{\mathrm{T}}, G = \left(f_1/\lambda_1, f_2/\lambda_2, \cdots, f_N/\lambda_N\right)^{\mathrm{T}} \end{aligned} \tag{43}$$

where superscript "T" represents transposition of a matrix. Then Eqs. (39) and (40) can be rewritten as

$$(\overline{w_2})_m = \overline{f}_m + \sum_{n=1}^{N} (B_1)_{mn} (w_1)_n \tag{44}$$

$$(\overline{w_1})_m = -\sum_{n=1}^{N} (B_2)_{mn} (w_2)_n \tag{45}$$

where $m = 1, 2, \dots, N$. They can be rewritten in a more compact matrix form.

$$\overline{W}_2 = \overline{F} + B_1 \cdot W_1 \tag{46}$$

$$\overline{W}_1 = -B_2 \cdot W_2 \tag{47}$$

Then

$$W_2 = \left(I + \overline{B}_1 B_2\right)^{-1} F \tag{48}$$

$$W_1 = -\overline{B}_2 \left(I + B_1 \overline{B}_2\right)^{-1} \overline{F} \tag{49}$$

where I is the $N \times N$ identity matrix. On the other hand, from (37) and (38), we know

$$U_N = \sum_{n=1}^{N} f_n w_{2n} = F^{\mathrm{T}} W_2 \tag{50}$$

$$V_N = 1 + \sum_{n=1}^{N} (f_n/\lambda_n) w_{1n} = 1 + G^{\mathrm{T}} W_1 \tag{51}$$

Substituting Eqs. (48), (49) into (50) and (51) and then substituting (50) and (51) into formula (36), we thus attain

$$\begin{aligned} \overline{u}_N &= -i2\frac{F^{\mathrm{T}}W_2}{1+G^{\mathrm{T}}W_1} = -i2\frac{F^{\mathrm{T}}(I+\overline{B}_1B_2)^{-1}F}{1-G^T\overline{B}_2(I+B_1\overline{B}_2)^{-1}\overline{F}} \\ &= -i2\frac{\det(I+\overline{B}_1B_2+FF^{\mathrm{T}})-\det(I+\overline{B}_1B_2)}{\det[I+(B_1-\overline{F}G^{\mathrm{T}})\overline{B}_2]}\cdot\frac{\det(I+B_1\overline{B}_2)}{\det(I+\overline{B}_1B_2)} \equiv -2i\frac{A\cdot D}{\overline{D}^2} \end{aligned} \tag{52}$$

where

$$A \equiv \det(I+\overline{B}_1B_2+FF^{\mathrm{T}}) - \det(I+\overline{B}_1B_2) \tag{53}$$

$$D \equiv \det(I+B_1\overline{B}_2) \tag{54}$$

In the subsequent chapter, we will prove that

$$\det[I+(B_1-\overline{F}G^{\mathrm{T}})\overline{B}_2] = \det(I+\overline{B}_1B_2) \tag{55}$$

It is obvious that formula (52) has the usual standard form of soliton solution. Here in formula (52), some algebra techniques have been used and can be found in Appendix A.1 in Part 2.

2.3 Explicit expression of N-soliton solution

2.3.1 Verification of standard form for the N-soliton solution

We only need to prove that Eq. (55) holds. Firstly, we define $N \times N$ matrices P_1, P_2, Q_1, Q_2, respectively, as

$$(P_1)_{nm} \equiv (B_1-\overline{F}G^{\mathrm{T}})_{nm} = \overline{f}_n\frac{\lambda_m}{\lambda_n^2-\lambda_m^2}f_m;\ (P_2)_{mn} \equiv (\overline{B}_2)_{mn} = f_m\frac{\lambda_m}{\overline{\lambda_m^2}-\overline{\lambda_n^2}}\overline{f}_n \tag{56}$$

$$(Q_1)_{nm} \equiv (\overline{B}_1)_{nm} = f_n\frac{\lambda_n^2}{\lambda_n^2-\overline{\lambda_m^2}}\left(\frac{\overline{f}_m}{\overline{\lambda}_m}\right);\ (Q_2)_{mn} \equiv (B_2)_{mn} = \overline{f}_m\frac{\overline{\lambda}_m}{\overline{\lambda_m^2}-\lambda_n^2}f_n \tag{57}$$

Then

$$\begin{aligned} \overline{D} &= \det(I+Q_1Q_2) = 1 + \sum_{r=1}^{N}\sum_{1\le n_1<n_2<\cdots<n_r\le N}\overline{D}_r(n_1,n_2,\cdots n_r) \\ &= 1+\sum_{r=1}^{N}\sum_{1\le n_1<\cdots<n_r\le N}\sum_{1\le m_1<\cdots<m_r\le N} Q_1(n_1,n_2,\cdots n_r;m_1,m_2,\cdots,m_r)Q_2(m_1,m_2,\cdots m_r;n_1,n_2\cdots,n_r) \end{aligned} \tag{58}$$

where $Q_1(n_1, n_2, \cdots n_r; m_1, m_2, \cdots, m_r)$ denotes a minor, which is the determinant of a submatrix of Q_1 consisting of elements belonging to not only rows $(n_1, n_2, \ldots, n_r)$ but also columns $(m_1, m_2, \ldots, m_r)$. Here use is made of Binet-Cauchy formula in the Appendices A.2–4 in Part 2. Then

$$\begin{aligned} &Q_1(n_1, n_2, \cdots n_r; m_1, m_2, \cdots, m_r)Q_2(m_1, m_2, \cdots m_r; n_1, n_2 \cdots, n_r) \\ &= \prod_{n,m} \frac{f_n \overline{f}_m}{\lambda_n^2 - \overline{\lambda}_m^2} \frac{\lambda_n^2}{\overline{\lambda}_m} \prod_{n<n', m<m'} \left(\lambda_n^2 - \lambda_{n'}^2\right)\left(\lambda_{m'}^2 - \lambda_m^2\right) \prod_{m,n} \frac{\overline{f}_m f_n}{\overline{\lambda}_m^2 - \lambda_n^2} \overline{\lambda}_m \prod_{n<n', m<m'} \left(\overline{\lambda}_m^2 - \lambda_{m'}^2\right)\left(\lambda_{n'}^2 - \lambda_n^2\right) \\ &= (-1)^r \prod_{m,n} \frac{\lambda_n^2 f_n^2 \overline{f_m^2}}{\left(\lambda_n^2 - \overline{\lambda_m^2}\right)^2} \prod_{n<n', m<m'} \left(\lambda_n^2 - \lambda_{n'}^2\right)^2 \left(\overline{\lambda_m^2} - \overline{\lambda_{m'}^2}\right)^2 \end{aligned} \tag{59}$$

where

$$n, n' \in \{n_1, n_2, \cdots, n_r\}, m, m' \in \{m_1, m_2, \cdots, m_r\} \tag{60}$$

Similarly,

$$\begin{aligned} &P_1(n_1, n_2, \cdots, n_r; m_1, m_2, \cdots, m_r)P_2(m_1, m_2, \cdots, m_r; n_1, n_2, \cdots, n_r) \\ &\quad = (-1)^r \prod_{n,m} \frac{\overline{f}_n^2 f_m^2 \lambda_m^2}{\left(\overline{\lambda}_n^2 - \lambda_m^2\right)^2} \prod_{n<n', m<m'} \left(\lambda_m^2 - \lambda_{m'}^2\right)^2 \left(\overline{\lambda_n^2} - \overline{\lambda_{n'}^2}\right)^2 \end{aligned} \tag{61}$$

where

$$\begin{aligned} &n, n' \in \{n_1, n_2, \cdots, n_r\}; m, m' \in \{m_1, m_2, \cdots, m_r\}, \text{and} \\ &\det\left[I + \left(B_1 - \overline{F}G^{\mathrm{T}}\right)\overline{B}_2\right] = \det(I + P_1P_2) \\ &= 1 + \sum_{r=1}^{N} \sum_{1 \le n_1 < \cdots < n_r \le N} \sum_{1 \le m_1 < \cdots < m_r \le N} P_1(n_1, \cdots, n_r; m_1, \cdots, m_r)P_2(m_1, \cdots, m_r; n_1, \cdots, n_r) \end{aligned} \tag{62}$$

It is easy to find a kind of permutation symmetry existed between expressions (59) and (61), that is,

$$\begin{aligned} &P_1(n_1, \cdots, n_r; m_1, \cdots, m_r)P_2(m_1, \cdots, m_r; n_1, \cdots, n_r) \\ &= Q_1(m_1, \cdots, m_r; n_1, \cdots, n_r)Q_2(n_1, \cdots, n_r; m_1, \cdots, m_r) \end{aligned} \tag{63}$$

Comparing (58) with (62) and making use of (63), we thus complete verification of Eq. (55). The soliton solution is surely of a typical form as that in NLS equation and can be expressed as formula (52).

2.3.2 Introduction of time evolution function

The time evolution factor of the scattering data can be introduced by standard procedure [21]. Due to the fact that the second Lax operator $M \to -i2\lambda^4\sigma_3$ in the limit of $|x| \to \infty$, it is easy to derive the time dependence of scattering date.

$$d\lambda_n/dt = 0, da(\lambda_n)/dt = 0; c_n(t) = c_{n0}e^{i4\lambda_n^4 t}, c_{n0} = b_{n0}/\dot{a}(\lambda_n), b_n(t) = b_{n0}e^{i4\lambda_n^4 t} \tag{64}$$

Then the typical soliton arguments θ_n and φ_n can be defined according to

$$f_n^2 = 2c_{n0}e^{i2\lambda_n^2 x}e^{i4\lambda_n^4 t} \equiv 2c_{n0}e^{-\theta_n}e^{i\varphi_n} \tag{65}$$

where $\lambda \equiv \mu_n + i\nu_n$, and $\theta_n = 4\mu_n\nu_n\left[x + 4\left(\mu_n^2 - \nu_n^2\right)t\right] = 4\kappa_n(x - V_n t)$; $\varphi_n = 2\left(\mu_n^2 - \nu_n^2\right)x + \left[4\left(\mu_n^2 - \nu_n^2\right)^2 - 16\mu_n^2\nu_n^2\right] \cdot t$

$$V_n = -4\left(\mu_n^2 - \nu_n^2\right), \kappa_n = 4\mu_n\nu_n \tag{66}$$

2.3.3 Calculation of determinant of $\overline{D}$ and A

Substituting expression (64) and (65) into formula (59) and then into (58), we have

$$Q_1(n_1, n_2, \cdots, n_r; m_1, m_2, \cdots, m_r)Q_2(m_1, m_2, \cdots, m_r; n_1, n_2, \cdots, n_r) = (-1)^r \prod_{n,m} (2c_n)(2\overline{c}_m)e^{-\theta_n}e^{i\varphi_n}e^{-\theta_m}e^{-i\varphi_m} \frac{\lambda_n^2}{\left(\lambda_n^2 - \overline{\lambda_m^2}\right)^2} \prod_{n<n', m<m'} \left(\lambda_n^2 - \lambda_{n'}^2\right)^2\left(\overline{\lambda_m^2} - \overline{\lambda_{m'}^2}\right)^2 \tag{67}$$

with $n, n' \in \{n_1, n_2, \cdots, n_r\}$ and $m, m' \in \{m_1, m_2, \cdots, m_r\}$. Where use is made of Binet-Cauchy formula which is numerated in Appendix A. 3–4 in Part 2. Substituting expression (67) into formula (58) thus complete the calculation of determinant $\overline{D}$.

About the calculation of the most complicate determinant A in (52), we introduce a $N\times(N+1)$ matrix Ω_1 and a $(N+1) \times N$ matrix Ω_2 defined as

$$\begin{aligned}(\Omega_1)_{nm} &= \left(\overline{B}_1\right)_{nm} = (Q_1)_{nm}, (\Omega_1)_{n0} = f_n,\\ (\Omega_2)_{mn} &= (B_2)_{mn} = (Q_2)_{mn}, (\Omega_2)_{0n} = f_n\end{aligned} \tag{68}$$

with $n, m = 1, 2, \cdots, N$. We thus have

$$\det\left(I + \overline{B}_1 B_2 + FF^{\mathrm{T}}\right) = \det(I + \Omega_1\Omega_2) = 1 + \sum_{r=1}^{N} \sum_{1\le n_1<\cdots<n_r\le N} \sum_{0\le m_1<\cdots<m_r\le N} \Omega_1(n_1, n_2, \cdots, n_r; m_1, m_2, \cdots, m_r)\Omega_2(m_1, m_2, \cdots, m_r; n_1, n_2, \cdots, n_r) \tag{69}$$

The above summation obviously can be decomposed into two parts: one is extended to $m_1 = 0$, the other is extended to $m_1 \ge 1$. Subtracted from (69), the part that is extended to $m_1 \ge 1$, the remaining parts of (69) is just A in (52) (with $m_1 = 0$ and $m_2 \ge 1$). Due to (68), we thus have

$$\begin{aligned}A &= \det(I + \Omega_1\Omega_2) - \det(I + Q_1Q_2)\\ &= \sum_{r=1}^{N} \sum_{1\le n_1<\cdots<n_r\le N} \sum_{1\le m_2<\cdots<m_r\le N} A_r(n_1, n_2, \cdots, n_r; 0, m_2, \cdots, m_r)\\ &= \sum_{r=1}^{N} \sum_{1\le n_1<n_2<n_r\le N}^{N} \sum_{1\le m_2<m_3<\cdots<m_r\le N} \Omega_1(n_1, n_2, \cdots, n_r; 0, m_2, \cdots, m_r)\Omega_2(0, m_2, \cdots, m_r; n_1, n_2, \cdots, n_r)\end{aligned} \tag{70}$$

with

$$\Omega_1(n_1, n_2, \cdots, n_r; 0, m_2, \cdots, m_r)\Omega_2(0, m_2, \cdots, m_r; n_1, n_2, \cdots, n_r)$$

$$= (-1)^{r-1}\prod_{n,m}\frac{f_n^2\overline{f_m^2\lambda_m^2}}{\left(\lambda_n^2 - \overline{\lambda_m^2}\right)^2}\prod_{n<n',m<m'}\left(\lambda_n^2 - \lambda_{n'}^2\right)^2\left(\overline{\lambda_m^2} - \overline{\lambda_{m'}^2}\right)^2$$

$$= (-1)^{r-1}\prod_{n,m}(2c_n)(2\overline{c_m})e^{-\theta_n}e^{i\varphi_n}e^{-\theta_m}e^{-i\varphi_m}\frac{\overline{\lambda_m^2}}{\left(\lambda_n^2 - \overline{\lambda_m^2}\right)^2}\prod_{n<n',m<m'}\left(\lambda_n^2 - \lambda_{n'}^2\right)^2\left(\overline{\lambda_m^2} - \overline{\lambda_{m'}^2}\right)^2 \tag{71}$$

Here $n, n' \in \{n_1, n_2, \cdots, n_r\}$ and especially $m, m' \in \{m_2, \cdots, m_r\}$, which completes the calculation of determinant A in formula (52). Substituting the explicit expressions of D, $\overline{D}$, and A into (52), we finally attain the explicit expression of N-soliton solution to the DNLS equation under VBC and reflectionless case, based upon a newly revised IST technique.

An interesting conclusion is found that, besides a permitted well-known constant global phase factor, there is also an undetermined constant complex parameter b_{n0} before each of the typical soliton factor $e^{-\theta_n}e^{i\varphi_n}$, $(n = 1,2, \dots, N)$. It can be absorbed into $e^{-\theta_n}e^{i\varphi_n}$ by redefinition of soliton center and its initial phase factor. This kind of arbitrariness is in correspondence with the unfixed initial conditions of the DNLS equation.

2.4 The typical examples for one- and two-soliton solutions

We give two concrete examples – the one- and two-soliton solutions as illustrations of the general explicit soliton solution.

In the case of one-soliton solution, $N = 1$, $\lambda_2 = -\lambda_1$, $\lambda_1 = \rho_1 e^{i\beta_1} = \mu_1 + i\nu_1$, and

$$A_1 = \Omega_1(n_1 = 1; m_1 = 0)\Omega_2(m_1 = 0; n_1 = 1) = f_1^2 \tag{72}$$

$$\overline{D}_1 = Q_1(n_1 = 1; m_1 = 1)Q_2(m_1 = 1; n_2 = 1) = 1 - \left|f_1\right|^4\lambda_1^2/\left(\lambda_1^2 - \overline{\lambda_1^2}\right)^2 \tag{73}$$

$$f_1^2 = 2c_{10}e^{i2\lambda_1^2x}e^{i4\lambda_1^4t}; c_{10} = b_{10}/\dot{a}(\lambda_1); b_{10} = e^{4\mu_1\nu_1x_{10}}e^{i\alpha_{10}}; b_{10}e^{i2\lambda_1^2x}e^{i4\lambda_1^4t} \equiv e^{-\theta_1}e^{i\varphi_1};$$

$$\theta_1 = 4\mu_1\nu_1\left[x - x_{10} + 4\left(\mu_1^2 - \nu_1^2\right)t\right]; \varphi_1 = 2\left(\mu_1^2 - \nu_1^2\right)x + \left[4\left(\mu_1^2 - \nu_1^2\right)^2 - 16\mu_1\nu_1^2\right]t + \alpha_{10} \tag{74}$$

It is different slightly from the definition in (66) for that here b_{10} has been absorbed into the soliton center and initial phase. Then

$$A_1 = \lambda_1\left(\lambda_1^2 - \overline{\lambda_1^2}\right)e^{-\theta_1}e^{i\varphi_1}/\overline{\lambda}_1^2 = i2\rho_1\sin 2\beta_1 e^{i3\beta_1}e^{-\theta_1}e^{i\varphi_1}$$

$$\overline{D}_1 = 1 - \frac{\left|\lambda_1^2 - \overline{\lambda_1^2}\right|^2}{\left|\lambda_1\right|^2}\frac{\lambda_1^2}{\left(\lambda_1^2 - \overline{\lambda_1^2}\right)^2}e^{-2\theta_1} = 1 + e^{i2\beta_1}e^{-2\theta_1}$$

and

$$\overline{u}_1(x,t) = -i2A_1D_1/\overline{D}_1^2 = \frac{4\rho_1\sin 2\beta_1 e^{i3\beta_1}\left(1 + e^{-i2\beta_1}e^{-2\theta_1}\right)}{\left(1 + e^{i2\beta_1}e^{-2\theta_1}\right)^2}e^{-\theta_1}e^{i\varphi_1} \tag{75}$$

The complex conjugate of one-soliton solution $\overline{u}_1(x,t)$ in (75) is $u_1(x,t)$, which is just in conformity with that gotten from pure Marchenko formalism [24] (see the next section), up to a permitted global constant phase factor. In the case of two-soliton solution, $N = 2$, $\lambda_3 = -\lambda_1$, $\lambda_4 = -\lambda_2$ and

$$\lambda_1 = \rho_1 e^{i\beta_1} = \mu_1 + i\nu_1;\ \lambda_2 = \rho_2 e^{i\beta_2} = \mu_2 + i\nu_2 \tag{76}$$

$$\begin{aligned} c_{10} &= \frac{b_{10}}{\dot{a}(\lambda_1)} = b_{10}\frac{\lambda_1^2 - \overline{\lambda_1^2}}{2\lambda_1}\cdot\frac{\lambda_1^2 - \overline{\lambda_2^2}}{\lambda_1^2 - \lambda_2^2}\cdot\frac{\lambda_1^2}{\overline{\lambda_1^2}}\cdot\frac{\lambda_2^2}{\overline{\lambda_2^2}} \\ c_{20} &= \frac{b_{20}}{\dot{a}(\lambda_2)} = b_{20}\frac{\lambda_2^2 - \overline{\lambda_2}}{2\lambda_2}\cdot\frac{\lambda_2^2 - \overline{\lambda_1^2}}{\lambda_2^2 - \lambda_1^2}\cdot\frac{\lambda_1^2}{\overline{\lambda_1^2}}\cdot\frac{\lambda_2^2}{\overline{\lambda_2^2}} \end{aligned} \tag{77}$$

$$f_j^2 = 2c_{j0}e^{i2\lambda_j^2 x}e^{i4\lambda_j^4 t}, j = 1,2\ (C.f.(1.62)), b_{j0}e^{i2\lambda_j^2 x + i4\lambda_j^4 t} \equiv e^{-\theta_j}e^{i\varphi_j}, j = 1,2 \tag{78}$$

where $\theta_j = 4\mu_j\nu_j\left[x - x_{j0} + 4\left(\mu_j^2 - \nu_j^2\right)t\right]$

$$\varphi_j = 2\left(\mu_j^2 - \nu_j^2\right)x + \left[4\left(\mu_j^2 - \nu_j^2\right)^2 - 16\mu_j^2\nu_j^2\right]\cdot t + \alpha_{j0} \tag{79}$$

and b_{j0} is absorbed into the soliton center and the initial phase by

$$b_{j0} = e^{4\mu_j\nu_j x_{j0}}e^{i\alpha_{j0}}; j = 1,2 \tag{80}$$

And we get

$$\begin{aligned} A_2 &= \sum_{\substack{n_1 = 1,2 \\ m_1 = 0}} \Omega_1(n_1,0)\Omega_2(0,n_1) + \sum_{\substack{n_1 = 1, n_2 = 2 \\ m_1 = 0, m_2 = 1,2}} \Omega_1(n_1,n_2;0,m_2)\Omega_2(0,m_2;n_1,n_2) \\ &= \Omega_1(1;0)\Omega_2(0;1) + \Omega_1(2;0)\Omega_2(0;2) + \Omega_1(1,2;0,1)\Omega_2(0,1;1,2) + \Omega_1(1,2;0,2)\Omega_2(0,2;1,2) \\ &= f_1^2 + f_2^2 - |f_1|^4 f_2^2\frac{\left(\lambda_1^2 - \lambda_2^2\right)^2\overline{\lambda_1^2}}{\left(\lambda_1^2 - \overline{\lambda_1^2}\right)^2\left(\overline{\lambda_1^2} - \lambda_2^2\right)^2} - |f_2|^4 f_1^2\frac{\left(\lambda_1^2 - \lambda_2^2\right)^2\overline{\lambda_2^2}}{\left(\overline{\lambda_2^2} - \lambda_1^2\right)^2\left(\lambda_2^2 - \overline{\lambda_2^2}\right)^2} \\ &= \lambda_1\left(1 - e^{-i4\beta_1}\right)\frac{\lambda_1^2 - \overline{\lambda_2^2}}{\lambda_1^2 - \lambda_2^2}e^{i4(\beta_1+\beta_2)}e^{-\theta_1}e^{i\varphi_1} + \lambda_2\left(1 - e^{-i4\beta_2}\right)\frac{\overline{\lambda_1^2} - \lambda_2^2}{\lambda_1^2 - \lambda_2^2}e^{i4(\beta_1+\beta_2)}e^{-\theta_2}e^{i\varphi_2} \\ &+ \left[\lambda_1\left(1 - e^{-i4\beta_1}\right)e^{-i2\beta_2}\frac{\overline{\lambda_1^2} - \lambda_2^2}{\overline{\lambda_1^2} - \overline{\lambda_2^2}}e^{-\theta_2 - i\varphi_2} + \lambda_2\left(1 - e^{-i4\beta_2}\right)e^{-i2\beta_1}\frac{\lambda_1^2 - \overline{\lambda_2^2}}{\overline{\lambda_1^2} - \overline{\lambda_2^2}}e^{-\theta_1 - i\varphi_1}\right]\bullet e^{-(\theta_1+\theta_2)}e^{i(\varphi_1+\varphi_2)}e^{i4(\beta_1+\beta_2)} \\ &= i2\left|\frac{\lambda_1^2 - \overline{\lambda_2^2}}{\lambda_1^2 - \lambda_2^2}\right|[\rho_1\sin 2\beta_1 e^{i(\varphi-\alpha)}e^{i(3\beta_1+4\beta_2)}e^{-\theta_1+i\varphi_1} + \rho_2\sin 2\beta_2 e^{-i(\varphi+\alpha)}e^{i(4\beta_1+3\beta_2)}e^{-\theta_2+i\phi_2} \\ &+ \rho_1\sin 2\beta_1 e^{-i(\varphi-\alpha)}e^{i(3\beta_1+2\beta_2)}e^{-2\theta_2-\theta_1}\cdot e^{i\phi_1} + \rho_2\sin 2\beta_2 e^{i(\varphi+\alpha)}e^{i(2\beta_1+3\beta_2)}e^{-2\theta_1-\theta_2}\cdot e^{i\phi_2}] \end{aligned} \tag{81}$$

where

$$\begin{aligned} \varphi &= \arg\left(\lambda_1^2 - \overline{\lambda_2^2}\right) = \arctan\left(\rho_1^2\sin 2\beta_1 + \rho_2^2\sin 2\beta_2\right)/\left(\rho_1^2\cos 2\beta_1 - \rho_2^2\cos 2\beta_2\right) \\ \alpha &= \arg\left(\lambda_1^2 - \lambda_2^2\right) = \arctan\left(\rho_1^2\sin 2\beta_1 - \rho_2^2\sin 2\beta_2\right)/\left(\rho_1^2\cos 2\beta_1 - \rho_2^2\cos 2\beta_2\right) \end{aligned} \tag{82}$$

and

$$\begin{aligned}
\overline{D}_2 &= 1+\sum_{r=1}^{2}\sum_{1\le n_1<n_2\le 2}\sum_{1\le m_1<m_2\le 2} Q_1(n_1,\cdots,n_r;m_1,\cdots,m_r)Q_2(m_1,\cdots,m_r;n_1,\cdots,n_r)\\
&= 1+Q_1(n_1=1;m_1=1)Q_2(m_1=1,n_1=1)+Q_1(n_1=1;m_1=2)Q_2(m_1=2,n_1=1)\\
&+Q_1(n_1=2;m_1=1)Q_2(m_1=1,n_1=2)+Q_1(n_1=2;m_1=2)Q_2(m_1=2,n_1=2)\\
&+Q_1(n_1=1,n_1=2;m_1=1,m_2=2)Q_2(m_1=1,m_2=2;n_1=1,n_2=2)\\
&= 1-|f_1|^4\frac{\lambda_1^2}{\left(\lambda_1^2-\overline{\lambda}_1^2\right)^2}-|f_2|^4\frac{\lambda_2^2}{\left(\lambda_2^2-\overline{\lambda}_2^2\right)^2}-f_1^2\overline{f}_2^2\frac{\lambda_1^2}{\left(\lambda_1^2-\overline{\lambda}_2^2\right)^2}-\overline{f}_1^2 f_2^2\frac{\lambda_2^2}{\left(\overline{\lambda}_1^2-\lambda_2^2\right)^2}\\
&+|f_1f_2|^4\cdot\frac{\lambda_1^2\lambda_2^2\left(\overline{\lambda}_1^2-\overline{\lambda}_2^2\right)^2\left(\lambda_1^2-\lambda_2^2\right)^2}{\left(\lambda_1^2-\overline{\lambda}_1^2\right)^2\left(\lambda_1^2-\overline{\lambda}_2^2\right)^2\left(\lambda_2^2-\overline{\lambda}_1^2\right)^2\left(\lambda_2^2-\overline{\lambda}_2^2\right)^2}
\end{aligned} \tag{83}$$

$$\begin{aligned}
D_2 &= 1+\left|\frac{\lambda_1^2-\overline{\lambda}_2^2}{\lambda_1^2-\lambda_2^2}\right|^2\left(e^{-i2\beta_1}e^{-2\theta_1}+e^{-i2\beta_2}e^{-2\theta_2}\right)\\
&+\left(1-\left|\frac{\lambda_1^2-\overline{\lambda}_2^2}{\lambda_1^2-\lambda_2^2}\right|^2\right)e^{-(\theta_1+\theta_2)}e^{-i(\beta_1+\beta_2)}\left[\frac{\rho_1}{\rho_2}e^{i(\varphi_2-\varphi_1)}+\frac{\rho_2}{\rho_1}e^{i(\varphi_1-\varphi_2)}\right]\\
&+e^{-i2(\beta_1+\beta_2)}e^{-2(\theta_1+\theta_2)}
\end{aligned} \tag{84}$$

where

$$\left|\frac{\lambda_1^2-\overline{\lambda}_2^2}{\lambda_1^2-\lambda_2^2}\right|^2=\frac{(\rho_1/\rho_2-\rho_2/\rho_1)^2+4\sin^2(\beta_1+\beta_2)}{(\rho_1/\rho_2-\rho_2/\rho_1)^2+4\sin^2(\beta_1-\beta_2)} \tag{85}$$

Substituting (81) and (84) into formula (52), we thus get the two-soliton solution to the DNLS equation with VBC

$$\overline{u}_2=-i2A_2D_2/\overline{D}_2^2 \tag{86}$$

Once again we find that, up to a permitted global constant phase factor, the above two-soliton solution is equivalent to that gotten in Ref. [23, 24], verifying the validity of our formula of N-soliton solution and the reliability of those linear algebra techniques. As a matter of fact, a general and strict demonstration of our revised IST for DNLS equation with VBC has been given in one paper by use of Liouville theorem [25].

2.5 The asymptotic behaviors of N-soliton solution

The complex conjugate of expression (52) gives the explicit expression of N-soliton solution as

$$u_N=i2\overline{A}_N\overline{D}_N/D_N^2 \tag{87}$$

Without the loss of generality, for $\lambda_n=\mu_n+i\nu_n$, $V_n=-4\left(\mu_n^2-\nu_n^2\right)$, $n=1,2,\cdots,N,$ we assume $V_1<V_2<\cdots<V_n<\cdots V_N$ and define the n'th vicinity area as $\Gamma_n: x-x_{no}-V_nt\sim 0, (n=1.2,\cdots,N)$.

As $t\to-\infty$, N vicinity areas $\Gamma_n, n=1,2\ldots,N$, queue up in a descending series

$$\Gamma_N,\Gamma_{N-1},\cdots,\Gamma_1 \tag{88}$$

and in the vicinity of Γ_n, we have (note that $\kappa_j>0$)

$$\theta_j = 4\kappa_j(x - x_{j0} - V_jt) \rightarrow \begin{cases} +\infty, & \text{for } j > n \\ -\infty, & \text{for } j < n \end{cases} \tag{89}$$

Here the complex constant $2c_{n0}$ in expression (65) has been absorbed into $e^{-\theta_n}e^{i\varphi_n}$ by redefinition of the soliton center x_{n0} and the initial phase α_{n0}.

Introducing a typical factor $F_n = -e^{-2\theta_n}/\left(\lambda_n^2 - \overline{\lambda}_n^2\right)^2 > 0,\ n = 1, 2, \cdots, N$; then

$$D_n(1, 2, \cdots, n) = \prod_{j=1}^{n} \overline{\lambda_j^2} F_j \prod_{l<m} \left| \frac{\lambda_l^2 - \lambda_m^2}{\lambda_l^2 - \overline{\lambda}_m^2} \right|^4 \tag{90}$$

where $l, m \in \{1, 2, \cdots, n\}$. Thus

$$D \simeq D_{n-1}(1, 2, \cdots, n-1) + D_n(1, 2, \cdots, n) = \left(1 + \overline{\lambda_n^2} F_n \prod_{j=1}^{n-1} \left| \frac{\lambda_j^2 - \lambda_n^2}{\lambda_j^2 - \overline{\lambda}_n^2} \right|^4 \right) D_{n-1}(1, 2, \cdots, n-1) \tag{91}$$

and

$$\overline{A} \simeq \overline{A}_n(1, 2, \cdots, n; 0, 1, 2, \cdots n-1) = D_{n-1} e^{-\theta_n} e^{-i\varphi_n} \prod_{j=1}^{n-1} \frac{\left(\overline{\lambda_j^2} - \overline{\lambda_n^2}\right)^2}{\left(\lambda_j^2 - \overline{\lambda_n^2}\right)^2} e^{i4\beta_j} \tag{92}$$

In the vicinity of Γ_n,

$$u(x, t) = i2\overline{A}D/D^2 \simeq u_1\left(\theta_n + \Delta\theta_n^{(-)}, \varphi_n + \Delta\varphi_n^{(-)}\right) \tag{93}$$

Here

$$\Delta\theta_n^{(-)} = 2\sum_{j=1}^{n-1} \ln \left| \frac{\lambda_j^2 - \overline{\lambda_n^2}}{\lambda_j^2 - \lambda_n^2} \right|^2 \tag{94}$$

$$\begin{aligned} \Delta\varphi_n^{(-)} &= -\sum_{j=1}^{n-1} \left\{ \arg \left[\frac{\left(\overline{\lambda_j^2} - \overline{\lambda_n^2}\right)^2}{\left(\lambda_j^2 - \overline{\lambda_n^2}\right)^2} \right]^2 + 4\beta_j \right\} \\ &= 2\sum_{j=1}^{n-1} \left[\arg\left(\lambda_j^2 - \overline{\lambda_n^2}\right) - \arg\left(\overline{\lambda_j^2} - \overline{\lambda_n^2}\right) - 2\beta_j \right] \end{aligned} \tag{95}$$

then

$$u_N \simeq \sum_{n=1}^{N} u_1\left(\theta_n + \Delta\theta_n^{(-)}, \varphi_n + \Delta\varphi_n^{(-)}\right) \tag{96}$$

Each $u_1(\theta_n, \varphi_n)$, $(1, 2, \cdots, n)$ is a one-soliton solution characterized by one parameter λ_n, moving along the positive direction of the x-axis, queuing up in a series with descending order number n as in series (88). As $t \rightarrow \infty$, in the vicinity of Γ_n, we have (note that $\kappa_j > 0$)

$$\theta_j = 4\kappa_j(x - x_{j0} - V_j t) \to \begin{cases} -\infty, & \text{for } j > n \\ +\infty, & \text{for } j < n \end{cases} \quad D \simeq D_{N-n}(n+1, n+2, \cdots, N) + D_{N-n+1}(n, n+1, \cdots, N)$$

$$= \left(1 + \overline{\lambda_n^2} F_n \prod_{j=n+1}^{N} \left|\frac{\lambda_j^2 - \lambda_n^2}{\lambda_j^2 - \overline{\lambda_n^2}}\right|^4\right) D_{N-n}(n+1, n+2, \cdots, N) \tag{97}$$

$$\begin{aligned}\overline{A} &\simeq \overline{A}_{N-n+1}(n, n+1, \cdots, N; 0, n+1, n+2, \cdots, N) \\ &= D_{N-n}(n+1, n+2, \cdots, N) e^{-\theta_n} e^{-i\varphi_n} \prod_{j=n+1}^{N} \frac{\left(\overline{\lambda_j^2} - \overline{\lambda_n^2}\right)^2 \lambda_j^2}{\left(\lambda_j^2 - \overline{\lambda_n^2}\right)^2 \overline{\lambda_j^2}},\end{aligned} \tag{98}$$

So as $t \to \infty$, in the vicinity of Γ_n,

$$u = i2\overline{AD}/D^2 \simeq u_1\left(\theta_n + \Delta\theta_n^{(+)}, \varphi_n + \Delta\varphi_n^{(+)}\right) \tag{99}$$

$$\Delta\theta_n^{(+)} = 2 \sum_{j=n+1}^{N} \ln \left|\frac{\lambda_j^2 - \overline{\lambda_n^2}}{\lambda_j^2 - \lambda_n^2}\right|^2 \tag{100}$$

$$\begin{aligned}\Delta\varphi_n^{(+)} &= -\sum_{j=n+1}^{N} \left\{ \arg\left[\frac{\left(\overline{\lambda_j^2} - \overline{\lambda_n^2}\right)^2}{\left(\lambda_j^2 - \overline{\lambda_n^2}\right)^2}\right] + 4\beta_j \right\} \\ &= 2 \sum_{j=n+1}^{N} \left[\arg\left(\lambda_j^2 - \overline{\lambda_n^2}\right) - \arg\left(\overline{\lambda_j^2} - \overline{\lambda_n^2}\right) - 2\beta_j\right],\end{aligned} \tag{101}$$

then as $t \to \infty$,

$$u_N \simeq \sum_{n=1}^{N} u_1\left(\theta_n + \Delta\theta_n^{(+)}, \varphi_n + \Delta\varphi_n^{(+)}\right) \tag{102}$$

That is to say, the N-soliton solution can be viewed as N well-separated exact one- solitons, queuing up in a series with ascending order number n: $\Gamma_1, \Gamma_2, \cdots, \Gamma_N$. In the course going from $t \to -\infty$ to $t \to \infty$, the n'th one-soliton overtakes the solitons from the first to $n-1$'th and is overtaken by the solitons from $n+1$'th to N'th. In the meantime, due to collisions, the n'th soliton got a total forward shift $\Delta\theta_n^{(-)}/\kappa_n$ from exceeding those slower soliton from the first to $n-1$'th, and got a total backward shift $\Delta\theta_n^{(+)}/\kappa_n$ from being exceeded by those faster solitons from $n+1$'th to N'th, and just equals to the summation of shifts due to each collision between two solitons, together with a total phase shift $\Delta\varphi_n$, that is,

$$\Delta x_n = \left|\Delta\theta_n^{(+)} - \Delta\theta_n^{(-)}\right|/\kappa_n \tag{103}$$

$$\Delta\varphi_n = \Delta\varphi_n^{(+)} - \Delta\varphi_n^{(-)} \tag{104}$$

2.6 N-soliton solution to MNLS equation

Finally, we indicate that the exact N-soliton solution to the DNLS equation can be converted to that of MNLS equation by a gauge-like transformation. A nonlinear Schrödinger equation including the nonlinear dispersion term expressed as

$$i\partial_t v + \partial_{xx} v + i\alpha\partial_x\left(|v|^2 v\right) + 2\beta|v|^2 v = 0 \tag{105}$$

is also integrable [23] and called modified nonlinear Schrödinger (MNLS for brevity) equation. It is well known that MNLS equation well describes transmission of femtosecond pulses in optical fibers [4–6] and is related to DNLS equation by a gauge-like transformation [23] formulated as

$$v(x,t) = u(X,T)e^{i2\rho X + i4\rho^2 T} \tag{106}$$

with $x = \alpha^{-1}(X + 4\rho T)$, $t = \alpha^{-2}T$; $X = \alpha x - 4\beta t$, $T = \alpha^2 t$; $\rho = \beta\alpha^{-2}$. Using a method that is analogous to reference [16], and applying above gauge-like transformation to Eq. (105), the MNLS equation with VBC can be transformed into DNLS equation with VBC.

$$i\partial_T u + \partial_{XX} u + i\partial_X\left(|u|^2 u\right) = 0 \tag{107}$$

with $u = u(X,T)$. So according to (106), the N-soliton solution to MNLS equation can also be attained by a gauge-like transformation from that of DNLS equation.

The N-soliton solution to the DNLS equation with VBC has been derived by means of a IST considered anew and some special linear algebra techniques. The one- and two-soliton solutions have been given as two typical examples in illustration of the general formula of the N-soliton solution. It is found to be perfectly in agreement with that gotten in the following section based on a pure Marchenko formalism or Hirota's Bilinear derivative transformation [24, 26, 27]. The demonstration of the revised IST considered anew for DNLS equation with VBC has also been given by use of Liouville theorem [25].

The newly revised IST technique for DNLS equation with VBC supplies substantial foundation for its direct perturbation theory.

3. A simple method to derive and solve Marchenko equation for DNLS equation

Gel'fand-Levitan-Marchenko (GLM for brevity) equations can be viewed as an integral-transformed version of IST for those integrable nonlinear equations [21, 24, 28].

In this section, a simple method is used to derive and solve Marchenko equation (or GLM equation) for DNLS E with VBC [28]. Firstly, starting from the first Lax equation, we derive two conditions to be satisfied by the kernel matrix $N(x,y)$ of GLM by applying the Lax operator $\partial_x - L$ upon the integral representation of Jost function for DNLSE. Secondly, based on Lax equation, a strict demonstration has been given for the validness of Marchenko formalism. At last, the Marchenko formalism is determined by choosing a suitable $F(x+y)$ and $G(x+y)$, and their relation (135) has been constructed. The one and multi-soliton solution in the reflectionless case is attained based upon a pure Marchenko formalism by avoiding direct use of inverse scattering data and verified by using direct substitution method with Mathematica.

3.1 The lax pair and its Jost functions of DNLS equation

DNLS equation is usual expressed as

$$iu_t + u_{xx} + i\left(|u|^2 u\right)_x = 0 \tag{108}$$

with vanishing boundary, $|x| \to \infty$, $u \to 0$. Here the subscript denotes partial derivative. Its Lax pair is given by

$$L = -i\lambda^2\sigma_3 + \lambda U, U = \begin{pmatrix} 0 & u \\ -\overline{u} & 0 \end{pmatrix} \tag{109}$$

$$M = -i2\lambda^4\sigma_3 + 2\lambda^3 U - i\lambda^2 U^2\sigma_3 - \lambda\left(-U^3 + iU_x\sigma_3\right) \tag{110}$$

The first Lax equation is

$$\partial_x f(x,\lambda) = L(x,\lambda) f(x,\lambda) \tag{111}$$

In the case of $|x| \to \infty$, $u \to 0$, $L \to L_0 = -i\lambda^2\sigma_3$, the free Jost solution is

$$E(x,\lambda) = e^{-i\lambda^2 x\sigma_3}, E_{\bullet 1}(x,\lambda) = \begin{pmatrix} 1 \\ 0 \end{pmatrix} e^{-i\lambda^2 x}; E_{\bullet 2}(x,\lambda) = \begin{pmatrix} 0 \\ 1 \end{pmatrix} e^{i\lambda^2 x} \tag{112}$$

where λ^2 is a real squared parameter, $E(x,\lambda)$ expresses two independent solutions with two components. The Jost solutions of (4) are defined by their asymptotic properties at $x \to \pm\infty$,

$$\Psi(x,\lambda) = (\tilde{\psi}(x,\lambda), \psi(x,\lambda)) \to E(x,\lambda), \quad as \ \ x \to \infty \tag{113}$$

$$\Phi(x,\lambda) = (\varphi(x,\lambda), \tilde{\varphi}(x,\lambda)) \to E(x,\lambda), \quad as \ \ x \to -\infty \tag{114}$$

3.2 The integral representation of Jost function

As usual, we introduce the integral representation,

$$\Psi(x,\lambda) = E(x,\lambda) + \int_x^\infty \mathrm{d}y \left\{\lambda^2 N^d(x,y) + \lambda N^o(x,y)\right\} E(y,\lambda) \tag{115}$$

where the superscripts d and o mean the diagonal and off-diagonal elements, respectively. According to the conventional operation in IST, the time variable is suppressed temporarily. Here

$$N^d(x,y) = \begin{pmatrix} N(x,y)_{11} & 0 \\ 0 & N(x,y)_{22} \end{pmatrix}, N^o(x,y) = \begin{pmatrix} 0 & N(x,y)_{12} \\ N(x,y)_{21} & 0 \end{pmatrix}$$

Due to the symmetry of the first Lax operator $\lambda^2(-i\sigma_3)_{11} = \lambda^2\overline{(-i\sigma_3)_{22}}$ and $\lambda U_{21} = -\lambda\overline{U}_{12}$, the kernel matrix $N(x,y)$ of the integral representation of Jost function should have the same symmetry as follows:

$$\lambda^2 N^d(x,y)_{11} = \lambda^2\overline{N^d(x,y)_{22}}; \lambda N^o(x,y)_{21} = -\lambda\overline{N^o(x,y)_{12}} \tag{116}$$

Substitute Eq. (115) into the first Lax Eq. (111). By simply partial integration, we have the following terms:

$$\{\partial_x - L\}E(x,\lambda) = -(L-L_0)E(x,\lambda) = -\lambda U(x)E(x,\lambda) \tag{117}$$

$$\begin{aligned}&\partial_x \int_x^\infty \mathrm{d}y\{\lambda^2 N^d(x,y)+\lambda N^o(x,y)\}E(y,\lambda)\\&= -\left[\lambda^2 N^d(x,x)+\lambda N^o(x,x)\right]E(x,\lambda)+\int_x^\infty \mathrm{d}y\left[\lambda^2 N_x^d(x,y)+\lambda N_x^o(x,y)\right]E(y,\lambda)\end{aligned} \tag{118}$$

$$\begin{aligned}&-(-i\lambda^2\sigma_3)\int_x \mathrm{d}y\left[\lambda^2 N^d(x,y)+\lambda N^\circ(x,y)\right]E(y,\lambda) = -\int_x \mathrm{d}y\sigma_3\left[\lambda^2 N^d(x,y)+\lambda N^o(x,y)\right]\sigma_3 E'(y,\lambda)\\&= \sigma_3\left[\lambda^2 N^d(x,x)+\lambda N^o(x,x)\right]\sigma_3 E(x,\lambda)+\int_x^\infty \mathrm{d}y\sigma_3\left[\lambda^2 N_y^d(x,y)+\lambda N_y^o(x,y)\right]\sigma_3 E(y,\lambda)\end{aligned} \tag{119}$$

and

$$\begin{aligned}&-\lambda U(x)\int_x^\infty \mathrm{d}y\{\lambda^2 N^d(x,y)\}E(y,\lambda) = -\int_x^\infty \mathrm{d}y\lambda U(x)N^d(x,y)i\sigma_3 E'(y,\lambda)\\&= i\lambda U(x)N^d(x,x)\sigma_3 E(x,\lambda)+i\int_x^\infty \mathrm{d}y\lambda U(x)N_y^d(x,y)\sigma_3 E(y,\lambda)\end{aligned} \tag{120}$$

Use is made of that $-i\lambda^2\sigma_3 E(y,\lambda) = E'(y,\lambda)$, then

$$-\lambda U(x)\int_x^\infty \mathrm{d}yN^o(x,y) \equiv -\int_x^\infty \mathrm{d}y\lambda U(x)N^o(x,y) \tag{121}$$

According to equation $(\partial_x - L)\Psi(x,\lambda) = 0$, adding up the l.h.s. and r.h.s., respectively, of Eq. (117)–(120), (121). We obtain two equations involving with terms λ^2 and λ outside of the integral $\int \mathrm{d}y\cdots$ as follows:

$$\lambda^2 : -N^d(x,x)+\sigma_3 N^d(x,x)\sigma_3 = 0 \tag{122}$$

$$\lambda^1 : -U(x)-N^o(x,x)+\sigma_3 N^o(x,x)\sigma_3+iU(x)N^d(x,x)\sigma_3 = 0 \tag{123}$$

Or

$$U_{12} = u(x) = -2\frac{N_{12}(x,x)}{1+i\overline{N}_{11}(x,x)}, \tag{124}$$

and the equations in the integral $\int \mathrm{d}y\{S\} = 0$, where $\{S\}$ is equal to

$$\left[\lambda^2 N_x^d(x,y)+\lambda N_x^o(x,y)\right]+\sigma_3\left[\lambda^2 N_y^d(x,y)+\lambda N_y^o(x,y)\right]\sigma_3-\lambda U(x)\left[-iN_y^d(x,y)\sigma_3+\lambda N^o(x,y)\right] = 0 \tag{125}$$

Therefore, Eq. (125) gives two conditions to be satisfied by the kernel matrix $N(x,y)$ in the integral representation of Jost solution

$$\lambda^2 \text{ terms} : A(x,y) \equiv N_x^d(x,y)+\sigma_3 N_y^d(x,y)\sigma_3-U(x)N^o(x,y) = 0 \tag{126}$$

$$\lambda^1 \text{ terms} : B(x,y) \equiv N_x^o(x,y)+\sigma_3 N_y^o(x,y)\sigma_3+iU(x)N_y^d(x,y)\sigma_3 = 0 \tag{127}$$

Since (122) is an identity, Eq. (123) or (124) gives the solution $U(x)$ or $u(x)$ in terms of $N(x,x)$, thus the first Lax equation gives two conditions (126) and (127) which should be satisfied by the integral kernel $N(x,y)$. Note that the time variable of $u(x)$ in (124) is suppressed temporarily.

3.3 Marchenko equation for DNLSE and its demonstration

In Eq. (115), the $N^d(x,y)$ and $N^o(x,y)$ appear in different manner, we assume the form of Marchenko equation for DNLSE with VBC is

$$N^d(x,y) + \int_x^\infty dz N^o(x,z)F(z+y) = 0 \tag{128}$$

$$N^o(x,y) + F(x+y) + \int_x^\infty dz N^d(x,z)G(z+y) = 0 \tag{129}$$

where $F(x+y)$ is only with off-diagonal terms. $G(x+y)$ is considered as another function with only off-diagonal terms. We notice that the Marchenko equation needn't involve obviously the function of spectral parameter λ.

We now show the kernel $N(x,y)$ determined by (128) and (129) indeed satisfy the conditions (126) and (127) as long as we choose a suitable form of expression for $G(x+y)$.

Making partial derivation in (128) with respect to x and y, respectively, we obtain

$$N_x^d(x,y) - N^o(x,x)F(x+y) + \int_x^\infty dz N_x^o(x,z)F(z+y) = 0 \tag{130}$$

$$N_y^d(x,y) + \int_x^\infty dz N^o(x,z)F'(z+y) = 0 \tag{131}$$

By partial integrating, Eq. (131) becomes

$$N_y^d(x,y) - N^o(x,x)F(x+y) - \int_x^\infty dz N_z^o(x,z)F(z+y) = 0 \tag{132}$$

Use is made of the fact that $F'(z+y) = F_y(z+y) = F_z(z+y)$ in (131). Making a weighing summation as follows:

$$\{l.h.s. \quad of \quad (20)\} + \sigma_3\{l.h.s. \quad of \quad (22)\}\sigma_3 - U(x)\cdot\{l.h.s. \quad of \quad (19)\} = 0$$

We find

$$\begin{aligned} &N_x^d(x,y) + \sigma_3 N_y^d(x,y)\sigma_3 - U(x)N^o(x,y) - N^o(x,x)F(x+y) - \sigma_3 N^o(x,x)F(x+y)\sigma_3 - U(x)F(x+y) \\ &+ \int_x^\infty dz\{N_x^o(x,z)F(z+y) - \sigma_3 N_z^o(x,z)F(z+y)\sigma_3 - U(x)N^d(x,z)G(z+y)\} = 0 \end{aligned} \tag{133}$$

Since $F(x)$ is off-diagonal, $F(x)\sigma_3 = -\sigma_3 F(x)$. Thus the terms involving with $F(x+y)$ outside of integral are equal to $-iU(x)N^d(x,x)\sigma_3 F(x+y)$ by use of Eq. (123). Then (133) can be rewritten as

$$\begin{aligned} &A(x,y) - iU(x)N^d(x,x)\sigma_3 F(x+y) \\ &+ \int_x^\infty dz\left[B(x,z)F(z+y) - iU(x)N_z^d(x,z)\sigma_3 F(z+y) - U(x)N^d(x,z)G(z+y)\right] = 0 \end{aligned} \tag{134}$$

If we choose

$$G(z+y) = i\sigma_3 F'(z+y), \tag{135}$$

then

$$iU(x)N^d(x,x)\sigma_3F(x+y)+\int_z^\infty dz\{iU(x)N_z^d(x,z)\sigma_3F(z+y)+U(x)N^d(x,z)i\sigma_3F'(z+y)\}=0 \tag{136}$$

Thus, Eq. (134) becomes

$$A(x,y)+\int_x^\infty dzB(x,z)F(z+y)=0 \tag{137}$$

Now substituting (135) into (129), we find

$$N^o(x,y)+F(x+y)+\int_x^\infty dzN^d(x,z)i\sigma_3F'(z+y)=0 \tag{138}$$

Making partial derivation with respect to x and y, respectively, on the l.h.s. of Eq. (138), we have

$$N_x^o(x,y)+F'(x+y)-N^d(x,x)i\sigma_3F'(x+y)+\int_x^\infty dzN_x^d(x,z)i\sigma_3F'(z+y)=0 \tag{139}$$

$$N_y^o(x,y)+F'(x+y)+\int_x^\infty dzN^d(x,z)i\sigma_3F''(z+y)=0 \tag{140}$$

or

$$N_y^o(x,y)+F'(x+y)-N^d(x,x)i\sigma_3F'(x+y)-\int_x^\infty dzN_z^d(x,x)i\sigma_3F'(z+y)=0 \tag{141}$$

Now we make a weighing summation as

$$\{l.h.s.\ of\ (29)\}+\sigma_3\{l.h.s.\ of\ (31)\}\sigma_3+iU(x)\{l.h.s.\ of\ (21)\}\sigma_3=0$$

Hence, we have

$$\begin{aligned}&N_x^o(x,y)+\sigma_3N_y^o(x,y)\sigma_3+iU(x)N_y^d(x,y)\sigma_3\\&+F'(x+y)+\sigma_3F'(x+y)\sigma_3-N^d(x,x)i\sigma_3F'(x+y)-\sigma_3N^d(x,x)i\sigma_3F'(x+y)\sigma_3\\&+\int_x^\infty dz\{N_x^d(x,z)i\sigma_3F'(z+y)-\sigma_3N_z^d(x,z)i\sigma_3F'(z+y)\sigma_3+iU(x)N^o(x,z)F'(z+y)\sigma_3\}=0\end{aligned} \tag{142}$$

Noticing $F(x)\sigma_3=-\sigma_3F(x)$, Eq. (142) becomes

$$B(x,y)+\int_x^\infty dzA(x,z)i\sigma_3F'(z+y)=0 \tag{143}$$

We find that, as long as we choose a suitable form for $G(x+y)$ as well as $F(x+y)$ according to Eq. (135), Eq. (128) and (129) will just satisfy the two conditions (126) and (127) derived from the first Lax Eq. (111). On the other hand, owing to the symmetry properties of $N^o(x,y)$ and $N^d(x,y)$, the function $f(x+y)$ in (128) and (129) can only has off-diagonal elements, we write

$$F(x+y) = \begin{pmatrix} 0 & -\overline{f(x+y)} \\ f(x+y) & 0 \end{pmatrix}; G(z+y) = \begin{pmatrix} 0 & -\overline{h(z+y)} \\ h(z+y) & 0 \end{pmatrix} = i\sigma_3 F'(z+y) \tag{144}$$

Considering the dependence of the Jost solutions on the squared spectral parameter λ^2, in the reflectionless case, we choose

$$f(x+y) = \sum_{n=1}^{N} C_n(t) e^{i\lambda_n^2(x+y)} \tag{145}$$

where $C_n(t)$ contains a time-dependent factor $e^{i4\lambda_n^4 t}$, which can be introduced by a standard procedure [29], due to a fact of the Lax operator $M \to -i2\lambda^4\sigma_3$ as $x \to \pm\infty$.

As is well known, Lax equations are linear equation so that a constant factor can be introduced in its solution, that is, $C_n = e^{\beta_n + i\alpha_n} e^{i4\lambda_n^4 t}$. It means that β_n is related to the center of soliton and α_n expresses the initial phase up to a constant factor. Thus, the time-independent part of C_n is inessential and can be absorbed or normalized only by redefinition of the soliton center and initial phase. On the other hand, notice the terms generated by partial integral in (133)–(142), in order to ensure the convergence of the partial integral, we must let $\lim_{x\to\infty} e^{i\lambda_n^2 x} = 0$, so we only consider the N zero points of $a(\lambda)$ in the first quadrant of complex plane of λ (also in the upper half part of the complex plane of λ^2), that is, the discrete spectrum for λ_1, λ_2, $\cdots\cdots\lambda_N$, although $-\lambda_n$, $(n = 12, \cdots, N)$ in the third quadrant of the complex plane of λ are also the zero points of $a(\lambda)$ due to symmetry of Lax operator and transition matrix. Then Eq. (145) corresponds to the N-soliton solution in the reflectionless case, and we have completed the derivation and manifestation of Marchenko equation (128) and (129), (144), and (145) for DNLSE with VBC.

3.4 A multi-soliton solution of the DNLS equation based upon pure Marchenko formalism

When there are N simple poles $\lambda_1, \lambda_2, \cdots, \lambda_N$ in the first quadrant of the complex plane of λ, the Marchenko equation will give a N-solition solution to the DNLS equation with VBC in the reflectionless case. We can assume that

$$f(x+y) = F_{21}(x+y) = \sum_{n=1}^{N} g_n(x,t) h_n(y) \equiv G(x,t) H(y)^{\mathrm{T}} \tag{146}$$

where $g_n(x,t) \equiv C_n(t) e^{i\lambda_n^2 x}$, $h_n(y) \equiv e^{i\lambda_n^2 y}$, $n = 1, 2, \cdots, N$, and

$$G(x,t) \equiv (g_1(x,t), g_2(x,t), \cdots, g_N(x,t))_; \; H(y)^{\mathrm{T}} \equiv (h_1(y), h_2(y), \cdots, h_N(y))^{\mathrm{T}} \tag{147}$$

Here and hereafter the superscript T represents transposing of a matrix. On the other hand, we assume that

$$N_{11}(x,y) = N_{11}(x) H(y)^{\mathrm{T}}, N_{12}(x,y) = N_{12}(x) \overline{H}(y)^{\mathrm{T}} \tag{148}$$

Then

$$F_{12}(x+y) = -\overline{F}_{12}(x+y) = -\overline{G}(x)\overline{H}(y)^{\mathrm{T}};\ F'_{12}(x+y) = i\overline{\lambda_n^2}\overline{C}_n e^{-i\overline{\lambda_n^2}(x+y)} = -\overline{G'(x)}\overline{H}(y)^{\mathrm{T}} \tag{149}$$

Substituting (146)–(149) into the Marchenko equation (128) and (129), we have

$$\begin{cases} N_{11}(x)\mathrm{H}(y)^{\mathrm{T}} + N_{12}(x)\int_x^{\infty}\mathrm{d}z\overline{\mathrm{H}}(z)^{\mathrm{T}}\mathrm{G}(z)\mathrm{H}(y)^{\mathrm{T}} = 0 \\ N_{12}(x)\overline{\mathrm{H}}(y)^{\mathrm{T}} - \overline{\mathrm{G}}(x)\overline{\mathrm{H}}(y)^{\mathrm{T}} - iN_{11}(x)\int_x^{\infty}\mathrm{d}z\mathrm{H}(z)^{\mathrm{T}}\overline{\mathrm{G}}'(z)\overline{\mathrm{H}}(y)^{\mathrm{T}} = 0 \end{cases} \tag{150}$$

or

$$N_{11}(x) + N_{12}(x)\Delta_1(x) = 0, \tag{151}$$

$$N_{12}(x) - iN_{11}(x)\Delta_2(x) = \overline{G}(x) \tag{152}$$

here

$$\Delta_1(x) = \int_x^{\infty}\overline{\mathrm{H}}(z)^{T}\mathrm{G}(z)\mathrm{d}z,\ \Delta_2(x) = \int_x^{\infty}\mathrm{H}(z)^{\mathrm{T}}\overline{\mathrm{G}}'(z)\mathrm{d}z \tag{153}$$

Both of them are $N \times N$ matrices and their matrix element are, respectively, expressed as

$$\Delta_1(x)_{mn} = \int_x^{\infty} e^{-i\overline{\lambda_m^2}z}C_n e^{i\lambda_n^2 z}\mathrm{d}z = \overline{h}_m(x)\frac{-i}{\overline{\lambda_m^2} - \lambda_n^2}g_n(x) \tag{154}$$

$$\Delta_2(x)_{mn} = \int_x^{\infty} e^{-i\lambda_m^2 z}\overline{C}_n\left(-i\overline{\lambda_n^2}\right)e^{-i\overline{\lambda_n^2}z}\mathrm{d}z = h_m(x)\frac{\overline{\lambda_n^2}}{\lambda_m^2 - \overline{\lambda_n^2}}\overline{g}_n(x) \tag{155}$$

From (151) and (152), we immediately get

$$N_{11}(x) = -\overline{\mathrm{G}}(x)[1 + i\Delta_1(x)\Delta_2(x)]^{-1}\Delta_1(x) \tag{156}$$

$$N_{12}(x) = \overline{\mathrm{G}}(x)[1 + i\Delta_1(x)\Delta_2(x)]^{-1} \tag{157}$$

from (148), (156), and (157), we have

$$N_{11}(x,y) = -\overline{G}(x)[\mathrm{I} + i\Delta_1(x)\Delta_2(x)]^{-1}\Delta_1(x)H(y)^{\mathrm{T}} \tag{158}$$

$$N_{12}(x,y) = \overline{\mathrm{G}}(x)[\mathrm{I} + i\Delta_1(x)\Delta_2(x)]^{-1}\overline{\mathrm{H}}(y)^{\mathrm{T}} \tag{159}$$

then

$$\begin{aligned} N_{11}(x,x) &= i\mathrm{Tr}\left\{i\Delta_1(x)\mathrm{H}(x)^{\mathrm{T}}\overline{\mathrm{G}}(x)[\mathrm{I} + i\Delta_1(x)\Delta_2(x)]^{-1}\right\} \\ &= i\left\{\frac{\det\left[\mathrm{I} + i\Delta_1(x)\Delta_2(x) + i\Delta_1(x)\mathrm{H}(x)^{\mathrm{T}}\overline{\mathrm{G}}(x)\right]}{\det[\mathrm{I} + i\Delta_1(x)\Delta_2(x)]} - 1\right\} \end{aligned} \tag{160}$$

and

$$N_{12}(x,x) = \mathrm{Tr}\left\{\overline{\mathrm{H}}(x)^{\mathrm{T}}\overline{\mathrm{G}}(x)[\mathrm{I} + i\Delta_1(x)\Delta_2(x)]^{-1}\right\} = \frac{\det\left[\mathrm{I} + \mathrm{i}\Delta_1(\mathrm{x})\Delta_2(\mathrm{x}) + \overline{\mathrm{H}}^{\mathrm{T}}(\mathrm{x})\overline{\mathrm{G}}(\mathrm{x})\right]}{\det[\mathrm{I} + \mathrm{i}\Delta_1(\mathrm{x})\Delta_2(\mathrm{x})]} - 1 \tag{161}$$

Substituting (160) and (161) into Eq. (124), we thus attain the N-soliton solution as follows in a pure Marchenko formalism.

$$u_N(x,t) = -2\frac{\det\left(\mathrm{I} + i\Delta_1\Delta_2 + \overline{\mathrm{H}}^{\mathrm{T}}\overline{\mathrm{G}}\right) - \det(\mathrm{I} + i\Delta_1\Delta_2)}{\det\left(\mathrm{I} - i\overline{\Delta}_1\overline{\Delta}_2 - i\overline{\Delta}_1\overline{\mathrm{H}}^{\mathrm{T}}\mathrm{G}\right)} \cdot \frac{\det(\mathrm{I} - i\overline{\Delta}_1\overline{\Delta}_2)}{\det(\mathrm{I} + i\Delta_1\Delta_2)} = -2C\overline{D}/D^2 \tag{162}$$

where

$$D \equiv \det(\mathrm{I} + i\Delta_1\Delta_2), C \equiv \det\left(\mathrm{I} + i\Delta_1\Delta_2 + \overline{\mathrm{H}}^{\mathrm{T}}\overline{\mathrm{G}}\right) - \det(\mathrm{I} + i\Delta_1\Delta_2), \tag{163}$$

and we will prove that in (136)

$$\det\left(\mathrm{I} - i\overline{\Delta}_1\overline{\Delta}_2 - i\overline{\Delta}_1\overline{\mathrm{H}}^{\mathrm{T}}\mathrm{G}\right) = \det(\mathrm{I} + i\Delta_1\Delta_2) \equiv D \tag{164}$$

By means of some linear algebraic techniques, especially the Binet-Cauchy formula for some special matrices (see the Appendices 2–3 in Part2), the determinant D and C can be expanded explicitly as a summation of all possible principal minors. Firstly, we can prove identity (164) by means of Binet-Cauchy formula.

$$\det\left(I - i\overline{\Delta}_1\overline{\Delta}_2 - i\overline{\Delta}_1\overline{H}^{\mathrm{T}}G\right) = \det\left[I + \left(-i\overline{\Delta}_1\right)\left(\overline{\Delta}_2 + \overline{H}^{\mathrm{T}}G\right)\right] := \det(I + \mathrm{M}_1\mathrm{M}_2) \tag{165}$$

where

$$(\mathrm{M}_1)_{nm} \equiv \left(-i\overline{\Delta}_1\right)_{mn} = h_n \frac{1}{\lambda_n^2 - \overline{\lambda_m^2}} \overline{g}_m, (\mathrm{M}_2)_{mn} \equiv \left(\overline{\Delta}_2 + \overline{H}^{\mathrm{T}}G\right)_{mn} = \overline{h}_m \frac{\overline{\lambda_m^2}}{\overline{\lambda_m^2} - \lambda_n^2} g_n \tag{166}$$

The complex constant factor c_{n0} can be absorbed into the soliton center and initial phase by redefining

$$g_n(x,t)h_n(x) = C_n(t)e^{i2\lambda_n^2 x} = c_{no}e^{i4\lambda_n^4 t}e^{i2\lambda_n^2 x} \equiv e^{-\theta_n}e^{i\varphi_n} \tag{167}$$

here $\lambda_n = \mu_n + i\nu_n$, and

$$\begin{aligned}\theta_n &\equiv 4\mu_n\nu_n\left[x - x_{n0} + 4\left(\mu_n^2 - \nu_n^2\right)t\right] = 4\kappa_n(x - x_{n0} - \upsilon_n t); \kappa_n = 4\mu_n\nu_n; \upsilon_{\mathrm{n}} \\ &= -4\left(\mu_n^2 - \nu_n^2\right); \ \varphi_n \equiv 2\left(\mu_n^2 - \nu_n^2\right)x + \left[4\left(\mu_n^2 - \nu_n^2\right)^2 - 16\mu_n^2\nu_n^2\right] \bullet t + \alpha_{no}; c_{n0} \\ &\equiv e^{4\kappa_n x_{no}}e^{i\alpha_{no}}; n = 1, 2, \cdots, N\end{aligned} \tag{168}$$

$$\det\left(I - i\overline{\Delta}_1\overline{\Delta}_2 - i\overline{\Delta}_1\overline{H}^{\mathrm{T}}G\right) = 1 + \sum_{r=1}^{N} \sum_{1 \le n_1 < n_2 < \cdots < n_r \le N} \sum_{1 \le m_1 < m_2 < \cdots < m_r \le N} \mathrm{M}_1(n_1, n_2, \cdots, n_r; m_1, m_2, \cdots m_r)\mathrm{M}_2(m_1, m_2, \cdots, m_r, n_1, n_2, \cdots, n_r) \tag{169}$$

where $\mathrm{M}_1(n_1, n_2, \cdots, n_r; m_1, m_2, \cdots m_r)$ denotes a minor, which is the determinant of a submatrix of M_1, consisting of elements belonging to not only $(n_1, n_2, \ldots, n_r)$ rows but also columns $(m_1, m_2, \ldots, m_r)$.

$$\mathrm{M}_1(n_1, n_2, \cdots, n_r; m_1, m_2, \cdots m_r) = \prod_{n,m} \frac{h_n \overline{g}_m}{\lambda_n^2 - \overline{\lambda_m^2}} \prod_{n<n', m<m'} \left(\lambda_n^2 - \lambda_{n'}^2\right)\left(\overline{\lambda_{m'}^2} - \overline{\lambda_m^2}\right) \tag{170}$$

$$\mathrm{M}_2(m_1, m_2, \cdots, m_r; n_1, n_2, \cdots n_r) = \prod_{n,m} \frac{\overline{h}_m g_n \overline{\lambda_m^2}}{\overline{\lambda_m^2} - \lambda_n^2} \prod_{n<n', m<m'} \left(\overline{\lambda_m^2} - \overline{\lambda_{m'}^2}\right)\left(\lambda_{n'}^2 - \lambda_n^2\right) \tag{171}$$

where $n, n' \in \{n_1, n_2, \cdots, n_r\}$, $m, m' \in \{m_1, m_2, \cdots, m_r\}$, then $\mathrm{M}_1(n_1, n_2, \cdots, n_r; m_1, m_2, \cdots m_r)\mathrm{M}_2(m_1, m_2, \cdots, m_r; n_1, n_2, \cdots n_r)$

$$= (-1)^r \prod_{n,m} \frac{e^{-\theta_n} e^{i\varphi_n} e^{-\theta_m} e^{-i\varphi_m} \overline{\lambda_m^2}}{\left(\lambda_n^2 - \overline{\lambda_m^2}\right)^2} \prod_{n<n', m<m'} \left(\overline{\lambda_m^2} - \overline{\lambda_{m'}^2}\right)^2 \left(\lambda_n^2 - \lambda_{n'}^2\right)^2 \tag{172}$$

If we define matrices $Q_1 = i\Delta_1$ and $Q_2 = \Delta_2$, then we can similarly attain

$$D = \det(\mathrm{I} + i\Delta_1\Delta_2) = \det(\mathrm{I} + \mathrm{Q}_1\mathrm{Q}_2) = 1 + \sum_{r=1}^{N} \sum_{1 \le n_1 < n_2 < \cdots < n_r \le N} \sum_{1 \le m_1 < m_2 < \cdots < m_r \le N} Q_1(n_1, n_2, \cdots, n_r; m_1, m_2, \cdots m_r) Q_2(m_1, m_2, \cdots, m_r; n_1, n_2, \cdots, n_r) \tag{173}$$

and

$$\mathrm{Q}_1(n_1, n_2, \cdots, n_r; m_1, m_2, \cdots m_r)\mathrm{Q}_2(m_1, m_2, \cdots, m_r; n_1, n_2, \cdots, n_r)$$
$$= (-1)^r \prod_{n,m} \frac{e^{-\theta_m} e^{i\varphi_m} e^{-\theta_n} e^{-i\varphi_n} \overline{\lambda_n^2}}{\left(\lambda_m^2 - \overline{\lambda_n^2}\right)^2} \prod_{n<n', m<m'} \left(\lambda_m^2 - \lambda_{m'}^2\right)^2 \left(\overline{\lambda_n^2} - \overline{\lambda_{n'}^2}\right)^2 \tag{174}$$

where $n, n' \in \{n_1, n_2, \cdots, n_r\}$, $m, m' \in \{m_1, m_2, \cdots, m_r\}$. Comparing (172) and (174), we find the following permutation symmetry between them

$$\mathrm{M}_1(n_1, n_2, \cdots, n_r; m_1, m_2, \cdots m_r)\mathrm{M}_2(m_1, m_2, \cdots, m_r; n_1, n_2, \cdots, n_r)$$
$$= \mathrm{Q}_1(m_1, m_2, \cdots, m_r; n_1, n_2, \cdots n_r)\mathrm{Q}_2(n_1, n_2, \cdots, n_r; m_1, m_2, \cdots, m_r)$$

Using above identity, comparing (169), (172), (173), and (174), we find that identity (164) holds and complete the computation of D.

Secondly, we compute the most complicate determinant C in (163). In order to calculate $\det\left(I + i\Delta_1\Delta_2 + \overline{H}^T\overline{G}\right)$, we introduce an $N \times (N+1)$ matrix Ω_1 and an $(N+1) \times N$ matrix Ω_2

$$(\Omega_1)_{nm} = (i\Delta_1)_{nm}, (\Omega_1)_{n0} = \overline{h}_n = \frac{\overline{h}_n \overline{\lambda_n^2}}{\lambda_n^2 - 0^2}; (\Omega_2)_{mn} = (\Delta_2)_{mn},$$
$$(\Omega_2)_{0n} = \overline{g}_n = \frac{-\overline{\lambda_n^2}\overline{g}_n}{0^2 - \overline{\lambda_n^2}} \tag{175}$$

with $n, m = 1, 2, \ldots, N$. We thus have

$$\det(I+\Omega_1\Omega_2) = 1 + \sum_{r=1}^{N} \sum_{1\le n_1\le n_2<\cdots<n_r\le N} \sum_{0\le m_1<m_2<\cdots<m_r\le N} \Omega_1(n_1,n_2,\cdots,n_r;m_1,m_2,\cdots m_r)\Omega_2(m_1,m_2,\cdots,m_r;n_1,n_2,\cdots,n_r) \tag{176}$$

The above summation obviously can be decomposed into two parts: one is extended to $m_1 = 0$ and the other extended to $m_1 \ge 1$. Subtracted from (176), the part that is extended to $m_1 \ge 1$, the remaining parts of (176) is just C in Eq. (163) (with $m_1 = 0, m_2 \ge 1$). Due to (175), we have

$$\begin{aligned} C &= \det(I+\Omega_1\Omega_2) - \det(I+i\Delta_1\Delta_2) \\ &= \sum_{r=1}^{N} \sum_{1\le n_1<n_2<\cdots<n_r\le N} \sum_{1\le m_2<m_3<\cdots<m_r\le N} \Omega_1(n_1,n_2,\cdots,n_r;0,m_2,\cdots,m_r)\Omega_2(0,m_2,\cdots,m_r;n_1,n_2,\cdots,n_r) \end{aligned} \tag{177}$$

$$\Omega_1(n_1,n_2,\cdots,n_r;0,m_2,\cdots m_r) = \prod_n \bar{h}_n \prod_m g_m \prod_{n<n',m<m'} \left(\overline{\lambda_n^2} - \overline{\lambda_{n'}^2}\right)\left(\lambda_{m'}^2 - \lambda_m^2\right) \prod_{n,m} \frac{1}{\overline{\lambda_n^2} - \lambda_m^2} \tag{178}$$

$$\begin{aligned} \Omega_2(0,m_2,\cdots,m_r;n_1,n_2,\cdots n_r) = &\prod_n \bar{g}_n \prod_m h_m \prod_{n<n',m<m'} \left(\overline{\lambda_n^2} - \overline{\lambda_{n'}^2}\right)\left(\lambda_{m'}^2 - \lambda_m^2\right) \prod_{n,m} \frac{1}{\lambda_m^2 - \overline{\lambda_n^2}} \\ &\cdot (-1)^{r+1} \prod_m \lambda_m^2, \end{aligned} \tag{179}$$

which leads to

$$\begin{aligned} &\Omega_1(n_1,n_2,\cdots,n_r;0,m_2,\cdots m_r)\Omega_2(0,m_2,\cdots,m_r;n_1,n_2,\cdots,n_r) \\ &= (-1)^{r+1} \prod_n e^{-\theta_n} e^{-i\varphi_n} \prod_m e^{-\theta_m} e^{i\varphi_m} \prod_{n<n',m<m'} \left(\overline{\lambda_n^2} - \overline{\lambda_{n'}^2}\right)^2 \left(\lambda_{m'}^2 - \lambda_m^2\right)^2 \prod_{n,m} \frac{1}{\left(\lambda_m^2 - \overline{\lambda_n^2}\right)^2} \prod_m \lambda_m^2, \end{aligned} \tag{180}$$

here $n, n' \in (n_1, n_2, \cdots, n_r)$, $m, m' \in (m_2, \cdots, m_r)$ in (178)–(180). Finally, substituting (174) into (173), (180) into (177), and (173 and 177) into (162), we thus attain the explicit N-soliton solution to the DNLS equation with VBC under the reflectionless case, based on a pure Marchenko formalism and in no need of the concrete spectrum expression of $a(\lambda)$. Obviously, the N-soliton solution permits uncertain complex constants c_{n0} $(n = 1, 2, \cdots, N)$ as well as an arbitrary global constant phase factor.

3.5 The special examples for one- and two-soliton solutions

In the case of one simple pole and one-soliton solution as $N = 1$, according to (173), (177), (174), and (180), we have

$$C_1 = \Omega_1(n_1 = 1; m_1 = 0)\Omega_2(m_1 = 0; n_1 = 1) = \bar{g}_1\bar{h}_1(-1)^{1+1} = \bar{g}_1\bar{h}_1 \tag{181}$$

$$D_1 = Q_1(n_1 = 1; m_1 = 1)Q_2(m_1 = 1; n_2 = 1) = 1 - \frac{g_1 h_1 \bar{g}_1 \bar{h}_1 \overline{\lambda_1^2}}{\left(\lambda_1^2 - \overline{\lambda_1^2}\right)^2} = 1 - |g_1 h_1|^2 \frac{\overline{\lambda_1^2}}{\left(\lambda_1^2 - \overline{\lambda_1^2}\right)^2} \tag{182}$$

From (167) and (168), we have (suppose $\lambda_1 = \rho_1 e^{i\beta_1} = \mu_1 + i\nu_1$ and $c_{10} = e^{4\kappa_1 x_{10}} e^{i\alpha_{10}}$)

$$g_1h_1 = C_1(t)e^{i2\lambda_1^2x} = c_{10}e^{i4\lambda_1^4t}e^{i2\lambda_1^2x} = e^{-\theta_1}e^{i\varphi_1} \tag{183}$$

$$\begin{aligned}\theta_1 &= 4\mu_1v_1\left[x - x_{10} + 4\left(\mu_1^2 - v_1^2\right)t\right]\\ \varphi_1 &= 2\left(\mu_1^2 - v_1^2\right)x + \left[4\left(\mu_1^2 - v_1^2\right)^2 - 16\mu_1^2v_1^2\right]\cdot t + \alpha_{10}\end{aligned} \tag{184}$$

Then from (181) and (182), we attain the one-soliton solution

$$u_1(x,t) = -2\frac{C_1\overline{D}_1}{D_1^2} = -2\left(1 - \frac{\lambda_1^2}{\left(\lambda_1^2 - \overline{\lambda_1^2}\right)^2}e^{-2\theta_1}\right)e^{-\theta_1}e^{-i\varphi_1}\Bigg/\left(1 - \frac{\overline{\lambda_1^2}}{\left(\lambda_1^2 - \overline{\lambda_1^2}\right)^2}e^{-2\theta_1}\right)^2 \tag{185}$$

By further redefinition of its soliton center and initial phase, the single soliton solution can be further rewritten as usual standard form. It is easy to find, up to a permitted well-known constant global phase factor, the one-soliton solution to DNLS equation gotten in the pure Marchenko formalism is in perfectly agreement with that gotten from other approaches [23, 24, 26, 27].

As $N = 2$ in the case of two-soliton solution corresponding to double simple poles, we have

$$u_2(x,t) = -2C_2\overline{D}_2/D_2^2 \tag{186}$$

$$\begin{aligned}
C_2 &= \sum_{\substack{n_1=1,2\\ m_1=0}} \Omega_1(n_1,0)\Omega_2(0;n_1) + \sum_{\substack{n_1=1,n_2=2\\ m_1=0,m_2=1,2}} \Omega_1(n_1,n_2;0,m_2)\Omega_2(0,m_2;\mathrm{n}_1,n_2)\\
&= \Omega_1(n_1=1;m_1=0)\Omega_2(m_1=0;n_1=1) + \Omega_1(n_1=2;m_1=0)\Omega_2(m_1=0;n_1=2)\\
&+\Omega_1(n_1=1,n_2=2;m_1=0,m_2=1)\Omega_2(m_1=0,m_2=1;n_2=1,n_2=2)\\
&+\Omega_1(n_1=1,n_2=2;m_1=0,m_2=2)\Omega_2(m_1=0,m_2=2;n_1=1,n_2=2)\\
&= \overline{g}_1\overline{h}_1 + \overline{g}_2\overline{h}_2 - |g_1\overline{h}_1|^2\overline{g}_2\overline{h}_2\frac{\left(\overline{\lambda_1^2} - \overline{\lambda_2^2}\right)^2\cdot\lambda_1^2}{\left(\lambda_1^2 - \overline{\lambda_1^2}\right)^2\left(\lambda_1^2 - \overline{\lambda_2^2}\right)^2} - |\overline{g}_2\overline{h}_2|^2\overline{g}_1\overline{h}_1\frac{\left(\overline{\lambda_1^2} - \overline{\lambda_2^2}\right)^2\cdot\lambda_2^2}{\left(\lambda_2^2 - \overline{\lambda_1^2}\right)^2\left(\lambda_2^2 - \overline{\lambda_2^2}\right)^2}\\
&= e^{-\theta}e^{-i\varphi_1} + e^{-\theta_2}e^{-i\varphi_2} - \frac{\left(\overline{\lambda_1^2} - \overline{\lambda_2^2}\right)^2\cdot\lambda_1^2}{\left(\lambda_1^2 - \overline{\lambda_1^2}\right)^2\left(\lambda_1^2 - \overline{\lambda_2^2}\right)^2}e^{-2\theta_1-\theta_2}e^{-i\varphi_2} - \frac{\left(\overline{\lambda_1^2} - \overline{\lambda_2^2}\right)^2\cdot\lambda_2^2}{\left(\lambda_2^2 - \overline{\lambda_1^2}\right)^2\left(\lambda_2^2 - \overline{\lambda_2^2}\right)^2}e^{-2\theta_2-\theta_1}e^{-i\varphi_1}
\end{aligned} \tag{187}$$

$$\begin{aligned}
D_2 &= 1 + \sum_{r=1}^{2}\ \sum_{1\le n_1<n_2\le 2}\ \sum_{1\le m_1<m_2\le 2} Q_1(n_1,n_2,\cdots,n_r;m_1,m_2,\cdots,m_r)Q_2(m_1,m_2,\cdots,m_r;n_1,n_2,\cdots,n_r)\\
&= 1 + Q_1(n_1=1;m_1=1)Q_2(m_1=1,n_1=1) + Q_1(n_1=1;m_1=2)Q_2(m_1=2;n_1=1)\\
&+Q_1(n_1=2;m_1=1)Q_2(m_1=1,n_1=2) + Q_1(n_1=2,m_1=2)Q_2(m_1=2;n_1=2)\\
&+Q_1(n_1=1,n_2=2;m_1=1,m_2=2)Q_2(m_1=1,m_2=2;n_1=1,n_2=2)\\
&= 1 - |g_1h_1|^2\frac{\overline{\lambda_1^2}}{\left(\lambda_1^2 - \overline{\lambda_1^2}\right)^2} - |g_2h_2|^2\frac{\overline{\lambda_2^2}}{\left(\lambda_2^2 - \overline{\lambda_2^2}\right)^2} - \overline{g}_1\overline{h}_1g_2h_2\frac{\overline{\lambda_1^2}}{\left(\overline{\lambda_1^2} - \lambda_2^2\right)^2}\\
&-\overline{g}_2\overline{h}_2g_1h_1\frac{\overline{\lambda_2^2}}{\left(\lambda_1^2 - \overline{\lambda_2^2}\right)^2} + |g_1h_1|^2|g_2h_2|^2\frac{\overline{\lambda_1^2\lambda_2^2}\left(\lambda_1^2 - \lambda_2^2\right)^2\left(\overline{\lambda_1^2} - \overline{\lambda_2^2}\right)^2}{\left(\lambda_1^2 - \overline{\lambda_1^2}\right)^2\left(\lambda_1^2 - \overline{\lambda_2^2}\right)^2\left(\lambda_2^2 - \overline{\lambda_1^2}\right)^2\left(\lambda_2^2 - \overline{\lambda_2^2}\right)^2}\\
&= 1 - \frac{\overline{\lambda_1^2}}{\left(\lambda_1^2 - \overline{\lambda_1^2}\right)^2}e^{-2\theta_1} - \frac{\overline{\lambda_2^2}}{\left(\lambda_2^2 - \overline{\lambda_2^2}\right)^2}e^{-2\theta_2} - \frac{\overline{\lambda_1^2}}{\left(\overline{\lambda_1^2} - \lambda_2^2\right)^2}e^{-\theta_1-\theta_2}e^{i(\varphi_2-\varphi_1)}\\
&- \frac{\overline{\lambda_2^2}}{\left(\lambda_1^2 - \overline{\lambda_2^2}\right)^2}e^{-\theta_1-\theta_2}e^{i(\varphi_1-\varphi_2)} + \frac{\overline{\lambda_1^2\lambda_2^2}\left(\lambda_1^2 - \lambda_2^2\right)^2\left(\overline{\lambda_1^2} - \overline{\lambda_2^2}\right)^2}{\left(\lambda_1^2 - \overline{\lambda_1^2}\right)^2\left(\lambda_1^2 - \overline{\lambda_2^2}\right)^2\left(\lambda_2^2 - \overline{\lambda_1^2}\right)^2\left(\lambda_2^2 - \overline{\lambda_2^2}\right)^2}e^{-2(\theta_1+\theta_2)}
\end{aligned} \tag{188}$$

Up to a permitted constant global phase factor, the two-soliton solution gotten above is actually equivalent to that gotten from both IST and Hirota's method [23, 24, 26, 27], verifying the validity of the algebraic techniques that is used and our formula of the generalized multi-soliton solution. Because Marchenko equations (128), (129), (144), and (145) had been strictly proved, the multi-soliton solution is certainly right as long as we correctly use the algebraic techniques, especially Binet-Cauchy formula for the principal minor expansion of some special matrices.

4. Soliton solution of the DNLS equation based on Hirota's bilinear derivative transform

Bilinear derivative operator D had been found and defined in the early 1970s by Hirota R., a Japanese mathematical scientist [30–33]. Hirota's bilinear-derivative transform (HBDT for brevity) can be used to deal with some partial differential equation and to find some special solutions, such as soliton solutions and rogue wave solutions [26, 27, 32]. In this section, we use HBDT to solve DNLS equation with VBC and search for its soliton solution. The DNLS equation with VBC, that is,

$$\mathrm{i}u_t + u_{xx} + \mathrm{i}\left(|u|^2 u\right)_x = 0, \tag{189}$$

is one of the typical integrable nonlinear models, which is of a different form from the following equation:

$$\mathrm{i}u_t + u_{xx} + \mathrm{i}2|u|^2 u_x = 0, \tag{190}$$

which had been solved in Ref. [14] by using HBDT. We have paid special attention to the following solution form in it [14]:

$$u = g/f, \tag{191}$$

where f, g are usually complex functions. Solution (191) is suitable for Eq. (190) and NLS equation, and so on, but not suitable for the DNLS equation. Just due to this fact, their work cannot deal with Eq. (189) at the same time. As is well known, rightly selecting an appropriate solution form is an important and key step to apply Hirota's bilinear derivative transform to an integrable equation like Eq. (189). Refs. [13, 16, 17, 23], etc., have proved the soliton solution of the DNLS equation must has following standard form

$$u = g\bar{f}/f^2 \tag{192}$$

here and henceforth a bar over a letter represents complex conjugation.

In view of the existing experiences of dealing with the DNLS equation, in the present section, we attempt to use the solution form (192) and HBDT to solve the DNLS equation. We demonstrate our solving approach step by step, and naturally extend our conclusion to the n-soliton case in the end.

4.1 Fundamental concepts and general properties of bilinear derivative transform

For two differentiable functions $A(x,t), B(x,t)$ of two variables x and t, Hirota's bilinear derivative operator, D, is defined as

$$D_t^n D_x^m A \cdot B = \left(\frac{\partial}{\partial t} - \frac{\partial}{\partial t'}\right)^n \left(\frac{\partial}{\partial x} - \frac{\partial}{\partial x'}\right)^m A\left(x,t\right) B(x',t')|_{t'=t,x'=x} \tag{193}$$

which is different from the usual derivative, for example,

$$\begin{aligned} D_x A \cdot B &= A_x B - AB_x \\ D_x^2 A \cdot B &= A_{xx} B - 2A_x B_x + AB_{xx} \\ D_x^3 A \cdot B &= A_{xxx} B - 3A_{xx} B_x + 3A_x B_{xx} - AB_{xxx} \end{aligned} \tag{194}$$

where $A(x,t), B(x,t)$ are two functions derivable for an arbitrary order, and the dot $\cdot$ between them represents a kind of ordered product. Hirota's bilinear derivative has many interesting properties. Some important properties to be used afterwards are listed as follows:

$$①D_t^n D_x^m A \cdot B = (-1)^{n+m} D_t^n D_x^m B \cdot A \tag{195}$$

for example, $D_x A \cdot B = -D_x B \cdot A$; $D_x A \cdot A = 0$; $D_x^2 A \cdot B = D_x^2 B \cdot A$; $D_x^n A \cdot 1 = \partial_x^n A$; $D_x^n 1 \cdot A = (-1)^n \partial_x^n A$

$$②D_x^n A \cdot B = D_x^{n-m} D_x^m A \cdot B, (m < n) \tag{196}$$

③ Suppose $\eta_i = \Omega_i t + \Lambda_i x + \eta_{0i}$, $i = 1, 2$, Ω_i, Λ_i, η_{0i} are complex constants, then

$$D_t^n D_x^m \exp(\eta_1) \cdot \exp(\eta_2) = (\Omega_1 - \Omega_2)^n (\Lambda_1 - \Lambda_2)^m \exp(\eta_1 + \eta_2) \tag{197}$$

Especially, we have $D_t^n D_x^m \exp(\eta_1) \cdot \exp(\eta_2) = 0$ as $\Omega_1 = \Omega_2$ or $\Lambda_1 = \Lambda_2$. Some other important properties are listed in the Appendix.

4.2 Bilinear derivative transform of DNLS equation

After a suitable solution form, for example, (192) has been selected, under the Hirota's bilinear derivative transform, a partial differential equation like (189) can be generally changed into [20, 26, 27].

$$F_1(D_t, D_x \cdots) g_1 \cdot f_1 + F_2(D_t, D_x \cdots) g_2 \cdot f_2 = 0 \tag{198}$$

where $F_i(D_t, D_x \cdots)$, $i = 1, 2$ are the polynomial functions of D_t, D_x $\cdots$; and g_i, f_i, $i = 1, 2$, are the differentiable functions of two variables x and t . Using formulae in the Appendix and properties ①–③ of bilinear derivative transform numerated in the last chapter, with respect to (192), we have

$$u_t = D_t g\bar{f} \cdot f^2/f^4 = \left[f\bar{f} D_t g \cdot f - gfD_t f \cdot \bar{f}\right]/f^4 \tag{199}$$

$$u_x = D_x g\bar{f} \cdot f^2/f^4 = \left[f\bar{f} D_x g \cdot f - gfD_x f \cdot \bar{f}\right]/f^4 \tag{200}$$

$$u_{xx} = \left[f\bar{f} D_x^2 g \cdot f - 2(D_x g \cdot f)\left(D_x f \cdot \bar{f}\right) + gfD_x^2 f \cdot \bar{f}\right]/f^4 - 2g\bar{f} D_x^2 f \cdot f/f^4 \tag{201}$$

$$\left(|u|^2 u\right)_x = [2g\bar{g} D_x g \cdot f - g^2 (\bar{g}f)_x]/f^4 \tag{202}$$

Substituting the above expressions (199)–(202) into Eq. (189), the latter can be reduced to [26, 27].

$$f\bar{f}(\mathrm{i}D_t + D_x^2)g \cdot f - gf(\mathrm{i}D_t + D_x^2)f \cdot \bar{f} + f^{-2}D_x f^3 \cdot \left[g\left(2D_x f \cdot \bar{f} - \mathrm{i}g\bar{g}\right)\right] = 0 \quad (203)$$

We can extract the needed bilinear derivative equations from Eq. (203) as follows:

$$(\mathrm{i}D_t + D_x^2)g \cdot f = 0 \quad (204)$$

$$(\mathrm{i}D_t + D_x^2)f \cdot \bar{f} = 0 \quad (205)$$

$$D_x f \cdot \bar{f} = \mathrm{i}g\bar{g}/2 \quad (206)$$

Functions $g(x,t), f(x,t)$ can be expanded, respectively, as series of a small parameter ε

$$g = \sum_i \varepsilon^i g^{(i)} \quad (207)$$

$$f = 1 + \sum_i \varepsilon^i f^{(i)} \quad (208)$$

Substituting (207) and (208) into (204)–(206) and equating the sum of the terms with the same orders of ε at two sides of (204)–(206), we attain

$$(\mathrm{i}\partial_t + \partial_x^2)g^{(1)} = 0 \quad (209)$$

$$(\mathrm{i}D_t + D_x^2)\left(f^{(1)} \cdot 1 + 1 \cdot \bar{f}^{(1)}\right) = 0 \quad (210)$$

$$D_x\left(f^{(1)} \cdot 1 + 1 \cdot \bar{f}^{(1)}\right) = 0 \quad (211)$$

$$(\mathrm{i}D_t + D_x^2)\left(g^{(2)} \cdot 1 + g^{(1)} \cdot f^{(1)}\right) = 0 \quad (212)$$

$$(\mathrm{i}D_t + D_x^2)\left(f^{(2)} \cdot 1 + 1 \cdot \bar{f}^{(2)} + f^{(1)} \cdot \bar{f}^{(1)}\right) = 0 \quad (213)$$

$$D_x\left(f^{(2)} \cdot 1 + 1 \cdot \bar{f}^{(2)} + f^{(1)} \cdot \bar{f}^{(1)}\right) = \mathrm{i}g^{(1)} \cdot \bar{g}^{(1)}/2 \quad (214)$$

$$(\mathrm{i}D_t + D_x^2)\left(g^{(3)} \cdot 1 + g^{(2)} \cdot f^{(1)} + g^{(1)} \cdot f^{(2)}\right) = 0 \quad (215)$$

$$(\mathrm{i}D_t + D_x^2)\left(f^{(3)} \cdot 1 + 1 \cdot \bar{f}^{(3)} + f^{(2)} \cdot \bar{f}^{(1)} + f^{(1)} \cdot \bar{f}^{(2)}\right) = 0 \quad (216)$$

$$D_x\left(f^{(3)} \cdot 1 + 1 \cdot \bar{f}^{(3)} + f^{(2)} \cdot \bar{f}^{(1)} + f^{(1)} \cdot \bar{f}^{(2)}\right) = \mathrm{i}\left(g^{(2)}\bar{g}^{(1)} + g^{(1)}\bar{g}^{(2)}\right)/2 \quad (217)$$

$$(\mathrm{i}D_t + D_x^2)\left(g^{(4)} \cdot 1 + g^{(3)} \cdot f^{(1)} + g^{(2)} \cdot f^{(2)} + g^{(1)} \cdot f^{(3)}\right) = 0 \quad (218)$$

$$(\mathrm{i}D_t + D_x^2)\left(f^{(4)} \cdot 1 + 1 \cdot \bar{f}^{(4)} + f^{(3)} \cdot \bar{f}^{(1)} + f^{(1)} \cdot \bar{f}^{(3)} + f^{(2)} \cdot \bar{f}^{(2)}\right) = 0 \quad (219)$$

$$\begin{aligned} &D_x\left(f^{(4)} \cdot 1 + 1 \cdot \bar{f}^{(4)} + f^{(3)} \cdot \bar{f}^{(1)} + f^{(1)} \cdot \bar{f}^{(3)} + f^{(2)} \cdot \bar{f}^{(2)}\right) \\ &\quad = \mathrm{i}\left(g^{(3)}\bar{g}^{(1)} + g^{(2)}\bar{g}^{(2)} + g^{(1)}\bar{g}^{(3)}\right)/2 \end{aligned} \quad (220)$$

$$(\mathrm{i}D_t + D_x^2)\left(g^{(5)} \cdot 1 + g^{(4)} \cdot f^{(1)} + g^{(3)} \cdot f^{(2)} + g^{(2)} \cdot f^{(3)}\right) = 0 \quad (221)$$

$$(\mathrm{i}D_t + D_x^2)\left(f^{(5)} \cdot 1 + 1 \cdot \bar{f}^{(5)} + f^{(4)} \cdot \bar{f}^{(1)} + f^{(1)} \cdot \bar{f}^{(4)} + f^{(3)} \cdot \bar{f}^{(2)} + f^{(2)} \cdot \bar{f}^{(3)} \right) = 0 \tag{222}$$

$$\begin{aligned} &D_x\left(f^{(5)} \cdot 1 + 1 \cdot \bar{f}^{(5)} + f^{(4)} \cdot \bar{f}^{(1)} + f^{(1)} \cdot \bar{f}^{(4)} + f^{(3)} \cdot \bar{f}^{(2)} + f^{(2)} \cdot \bar{f}^{(3)} \right) \\ &\quad = \mathrm{i}\left(g^{(4)}\bar{g}^{(1)} + g^{(3)}\bar{g}^{(2)} + g^{(2)}\bar{g}^{(3)} + g^{(1)}\bar{g}^{(4)} \right)/2 \end{aligned} \tag{223}$$

The above equations, (209)–(223), contain the whole information needed to search for a soliton solution of the DNLS equation with VBC.

4.3 Soliton solution of the DNLS equation with VBC based on HBDT

4.3.1 One-soliton solution

For the one-soliton case, due to (209)–(211) and considering the transform property ③, we can select $g^{(1)}$ and $f^{(1)}$ respectively as

$$g^{(1)} = e^{\eta_1}, \eta_1 = \Omega_1 t + \Lambda_1 x + \eta_{10}, \Omega_1 = \mathrm{i}\Lambda_1^2, \tag{224}$$

$$f^{(1)} = 0 \tag{225}$$

From (212), one can select $g^{(2)} = 0$. From (214), we can attain

$$f^{(2)} - \bar{f}^{(2)} = \frac{\mathrm{i}}{2}\frac{1}{\Lambda_1 + \bar{\Lambda}_1} \mathrm{e}^{\eta_1 + \bar{\eta}_1} \tag{226}$$

where the vanishing boundary condition, $u \to 0$ as $|x| \to \infty$, is used. Then

$$\partial_t\left(f^{(2)} - \bar{f}^{(2)} \right) = -\frac{1}{2}\frac{\Lambda_1^2 - \bar{\Lambda}_1^2}{\Lambda_1 + \bar{\Lambda}_1} \mathrm{e}^{\eta_1 + \bar{\eta}_1} \tag{227}$$

Substituting (226) and (227) into Eq. (213), we can attain

$$f^{(2)} + \bar{f}^{(2)} = \frac{\mathrm{i}}{2}\frac{\Lambda_1 - \bar{\Lambda}_1}{\left(\Lambda_1 + \bar{\Lambda}_1\right)^2} \mathrm{e}^{\eta_1 + \bar{\eta}_1} \tag{228}$$

From (226) and (228), we can get an expression of $f^{(2)}$

$$f^{(2)} = \frac{\mathrm{i}}{2}\frac{\Lambda_1}{\left(\Lambda_1 + \bar{\Lambda}_1\right)^2} \mathrm{e}^{\eta_1 + \bar{\eta}_1} \tag{229}$$

Due to (224) and (229), we can also easily verify that

$$(\mathrm{i}D_t + D_x^2) g^{(1)} \cdot f^{(2)} = 0 \tag{230}$$

which immediately leads to

$$(\mathrm{i}\partial_t + \partial_x^2) g^{(3)} = 0 \tag{231}$$

in Eq. (215). Then from (215), we can select $g^{(3)} = 0$. For the same reason, from (216)–(223), we can select $f^{(3)}$, $g^{(4)}, g^{(5)}$, ... ; $f^{(4)}, f^{(5)}$, ... all to be zero. Thus the series (207) and (208) have been successfully cut off to have limited terms as follows:

$$g_1 = e^{\eta_1} \tag{232}$$

$$f_1 = 1 + \frac{i}{2}\frac{\Lambda_1}{\left(\Lambda_1 + \overline{\Lambda}_1\right)^2} e^{\eta_1 + \overline{\eta}_1} \tag{233}$$

where ε^i has been absorbed into the constant $e^{\eta_{10}}$ by redefiniing η_{10}. In the end, we attain the one-soliton solution to the DNLS equation with VBC

$$u_1(x,t) = g_1 \overline{f}_1 / f_1^2 \tag{234}$$

which is characterized with two complex parameters Λ_1 and η_{10} and shown in **Figure 1**. If we redefine the parameter Λ_1 as $\Lambda_1 \equiv -i2\overline{\lambda}_1^2$ and $\lambda_1 \equiv \mu_1 + iv_1$, then

$$\begin{aligned} e^{i2\lambda_1^2 x + i4\lambda_1^4 t + \overline{\eta}_{10}}/2 &\equiv e^{-\Theta_1} e^{i\Phi_1};\, e^{\overline{\eta}_{10}}/2 \equiv e^{4\mu_1 v_1 x_{10}} e^{i\alpha_{10}};\, \Theta_1 \\ &\equiv 4\mu_1 v_1 \left[x - x_{10} + 4\left(\mu_1^2 - v_1^2\right)t\right];\, \Phi_1 \\ &\equiv 2\left(\mu_1^2 - v_1^2\right)x + \left[4\left(\mu_1^2 - v_1^2\right)^2 - 16\mu_1^2 v_1^2\right] t + \alpha_{10} \end{aligned} \tag{235}$$

Then

$$\begin{aligned} g_1 &= e^{\eta_1} \equiv e^{-i2\overline{\lambda}^2 x - i4\overline{\lambda}^4 t + \eta_{10}} = 2e^{-\Theta_1} e^{-i\Phi_1} \\ f_1 &= 1 + \frac{i}{2}\frac{\Lambda_1}{\left(\Lambda_1 + \overline{\Lambda}_1\right)^2} e^{\eta_1 + \overline{\eta}_1} \\ &= 1 - \frac{\overline{\lambda}^2}{\left(\lambda^2 - \overline{\lambda}^2\right)^2} e^{-2\Theta} \end{aligned} \tag{236}$$

$$u_1(x,t) = \frac{g_1 \overline{f}_1}{f_1^2} = 2\left(1 - \frac{\lambda_1^2}{\left(\lambda_1^2 - \overline{\lambda}_1^2\right)^2} e^{-2\Theta_1}\right) e^{-\Theta_1} e^{-i\Phi_1} \Bigg/ \left(1 - \frac{\overline{\lambda}_1^2}{\left(\lambda_1^2 - \overline{\lambda}_1^2\right)^2} e^{-2\Theta_1}\right)^2 \tag{237}$$

It is easy to find, up to a permitted constant global phase factor $e^{i\pi} = -1$, the one-soliton solution (234) or (237) gotten in this paper is in perfect agreement with that gotten from other approaches [16]. By further redefining its soliton center, initial phase and $\lambda_1 = \rho_1 e^{i\beta_1}$, the one-soliton solution can be changed into the usual typical form [16, 23, 26, 27].

$$u_1(x,t) = 4|\lambda_1| \sin 2\beta_1 \left(1 + e^{i2\beta_1} e^{-2\Theta_1}\right) e^{-\Theta_1} e^{-i\Phi_1} / \left(1 + e^{-i2\beta_1} e^{-2\Theta_1}\right)^2 \tag{238}$$

On the other hand, just like in Ref. [13], we can rewrite g_1 and f_1 in a more appropriate or "standard" form

$$g_1 = e^{\eta_1 + \varphi_1} \tag{239}$$

$$f_1 = 1 + e^{(\eta_1 + \varphi_1) + (\overline{\eta}_1 + \varphi_{1'}) + \theta_{11'}} \tag{240}$$

Here

$$e^{\varphi_1} = 1, e^{\varphi_{1'}} = i/\overline{\Lambda}_1, e^{\theta_{11'}} = i\Lambda_1(-i\overline{\Lambda}_1)/2(\Lambda_1 + \overline{\Lambda}_1)^2, \tag{241}$$

which makes us easily extend the solution form to the case of n-soliton solution.

4.3.2 The two-soliton solution

For the two-soliton case, again from (209), we can select $g_2^{(1)}$ as

$$g_2^{(1)} = e^{\eta_1} + e^{\eta_2}, \eta_i = \Omega_i t + \Lambda_i x + \eta_{i0}, \Omega_i = i\Lambda_i^2, i = 1, 2. \tag{242}$$

The similar procedures to that used in the one-soliton case can be used to deduce g_2 and f_2. From (210) and (211), we can select $f_2^{(1)} = 0$, then from (212), we has to select $g_2^{(2)} = 0$. From (213) and (214), we can get the expressions of $f_2^{(2)} - \overline{f}_2^{(2)}$ and $f_2^{(2)} + \overline{f}_2^{(2)}$, then attain $f_2^{(2)}$ to be

$$f_2^{(2)} = \frac{i\Lambda_1}{2(\Lambda_1 + \overline{\Lambda}_1)^2} e^{\eta_1 + \overline{\eta}_1} + \frac{i\Lambda_2}{2(\Lambda_2 + \overline{\Lambda}_2)^2} e^{\eta_2 + \overline{\eta}_2} + \frac{i\Lambda_1}{2(\Lambda_1 + \overline{\Lambda}_2)^2} e^{\eta_1 + \overline{\eta}_2} + \frac{i\Lambda_2}{2(\Lambda_2 + \overline{\Lambda}_1)^2} e^{\eta_2 + \overline{\eta}_1} \tag{243}$$

Substituting (242) and (243) into (215), one can attain $g_2^{(3)}$ to be

$$g_2^{(3)} = \frac{-i(\Lambda_1 - \Lambda_2)^2 e^{\eta_1 + \eta_2}}{2} \left[\frac{\overline{\Lambda}_1 e^{\overline{\eta}_1}}{(\Lambda_1 + \overline{\Lambda}_1)^2 (\Lambda_2 + \overline{\Lambda}_1)^2} + \frac{\overline{\Lambda}_2 e^{\overline{\eta}_2}}{(\Lambda_2 + \overline{\Lambda}_2)^2 (\Lambda_1 + \overline{\Lambda}_2)^2} \right] \tag{244}$$

Substituting the expressions of $g_2^{(1)}, g_2^{(2)}, g_2^{(3)}, f_2^{(1)}, f_2^{(2)}$ into (216) and (217), we can select that $f_2^{(3)} = 0$. Then from the expressions of $g_2^{(1)}, g_2^{(2)}, g_2^{(3)}, f_2^{(1)}, f_2^{(2)}, f_2^{(3)}$ and (218), we can select $g_2^{(4)} = 0$. From (219) and (220), we can get the expressions of $f_2^{(4)} - \overline{f}_2^{(4)}$ and $f_2^{(4)} + \overline{f}_2^{(4)}$, then get $f_2^{(4)}$ to be

$$f_2^{(4)} = -\frac{\Lambda_1 \Lambda_2 |\Lambda_1 - \Lambda_2|^4}{4(\Lambda_1 + \overline{\Lambda}_1)^2 (\Lambda_2 + \overline{\Lambda}_2)^2 (\Lambda_1 + \overline{\Lambda}_2)^2 (\Lambda_2 + \overline{\Lambda}_1)^2} e^{\eta_1 + \eta_2 + \overline{\eta}_1 + \overline{\eta}_2} \tag{245}$$

Due to (243) and (244), we can also easily verify that

$$(iD_t + D_x^2) g_2^{(3)} \cdot f_2^{(2)} = 0 \tag{246}$$

Then from (244), (245), (246), and (221), we can select $g_2^{(5)} = 0$. From (222)–(223) and so on, we find that the series of (207) and (208) can be cut off by selecting $g_2^{(5)}, f_2^{(5)}; g_2^{(6)}, f_2^{(6)} \cdots$, all to be zero. We thus attain the last result of g_2, f_2 to be

$$g_2 = e^{\eta_1} + e^{\eta_2} - \frac{i(\Lambda_1 - \Lambda_2)^2}{2} e^{\eta_1 + \eta_2} \bullet \left[\frac{\overline{\Lambda}_1}{(\Lambda_1 + \overline{\Lambda}_1)^2 (\Lambda_2 + \overline{\Lambda}_1)^2} e^{\overline{\eta}_1} + \frac{\overline{\Lambda}_2}{(\Lambda_2 + \overline{\Lambda}_2)^2 (\Lambda_1 + \overline{\Lambda}_2)^2} e^{\overline{\eta}_2} \right] \tag{247}$$

$$f_2 = 1 + \frac{i\Lambda_1}{2(\Lambda_1+\overline{\Lambda}_1)^2}e^{\eta_1+\overline{\eta}_1} + \frac{i\Lambda_2}{2(\Lambda_2+\overline{\Lambda}_2)^2}e^{\eta_2+\overline{\eta}_2} + \frac{i\Lambda_1}{2(\Lambda_1+\overline{\Lambda}_2)^2}e^{\eta_1+\overline{\eta}_2} + \frac{i\Lambda_2}{2(\Lambda_2+\overline{\Lambda}_1)^2}e^{\eta_2+\overline{\eta}_1}$$
$$- \frac{\Lambda_1\Lambda_2|\Lambda_1-\Lambda_2|^4}{4(\Lambda_1+\overline{\Lambda}_1)^2(\Lambda_2+\overline{\Lambda}_2)^2(\Lambda_1+\overline{\Lambda}_2)^2(\Lambda_2+\overline{\Lambda}_1)^2}e^{\eta_1+\eta_2+\overline{\eta}_1+\overline{\eta}_2} \tag{248}$$

It can also be rewritten in a standard form as follows:

$$\begin{aligned} g_2 &= e^{\eta_1+\varphi_1} + e^{\eta_2+\varphi_2} + e^{(\eta_1+\varphi_1)+(\eta_2+\varphi_2)+(\overline{\eta}_1+\varphi_{1'})+\theta_{12}+\theta_{11'}+\theta_{21'}} \\ &+ e^{(\eta_1+\varphi_1)+(\eta_2+\varphi_2)+(\overline{\eta}_2+\varphi_{2'})+\theta_{12}+\theta_{12'}+\theta_{22'}} \end{aligned} \tag{249}$$

$$\begin{aligned} f_2 = 1 &+ e^{(\eta_1+\varphi_1)+(\overline{\eta}_1+\varphi_{1'})+\theta_{11'}} + e^{(\eta_1+\varphi_1)+(\overline{\eta}_2+\varphi_{2'})+\theta_{12'}} + e^{(\eta_2+\varphi_2)+(\overline{\eta}_1+\varphi_{1'})+\theta_{21'}} \\ &+ e^{(\eta_2+\varphi_2)+(\overline{\eta}_2+\varphi_{2'})+\theta_{22'}} + e^{(\eta_1+\varphi_1)+(\eta_2+\varphi_2)+(\overline{\eta}_1+\varphi_{1'})+(\overline{\eta}_2+\varphi_{2'})+\theta_{12}+\theta_{11'}+\theta_{12'}+\theta_{21'}+\theta_{22'}+\theta_{1'2'}} \end{aligned} \tag{250}$$

where φ_i, θ_{ij} in (249) and (250) are defined afterwards in (256). We then attain the two-soliton solution as

$$u_2(x,t) = g_2\overline{f}_2/f_2^2 \tag{251}$$

which is characterized with four complex parameters Λ_1, Λ_2, η_{10}, and η_{20} and shown in **Figure 3**. By redefining parameters η_{i0} and

$$\Lambda_k = -i2\overline{\lambda}_k^{-2}, k = 1, 2, \tag{252}$$

we can easily transform it to a two-soliton form given in Ref. [23], up to a permitted constant global phase factor.

4.3.3 Extension to the N-soliton solution

Generally for the case of N-soliton solution, if we select $g^{(1)}$ in (209) to be

$$\begin{aligned} g_N^{(1)} &= e^{\eta_1} + e^{\eta_2} + \ldots + e^{\eta_N}, \\ \eta_i &= \Omega_i t + \Lambda_i x + \eta_{0i}, \Omega_i = i\Lambda_i^2, i = 1, 2, \cdots, N \end{aligned} \tag{253}$$

then using an induction method, we can write the N-soliton solution as

$$g_N = \sum_{\kappa_j=0,1}^{(1)} \exp\left\{\sum_{j=1}^{2N}\kappa_j\left(\eta_j+\varphi_j\right) + \sum_{1\le j<k}^{2N}\kappa_j\kappa_k\theta_{jk}\right\} \tag{254}$$

$$f_N = \sum_{\kappa_j=0,1}^{(0)} \exp\left\{\sum_{j=1}^{2N}\kappa_j\left(\eta_j+\varphi_j\right) + \sum_{1\le j<k}^{2N}\kappa_j\kappa_k\theta_{jk}\right\} \tag{255}$$

where $\eta_{N+j} = \overline{\eta}_j$ $(j \le N)$

$$\begin{aligned} &e^{\phi_j} = 1, e^{\varphi_{j+N}} = e^{\varphi_{j'}} = \left(-i\overline{\Lambda}_j\right)^{-1} \\ &e^{\theta_{jk}} = 2(\Lambda_j-\Lambda_k)^2/i\Lambda_j \cdot i\Lambda_k, e^{\theta_{j,k+N}} = e^{\theta_{j,k'}} = i\Lambda_j\left(-i\overline{\Lambda}_k\right)/2\left(\Lambda_j+\overline{\Lambda}_k\right)^2 \\ &e^{\theta_{j+N,k+N}} = e^{\theta_{j'k'}} = 2\left(\overline{\Lambda}_j-\overline{\Lambda}_k\right)^2/\left(-i\overline{\Lambda}_j\right)\left(-i\overline{\Lambda}_k\right), (1\le j\le N, 1\le k\le N), \end{aligned} \tag{256}$$

therein $\sum_{\kappa_j=0,1}^{(l)}$ represents a summation over $\kappa_j = 0, 1$ under the condition $\sum_{j=1}^{N}\kappa_j = l + \sum_{j=1}^{N}\kappa_{j+N}$.

Here, we have some discussion in order. Because what concerns us only is the soliton solutions, our soliton solution of DNLS equation with VBC is only a subset of the whole solution set. Actually in the whole process of deriving the bilinear-form equations and searching for the one and two-soliton solutions, some of the latter results are only the sufficient but not the necessary conditions of the former equations. Thereby some possible modes might have been missing. For example, the solutions of Eqs. (209)–(211) are not as unique as in (224) and (225), some other possibilities thus get lost here. This is also why we use a term "select" to determine a solution of an equation. In another word, we have selected a soliton solution. Meanwhile, we have demonstrated in **Figures 2** and **3**, the three-dimensional evolution of the one- and two-soliton amplitude with time and space, respectively. The elastic collision of two solitons in the two-soliton case has been demonstrated in **Figure 4(a–d)** too. It can be found that each soliton keeps the same form and characteristic after the collision as that before the collision. In this section, by means of introducing HBDT and employing an appropriate solution form (192),

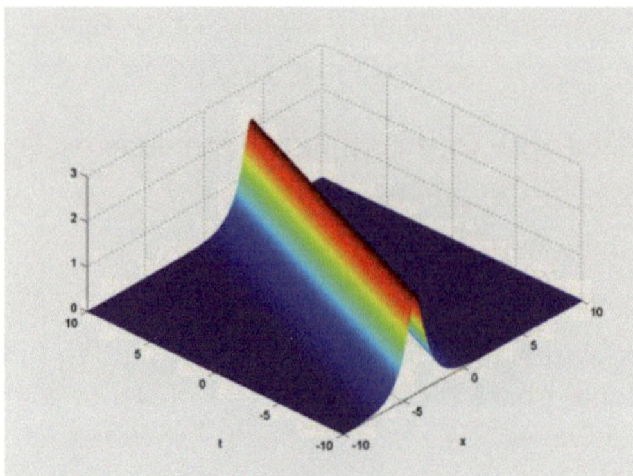

Figure 2.
The evolution of one-soliton solution with time and space under parameter $\Lambda_1 = -1 + 0.2i$, $\eta_{10} = 1$ *in (234).*

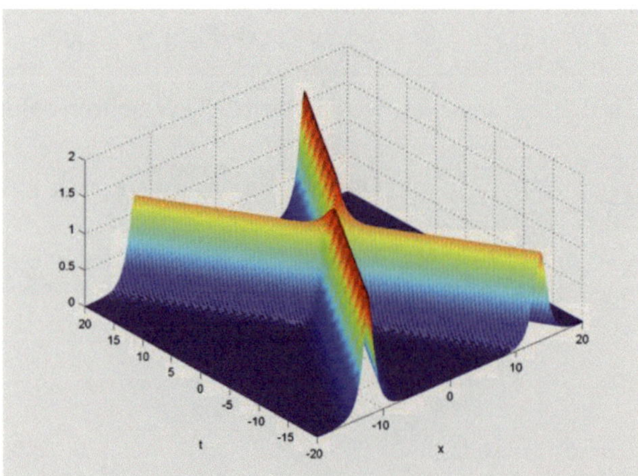

Figure 3.
The evolution of two-soliton solution with time and space under parameter $\Lambda_1 = 1 + 0.3i$, $\Lambda_2 = 1 - 0.3i$, $\eta_{10} = \eta_{20} = 1$ *in (251).*

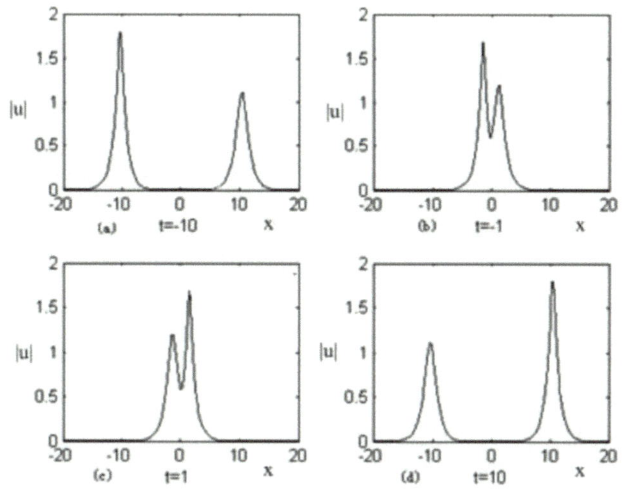

Figure 4.
The elastic collision between two solitons at 4 typical moments: (a) $t = -10$(normalized time); (b) $t = -1$; (c) $t = 1$; (d) $t = 10$, from -10 before collision to 10 after collision.

we successfully solve the derivative nonlinear Schrödinger equation with VBC. The one- and two-soliton solutions are derived and their equivalence to the existing results is manifested. The *N*-soliton solution has been given by an induction method. On the other hand, by using simple parameter transformations (e.g., (235) and (252)), the soliton solutions attained here can be changed into or equivalent to that gotten based on IST, up to a permitted global constant phase factor. This section impresses us so greatly for a fact that, ranked with the extensively used IST [23] and other methods, the HBDT is another effective and important tool to deal with a partial differential equation. It is especially suitable for some integrable nonlinear models.

Author details

Zhou Guo-Quan
Department of Physics, Wuhan University, Wuhan, P.R. China

*Address all correspondence to: zgq@whu.edu.cn

References

[1] Rogister A. Parallel propagation of nonlinear low-frequency waves in high-β plasma. Physics of Fluids. 1971;**14**:2733

[2] Ruderman MS. DNLS equation for large-amplitude solitons propagating in an arbitrary direction in a high-β hall plasma. Journal of Plasma Physics. 2002; **67**:271

[3] Hada T. In: Hada T, Matsumoto T, editors. Nonlinear Waves and Chaos in Space Plasmas. Tokyo: Terrapub; 1997. p. 121

[4] Tzoar N, Jain M. Self-phase modulation in long-geometry optical waveguides. Physical Review A. 1981;**23**: 1266

[5] Anderson D, Lisak M. Nonlinear asymmetric self-phase modulation and self-steepening of pulses in long optical waveguides. Physical Review A. 1983;**27**: 1393

[6] Govind PA. Nonlinear Fiber Optics. 3rd ed. New York: Academic Press; 2001

[7] Nakata I. Weak nonlinear electromagnetic waves in a ferromagnet propagating parallel to an external magnetic field. Journal of the Physical Society of Japan. 1991;**60**(11):3976

[8] Nakata I, Ono H, Yosida M. Solitons in a dielectric medium under an external magnetic field. Progress in Theoretical Physics. 1993;**90**(3):739

[9] Daniel M, Veerakumar V. Propagation of electromagnetic soliton in anti-ferromagnetic medium. Physics Letters A. 2002;**302**:77-86

[10] Kaup DJ, Newell AC. Solitons as particles, oscillators, and in slowly changing media: A singular perturbation theory. Proceedings of the Royal Society of London, Series A. 1978;**361**:413

[11] Kaup DJ, Newell AC. An exact solution for a derivative nonlinear Schrödinger equation. Journal of Mathematical Physics. 1978;**19**:798

[12] Kawata T, Inoue H. Exact solution of derivative nonlinear Schrödinger equation under nonvanishing conditions. Journal of the Physical Society of Japan. 1978;**44**:1968

[13] Steudel H. The hierarchy of multi-soliton solutions of derivative nonlinear Schrödinger equation. Journal of Physics A. 2003;**36**:1931

[14] Nakamura A, Chen HH. Multi-soliton solutions of derivative nonlinear Schrödinger equation. Journal of the Physical Society of Japan. 1980;**49**:813

[15] Huang NN, Chen ZY. Alfven solitons. Journal of Physics A: Mathematical and General. 1990;**23**:439

[16] Chen XJ, Yang JK, Lam WK. N-soliton solution for the derivative nonlinear Schrödinger equation with nonvanishing boundary conditions. Journal of Physics A: Mathematical and General. 2006;**39**(13):3263

[17] Lashkin VM. N-soliton solutions and perturbation theory for DNLS with nonvanishing condition. Journal of Physics A. 2007;**40**:6119-6132

[18] Ablowitz MJ. Solitons, Nonlinear Evolution Equations and Inverse Scattering. London: Cambridge University Press; 1991. p. P1381

[19] Wadati M, Konno K, Ichikawa YH. A generalization of inverse scattering method. Journal of the Physical Society of Japan. 1979;**46**:1965

[20] Chen HH, Lee YC, Liu CS. Integrability of nonlinear Hamiltonian systems by inverse scattering method. Physica Scripta. 1979;**20**(3–4):490

[21] Huang N-N. Theory of Solitons and Method of Perturbations. Shanghai: Shanghai Scientific and Technological Education Publishing House; 1996

[22] Huang N-N et al. Hamilton Theory about Nonlinear Integrable Equations. Beijing: Science Press; 2005. pp. P93-P95

[23] Zhou G-Q, Huang N-N. An *N*-soliton solution to the DNLS equation based on revised inverse scattering transform. Journal of Physics A: Mathematical and Theoretical. 2007; **40**(45):13607

[24] Zhou G-Q. A multi-soliton solution of the DNLS equation based on pure Marchenko formalism. Wuhan University. 2010;**15**(1):36-42

[25] Yang CN et al. Demonstration of inverse scattering transform for DNLS equation. Theoretical Physics. 2007;**48**: 299-303

[26] Zhou G-Q, Bi X-T. Soliton solution of the DNLS equation based on Hirota's bilinear derivative transform. Wuhan University. 2009;**14**(6):505-510

[27] Zhou G-Q. Hirota's bilinear derivative transform and soliton solution of the DNLS equation. Physics Bulletin. 2014;**4**:93-97

[28] Huang N-N. Marchenko equation for the derivative nonlinear schrödinger equation. Chinese Physics Letters. 2007; **24**(4):894-897

[29] Chen X-J et al. An inverse scattering transform for the derivative nonlinear Schrödinger equation. Physical Review E. 2004;**69**(6):066604

[30] Hirota R. Direct method of finding exact solutions of nonlinear evolution equations. In: Miura RM, editors. Bäcklund Transformations, the Inverse Scattering Method, Solitons, and Their Applications. Lecture Notes in Mathematics. Berlin, Heidelberg: Springer; 1976. pp. 40-68

[31] Hirota R. The Direct Method in Soliton Theory. Cambridge: Cambridge University Press; 2004

[32] Zhou G-Q, Li X-J. Space periodic solutions and rogue wave solution of the derivative nonlinear Schrödinger equation. Wuhan University Journal of Natural Sciences. 2017;**22**(5):9-15

[33] Hirota R. Exact solution of the Korteweg-de Vries equation for multiple collisions of solitons. Physical Review Letters. 1971;**27**(18):1192

Chapter 3

Soliton and Rogue-Wave Solutions of Derivative Nonlinear Schrödinger Equation - Part 2

Zhou Guo-Quan

Abstract

A revised and rigorously proved inverse scattering transform (IST for brevity) for DNLS$^+$ equation, with a constant nonvanishing boundary condition (NVBC) and normal group velocity dispersion, is proposed by introducing a suitable affine parameter in the Zakharov-Shabat IST integral; the explicit breather-type and pure N-soliton solutions had been derived by some algebra techniques. On the other hand, DNLS equation with a non-vanishing background of harmonic plane wave is also solved by means of Hirota's bilinear formalism. Its space periodic solutions are determined, and its rogue wave solution is derived as a long-wave limit of this space periodic solution.

Keywords: soliton, nonlinear equation, derivative nonlinear Schrödinger equation, inverse scattering transform, Zakharov-Shabat equation, Hirota's bilinear derivative method, DNLS equation, space periodic solution, rogue wave

1. Breather-type and pure N-soliton solution to DNLS$^+$ equation with NVBC based on revised IST

DNLS$^+$ equation with NVBC, the concerned theme of this section, is only a transformed version of modified nonlinear Schrödinger equation with normal group velocity dispersion and a nonlinear self-steepen term and can be expressed as

$$iu_t - u_{xx} + i\left(|u|^2 u\right)_x = 0, \tag{1}$$

here the subscripts represent partial derivatives.

Some progress have been made by several researchers to solve the DNLS equation for DNLS equation with NVBC, many heuristic and interesting results have been attained [1–14]. An early proposed IST worked on the Riemann sheets can only determine the modulus of the one-soliton solution [3, 15]. References [4, 5, 16] had attained a pure single dark/bright soliton solution. Reference [6] had derived a formula for $N-$ soliton solution in terms of Vandermonde-like determinants by means of Bäcklund transformation; but just as reference [7, 9] pointed out, this multi-soliton solution is still difficult to exhibit the internal elastic collisions among solitons and the typical asymptotic behaviors of multi-soliton of DNLS equation. By introducing an affine parameter in the integral of Zakharov-Shabat IST, reference [7] had found their pure $N-$ soliton solution for a special case that all the simple poles (zeros of $a(\lambda)$) were located on a circle of radius ρ centered at the origin, while reference [8] also found its multi-soliton solution for some extended case with N poles on a circle and M poles out of the circle, and further developed its perturbation theory based on IST. Reference [7] constructed their theory by introducing an damping factor in the integral of Zakharov-Shabat IST, to make it convergent, and further adopted a good idea of introducing an affine parameter to avoid the trouble of multi-value problem in Riemann sheets, but both of their results are assumed $N-$ soliton solutions and the soliton solution gotten from their IST had a self-dependent and complicated phase factor [7–9], hence reference [8] had to verify an identity demanded by the standard form of a soliton solution (see expression (52) in Ref. [8]). Such kind of an identity is rather difficult to prove for $N \geq 2$ case even by the use of computer techniques and Mathematica. On the other hand, author of reference [8] also admits his soliton solution is short of a rigorous verification of standard form. Then questions naturally generates – whether the traditional IST for DNLS equation with NVBC can be avoided and further improved? And whether a rigorous manifestation of soliton standard form can be given and a more reasonable and natural IST can be constructed?

A newly revised IST is thus proposed in this section to avoid the dual difficulty and the excessive complexity. An additional affine factor $1/\lambda$, $\lambda = (z + \rho^2 z^{-1})/2$, is introduced in the Z-S IST integral to make the contour integral convergent in the big circle [7, 10–13]. Meanwhile, the additional two poles on the imaginary axis caused by $\lambda = 0$ are removable poles due to the fact that the first Lax operator $L(\lambda) \to 0$, as $\lambda \to 0$. What is more different from reference [7] is that we locate the N simple poles off the circle of radius ρ centered at origin O, which corresponds to the general case of N breather-type solitons. When part of the poles approach the circle, the corresponding part of the breathers must tend to the pure solitons, which is the case described in Ref. [8]. The resulted one soliton solution can naturally tend to the well-established conclusion of VBC case as $\rho \to 0$ [17–20] and the pure one soliton solution in the degenerate case. The result of this section appears to be strict and reliable.

1.1 The fundamental concepts for the IST theory of DNLS equation

Under a Galileo transformation $(x,t) \to (x + \rho^2 t, t)$, DNLS $^+$ Eq. (1) can be changed into

$$iu_t - u_{xx} + i\left[\left(|u|^2 - \rho^2\right)u\right]_x = 0, \tag{2}$$

with nonvanishing boundary condition:

$$u \to \rho, \text{ as } |\mathrm{x}| \to \infty. \tag{3}$$

According to references [7–9], the phase difference between the two infinite ends should be zero. The Lax pair of DNLS$^+$ Eq. (3) is

$$\boldsymbol{L} = -i\lambda^2\sigma_3 + \lambda \boldsymbol{U}, \ \boldsymbol{U} = \begin{pmatrix} 0 & u \\ \overline{u} & 0 \end{pmatrix} = u\boldsymbol{\sigma}_+ + \overline{u}\sigma_- \tag{4}$$

$$\boldsymbol{M} = i2\lambda^4\sigma_3 - 2\lambda^3\boldsymbol{U} + i\lambda^2(U^2 - \rho^2)\sigma_3 - \lambda(U^2 - \rho^2)\boldsymbol{U} + i\lambda U_x\sigma_3 \tag{5}$$

where σ_3, σ_+, and $\boldsymbol{\sigma}_-$ involve in standard Pauli's matrices and their linear combination. Here and hereafter a bar over a variable represents complex conjugate. An affine parameter z and two aided parameters η, λ are introduced to avoid the trouble of dealing with double-valued functions on Riemann sheets

$$\lambda \equiv \left(z + \rho^2 z^{-1}\right)/2, \eta \equiv \left(z - \rho^2 z^{-1}\right)/2 \tag{6}$$

The Jost functions satisfy first Lax equation

$$\partial_x \boldsymbol{F}(x,\ z) = \boldsymbol{L}(x,\ z)\boldsymbol{F}(x,\ z), \tag{7}$$

here Jost functions $\boldsymbol{F}(x,\ z) \in \{\boldsymbol{\Psi}(x,z), \boldsymbol{\Phi}(x,z)\}$.

$$\boldsymbol{\Psi}(x,z) = (\tilde{\boldsymbol{\psi}}(x,z),\ \boldsymbol{\psi}(x,z)) \to \boldsymbol{E}(x,z), \text{as } x \to \infty \tag{8}$$

$$\boldsymbol{\Phi}(x,z) = \left(\boldsymbol{\phi}(x,z), \tilde{\boldsymbol{\phi}}(x,z)\right) \to \boldsymbol{E}(x,z), \text{as } x \to -\infty \tag{9}$$

The free Jost function $\boldsymbol{E}(x,z)$ can be easily attained as follows:

$$\boldsymbol{E}(x,z) = \left(\boldsymbol{I} + \rho z^{-1}\boldsymbol{\sigma}_2\right) \exp\left(-i\lambda\eta x\sigma_3\right), \tag{10}$$

which can be verified satisfying Eq. (7). The monodramy matrix is

$$\boldsymbol{T}(z) = \begin{pmatrix} a(z) & \tilde{b}(z) \\ b(z) & \tilde{a}(z) \end{pmatrix}, \tag{11}$$

which is defined by

$$\boldsymbol{\Phi}(x,z) = \boldsymbol{\Psi}(x,z)\ \boldsymbol{T}(z) \tag{12}$$

Some useful and important symmetry properties can be found

$$\boldsymbol{\sigma}_1\overline{\boldsymbol{L}}(z)\boldsymbol{\sigma}_1 = \boldsymbol{L}(\overline{z}), \boldsymbol{\sigma}_3\boldsymbol{L}(-z)\boldsymbol{\sigma}_3 = \boldsymbol{L}(z) \tag{13}$$

Symmetry relations in (13) lead to

$$\tilde{\boldsymbol{\psi}}(x,z) = \sigma_1\overline{\boldsymbol{\psi}(x,\overline{z})}, \tilde{\boldsymbol{\phi}}(x,z) = \sigma_1\overline{\boldsymbol{\phi}(x,\overline{z})} \tag{14}$$

$$\tilde{\boldsymbol{\psi}}(x,-z) = \boldsymbol{\sigma}_3\tilde{\boldsymbol{\psi}}(x,z), \boldsymbol{\psi}(x,-z) = -\boldsymbol{\sigma}_3\boldsymbol{\psi}(x,z) \tag{15}$$

$$\boldsymbol{\sigma}_1\overline{\boldsymbol{T}(x,\overline{z})}\boldsymbol{\sigma}_1 = \boldsymbol{T}(z), \boldsymbol{\sigma}_3\boldsymbol{T}(-z)\boldsymbol{\sigma}_3 = \boldsymbol{T}(z) \tag{16}$$

The above symmetry relations further result in

$$\tilde{a}(\overline{z}) = \overline{a(z)},\quad \tilde{b}(z) = \overline{b(z)} \tag{17}$$

$$a(-z) = a(z), b(-z) = -b(z) \tag{18}$$

Other important symmetry properties called reduction relations can also be easily found

$$\lambda\left(\rho^2 z^{-1}\right) = \lambda(z),\quad \eta\left(\rho^2 z^{-1}\right) = -\eta(z), \boldsymbol{L}\left(x, \rho^2 z^{-1}\right) = \boldsymbol{L}(x,z) \tag{19}$$

$$\boldsymbol{E}\left(x, \rho^2 z^{-1}\right) = \rho^{-1} z\left(\boldsymbol{I} + \rho z^{-1}\boldsymbol{\sigma}_2\right)e^{-i\eta\lambda x\sigma_3}\boldsymbol{\sigma}_2 \tag{20}$$

The above symmetry properties lead to following reduction relations among Jost functions

$$\tilde{\psi}\left(x,\rho^2 z^{-1}\right) = i\rho^{-1} z\psi(x,z), \psi\left(x,\rho^2 z^{-1}\right) = -i\rho^{-1} z\tilde{\psi}(x,z) \quad (21)$$

$$\phi\left(x,\rho^2 z^{-1}\right) = i\rho^{-1} z\tilde{\phi}(x,z), \tilde{\phi}\left(x,\rho^2 z^{-1}\right) = -i\rho^{-1} z\phi(x,z) \quad (22)$$

The important symmetries among the transition coefficients further resulted from (12), (21), and (22):

$$\sigma_2 T\left(\rho^2 z^{-1}\right)\sigma_2 = T(z), \tilde{a}\left(\rho^2 z^{-1}\right) = a(z), \tilde{b}\left(\rho^2 z^{-1}\right) = -b(z) \quad (23)$$

On the other hand, the simple poles, or zeros of $a(z)$, appear in quadruplet, and can be designated by z_n, $(n = 1, 2, \cdots, 2N)$, in the I quadrant, and $z_{n+2N} = -z_n$ in the III quadrant. Due to symmetry (17), (18) and (23), the n' th subset of zero points is

$$\left\{z_{2n-1}, z_{2n} = \rho^2 \bar{z}_{2n-1}^{-1}, -z_{2n-1}, -z_{2n} = -\rho^2 \bar{z}_{2n-1}^{-1}\right\} \quad (24)$$

And we arrange the $2N$ zeros in the first quadrant in following sequence

$$z_1, z_2; z_3, z_4; \cdots; z_{2N-1}, z_{2N} \quad (25)$$

According to the standard procedure [21, 22], the discrete part of $a(z)$ can be deduced

$$a(z) = \prod_{n=1}^{2N} \frac{z^2 - z_n^2}{z^2 - \bar{z}_n^2} \frac{\bar{z}_n}{z_n} \quad (26)$$

At the zeros of $a(z)$, or poles z_n, $(n = 1, 2, \cdots, 2N - 1, 2N)$, we have

$$\phi(x,z_n) = b_n \psi(x,z_n), \dot{a}(-z_n) = -\dot{a}(z_n) \quad (27)$$

Using symmetry relation in (14), (17), (21)–(24), (27), we can prove that

$$\bar{b}_{2n} = -b_{2n-1}, \bar{c}_{2n-1} = \rho^2 z_{2n}^{-2} c_{2n} \quad (28)$$

2. Relation between the solution and Jost functions of DNLS⁺ equation

The asymptotic behaviors of the Jost solutions in the limit of $|\lambda| \to \infty$ can be obtained by simple derivation. Let $F_{\cdot 1} = \tilde{\psi}(x,\lambda)$, then Eq. (7) can be rewritten as

$$\tilde{\psi}_{1x} + i\lambda^2 \tilde{\psi}_1 = \lambda u \tilde{\psi}_2, \tilde{\psi}_{2x} - i\lambda^2 \tilde{\psi}_2 = -\lambda \bar{u} \tilde{\psi}_1, \quad (29)$$

then we have

$$\tilde{\psi}_{1xx} - \left(\tilde{\psi}_{1x} + i\lambda^2 \tilde{\psi}_1\right) u_x / u + \lambda^4 \tilde{\psi}_1 - \lambda^2 |u|^2 \tilde{\psi}_1 = 0 \quad (30)$$

We assume a function g to satisfy the following equation

$$\tilde{\psi}_1(x,\lambda) = e^{-i\lambda\eta x + g} \quad (31)$$

$$\tilde{\psi}_{1x} = \left(-i\lambda\eta + g_x\right)\tilde{\psi}_1, \tilde{\psi}_{1xx} = \left[\left(-i\lambda\eta + g_x\right)^2 + g_{xx}\right]\tilde{\psi}_1 \quad (32)$$

Substituting (31)–(32) into Eq. (30), we have

$$g_{xx} + g_x^2 - (2i\lambda\eta + u_x/u)g_x - i\lambda(\lambda - \eta)u_x/u + \lambda^2(\lambda^2 - \eta^2) - \lambda^2|u|^2 = 0 \quad (33)$$

In the limit of $|z| \to \infty, g_x$ can be expanded as sum of series of $(z^{-2})^j, j = 0, 1, 2, \cdots$

$$g_x \equiv \mu = \mu_0 + \mu_2 z^{-2} + \mu_4 z^{-4} + \cdots, \quad (34)$$

where

$$\mu_0 = -i\left(\rho^2 - |u|^2\right)/2 = O(1) \quad (35)$$

Inserting formula (34) and (35) in Eq. (33) at $|z| \to \infty$ leads to
$u = \frac{\tilde{\psi}_1}{\lambda\tilde{\psi}_2}\left[i\lambda(\lambda - \eta) - \frac{1}{2}i\left(\rho^2 - |u|^2\right)\right] \to i\frac{\tilde{\psi}_1}{\tilde{\psi}_2}\frac{|u|^2}{z}, \quad (as \ |z| \to \infty)$
Then we find a useful formula

$$\bar{u} = -i \lim_{|z|\to\infty} z\tilde{\psi}_2(x,z)/\tilde{\psi}_1(x,z), \quad (36)$$

which expresses the conjugate of the solution u in terms of Jost functions as $|z| \to \infty$.

2.1 Introduction of time evolution factor

In order to make the Jost functions satisfy the second Lax equation, a time evolution factor $h(t,z)$ should be introduced by a standard procedure [21, 22] in the Jost functions and the scattering data. Considering the asymptotic behavior of the second Lax operator

$$\boldsymbol{M}(x,t;z) \to i2\lambda^4\boldsymbol{\sigma}_3 - 2\lambda^3\rho\boldsymbol{\sigma}_1, \quad as \ x \to \infty, \quad (37)$$

we let $h(t,k)$ to satisfy

$$[\partial/\partial t - \boldsymbol{M}(x,t;z)]h^{-1}(t,z)\boldsymbol{\psi}(x,t;z) = 0, \text{as } x \to \infty \quad (38)$$

then

$$\left[\partial/\partial t - i2\lambda^4\boldsymbol{\sigma}_3 - 2\lambda^3\rho\boldsymbol{\sigma}_1\right]h^{-1}(t,z)\boldsymbol{E}_{\bullet 2}(x,z) = 0. \quad (39)$$

Due to

$$\boldsymbol{E}_{\bullet 2}(x,z) = \left(-i\rho z^{-1}e^{i\lambda\eta x}, \ e^{i\lambda\eta x}\right)^{\mathrm{T}}, \quad (40)$$

from (39) and (40), we have

$$h(t,z) = e^{i2\lambda^3\eta t}. \quad (41)$$

Therefore, the complete Jost functions should depend on time as follows

$$h(t,z)\tilde{\boldsymbol{\psi}}(x,z), \quad h^{-1}(t,z)\boldsymbol{\psi}(x,z); h(t,z)\boldsymbol{\phi}(x,z), \quad h^{-1}(t,z)\tilde{\boldsymbol{\phi}}(x,z) \quad (42)$$

Nevertheless, hereafter the time variable in Jost functions will be suppressed because it has no influence on the treatment of Z-S equation. By a similar procedure [9, 15], the scattering data has following time dependences

$$a(z,t) = a(z,0), b(t,z) = b(0)e^{-i4\lambda^3\eta t} \tag{43}$$

2.2 Zakharov-Shabat equations and breather-type N-soliton solution

A 2×1 column function $\boldsymbol{\Pi}(x,z)$ is introduced as usual

$$\boldsymbol{\Pi}(x,z) \equiv \begin{cases} \boldsymbol{\phi}(x,z)/a(z), & as\ z\ in\ \text{I, III}\ quadrants. \\ \tilde{\boldsymbol{\psi}}(x,z), & as\ z\ in\ \text{II, IV}\ quadrants. \end{cases} \tag{44}$$

here and hereafter note "$\equiv$" represents definition. There is an abrupt jump for $\boldsymbol{\Pi}(x,z)$ across both real and imaginary axes

$$\boldsymbol{\phi}(x,z)/a(z) - \tilde{\boldsymbol{\psi}}(x,z) = r(z)\boldsymbol{\psi}(x,z), \tag{45}$$

where

$$r(z) = b(z)/a(z) \tag{46}$$

is called the reflection coefficient. Due to $\mu_0 \neq 0$ in (34), Jost solutions do not tend to the free Jost solutions $\boldsymbol{E}(x,z)$ in the limit of $|z| \to \infty$. This is their most typical property which means that the usual procedure of constructing the equation of IST by a Cauchy contour integral must be invalid. In view of these abortive experiences, we proposed a revised method to derive a suitable IST and the corresponding Z-S equation by multiplying an inverse spectral parameter $1/\lambda$, $\lambda = (z + \rho^2 z^{-1})/2$, before the Z-S integrand. Meanwhile, our modification produces no new poles since the Lax operator $\boldsymbol{L}(\lambda) \to 0$, as $\lambda \to 0$. In another word, the both additional poles $z_0 = \pm i\rho$ generated by $\lambda = 0$ are removable. Under reflectionless case, that is, $r(z) = 0$, the Cauchy integral along with contour Γ shown in **Figure 1** gives

$$\frac{1}{\lambda}\{\boldsymbol{\Pi}(x,z) - \boldsymbol{E}_{\cdot 1}(x,z)\}e^{i\lambda\eta x} = \frac{1}{2\pi i}\oint_{\Gamma} dz' \frac{1}{z'-z}\frac{1}{\lambda'}\{\boldsymbol{\Pi}(x,z') - \boldsymbol{E}_{\cdot 1}(x,z')\}e^{i\lambda'\eta' x} \tag{47}$$

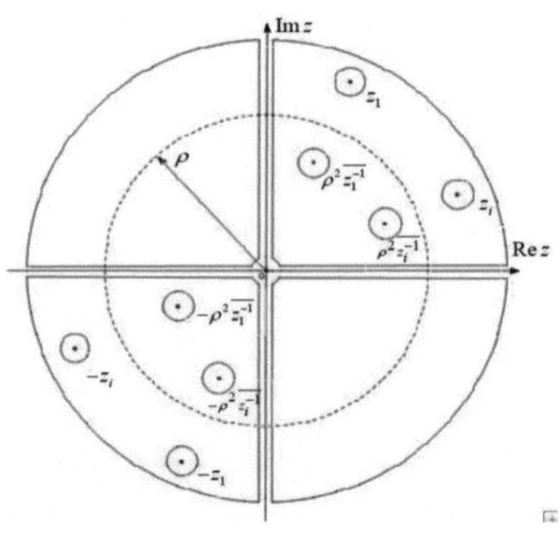

Figure 1.
The integral path for IST of the DNLS$^+$.

or

$$\tilde{\psi}(x,z) = e^{-i\lambda\eta x} + \lambda\left\{-\sum_{n=1}^{4N}\frac{1}{\lambda_n}\frac{1}{z_n - z}c_n\psi(x,z_n)e^{i\lambda_n\eta_n x}\right\}e^{-i\lambda\eta x} \tag{48}$$

where

$$c_n \equiv b_n/\dot{a}(z_n), \dot{a}(z_n) = da(z)/dz|_{z=z_n}; n = 1,2,\cdots,4N \tag{49}$$

Note that (27), (45), and (46) have been used in (48). The minus sign before the sum of residue number in (48) comes from the clock-wise contour integrals around the 4 N simple poles when the residue theorem is used, shown in **Figure 1**. By a standard procedure, the time dependences of b_n and c_n similar to (43) can be derived

$$b_n(t) = b_{n0}e^{-i4\lambda_n^3\eta_n t}, c_n = c_n(0)e^{-i4\lambda_n^3\eta t} \tag{50}$$

$$c_n(0) = b_{n0}/\dot{a}_n; n = 1,2,\cdots,4N \tag{51}$$

In the reflectionless case, the Zakharov-Shabat equations for DNLS$^+$ equation can be derived immediately from (48) as follows

$$\tilde{\psi}_1(x,z) = e^{-i\Lambda x} + \lambda\left[\sum_{n=1}^{2N}\frac{2z}{\lambda_n}\frac{1}{z^2 - z_n^2}c_n\psi_1(x,z_n)e^{i\Lambda_n x}\right]e^{-i\Lambda x} \tag{52}$$

$$\tilde{\psi}_2(x,z) = i\rho z^{-1}e^{-i\Lambda x} + \lambda\left[\sum_{n=1}^{2N}\frac{2z_n}{\lambda_n}\frac{1}{z^2 - z_n^2}c_n\psi_2(x,z_n)e^{i\Lambda_n x}\right]e^{-i\Lambda x} \tag{53}$$

here $\Lambda \equiv \lambda\eta, \Lambda_n \equiv \lambda(z_n)\eta(z_n) = \lambda_n\eta_n$, and in Eqs. (52) and (53), the terms corresponding to poles z_n, $(n = 1,2,\cdots,2N)$, have been combined with the terms corresponding to poles $z_{n+2N} = -z_n$. Substituting Eqs. (52) and (53) into formula (36) and letting $z \to \infty$, we attain the conjugate of the raw N-soliton solution (the time dependence is suppressed).

$$\overline{u}_N = U_N/V_N \tag{54}$$

$$U_N \equiv \rho\left[1 - \sum_{n=1}^{2N} i\frac{z_n}{\rho\lambda_n}c_n\psi_2(x,z_n)e^{i\Lambda_n x}\right] \tag{55}$$

$$V_N \equiv 1 + \sum_{n=1}^{2N}\frac{c_n}{\lambda_n}\psi_1(x,z_n)e^{i\Lambda_n x} \tag{56}$$

Letting $z = \rho^2 z_m^{-1}$, $(m = 1,2,\cdots,2N)$, respectively, in Eqs. (52) and (53), by use of reduction relations (19), (21), and (22), we can further change Eqs. (52) and (53) into the following form

$$\psi_1(x,z_m) = -i\rho z_m^{-1}e^{i\Lambda_m x} + \sum_{n=1}^{2N}\frac{\lambda_m c_n}{\lambda_n z_m^2}\frac{2\rho^3}{i\left(\rho^4 z_m^{-2} - z_n^2\right)}\psi_1(x,z_n)e^{i(\Lambda_n+\Lambda_m)x} \tag{57}$$

$$\psi_2(x,z_m) = e^{i\Lambda_m x} + \sum_{n=1}^{2N}\frac{\lambda_m z_n c_n}{\lambda_n z_m}\cdot\frac{2\rho}{i\left(\rho^4 z_m^{-2} - z_n^2\right)}\psi_2(x,z_n)e^{i(\Lambda_n+\Lambda_m)x} \tag{58}$$

$m = 1,2,\cdots,2N$. An implicit time dependence of the complete Jost functions ψ_1 and ψ_2 besides c_n should be understood. To solve Eq. (58), we define that

$$(\boldsymbol{\varphi}_2)_n \equiv i\frac{z_n}{\rho\lambda_n}\sqrt{\frac{c_n}{2}}\psi_2(x,z_n), \boldsymbol{\varphi_2} \equiv ((\varphi_2)_1, (\varphi_2)_2, \ldots, (\varphi_2)_{2N}) \tag{59}$$

$$f_n \equiv \sqrt{2c_n}e^{i\Lambda_n x}, g_n \equiv i\sqrt{\frac{c_n}{2}}\frac{z_n}{\rho\lambda_n}e^{i\Lambda_n x} = \frac{iz_n}{2\rho\lambda_n}f_n, n = 1, 2, \cdots, 2N \tag{60}$$

$$\boldsymbol{f} \equiv (f_1, f_2, \ldots f_{2N}), \boldsymbol{g} \equiv (g_1, g_2, \ldots g_{2N}) \tag{61}$$

$\boldsymbol{B} \equiv$ Matrix $(B_{nm})_{2N\times 2N}$, with

$$B_{nm} \equiv f_n\frac{\rho}{i\left(z_n^2 - \rho^4 z_m^{-2}\right)}f_m, m, n = 1, 2, \cdots, 2N. \tag{62}$$

Then Eq. (58) can be rewritten as

$$(\boldsymbol{\varphi}_2)_m = g_m - \sum_{n=1}^{2N}(\boldsymbol{\varphi}_2)_n B_{nm}, m = 1, 2, \cdots, 2N \tag{63}$$

or in a more compact form

$$\boldsymbol{\varphi}_2 = \boldsymbol{g} - \boldsymbol{\varphi}_2\boldsymbol{B}. \tag{64}$$

The above equation gives

$$\boldsymbol{\varphi}_2 = \boldsymbol{g}(\boldsymbol{I} + \boldsymbol{B})^{-1}. \tag{65}$$

Note that the choice of poles, z_n, $(n = 1, 2, \cdots, N)$, should make $\det(\boldsymbol{I} + \boldsymbol{B})$ nonzero and $(\boldsymbol{I} + \boldsymbol{B})$ an invertible matrix. On the other hand, Eq. (55) can be rewritten as

$$U_N = \rho\left[1 - \boldsymbol{\varphi}_2\boldsymbol{f}^{\mathrm{T}}\right], \tag{66}$$

hereafter a superscript "T" represents transposing of a matrix. Substituting Eq. (65) into (66) leads to

$$U_N = \rho\left[1 - \boldsymbol{g}(\boldsymbol{I} + \boldsymbol{B})^{-1}\boldsymbol{f}^{\mathrm{T}}\right] = \rho\frac{\det\left(\boldsymbol{I} + \boldsymbol{B} - \boldsymbol{f}^{\mathrm{T}}\boldsymbol{g}\right)}{\det(\boldsymbol{I} + \boldsymbol{B})} = \rho\frac{\det(\boldsymbol{I} + \boldsymbol{A})}{\det(\boldsymbol{I} + \boldsymbol{B})} \tag{67}$$

where

$$\boldsymbol{A} \equiv \boldsymbol{B} - \boldsymbol{f}^{\mathrm{T}}\boldsymbol{g} \tag{68}$$

with

$$A_{nm} \equiv B_{nm} - f_n g_m = \left(z_m z_n \lambda_n / \rho^2 \lambda_m\right) B_{nm}. \tag{69}$$

To solve Eq. (57), we define that

$$(\boldsymbol{\varphi_1})_m \equiv i\sqrt{\frac{c_m}{2}}\frac{z_m}{\rho\lambda_m}\psi_1(x,z_m), \boldsymbol{\varphi_1} \equiv ((\varphi_1)_1, (\varphi_1)_2, \cdots, (\varphi_1)_{2N}) \tag{70}$$

$$f'_m \equiv i\sqrt{2c_m}\cdot\frac{\rho}{z_m}e^{i\Lambda_m x} = i\frac{\rho}{z_m}f_m; g'_m = \sqrt{\frac{c_m}{2}}\cdot\frac{1}{\lambda_m}e^{i\Lambda_m x} = \frac{-iz_m}{2\rho\lambda_m}f'_m \tag{71}$$

$$f' = (f'_1, f'_2, \dots f'_{2N}); \boldsymbol{g}' = (g'_1, g'_2, \dots g'_{2N}); \tag{72}$$

$$D'_{nm} \equiv f'_n \left[\frac{-\rho}{i\left(z_n^2 - \rho^4 z_m^{-2}\right)} \right] f'_m = \frac{\rho^2}{z_n z_m} B_{nm} \text{ (c.f.(1.70) and (1.61))} \tag{73}$$

with $n, m = 1, 2, \cdots, 2N$. Then Eq. (57) can be rewritten as

$$(\boldsymbol{\varphi_1})_m = g'_m - \sum_{n=1}^{2N} (\boldsymbol{\varphi_1})_n D'_{nm}, m = 1, 2, \cdots, 2N \tag{74}$$

or in a more compact form

$$\boldsymbol{\varphi}_1 = g' - \boldsymbol{\varphi_1} \boldsymbol{D}' \tag{75}$$

The above equation givess

$$\boldsymbol{\varphi_1} = g'(\boldsymbol{I} + \boldsymbol{D}')^{-1} \tag{76}$$

Note that the choice of poles, z_n, $(\mathrm{n} = 1, 2, \cdots, N)$, should make $\det(\boldsymbol{I} + \boldsymbol{D}')$ nonzero and $(\boldsymbol{I} + \boldsymbol{D}')$ an invertible matrix. On the other hand, Eq. (56) can be rewritten as

$$V_N = 1 - \sum_{n=1}^{2N} (\boldsymbol{\varphi_1})_n f'_n = 1 - \boldsymbol{\varphi_1} f'^{\mathrm{T}} \tag{77}$$

Substituting Eq. (76) into (77), we thus attain

$$V_N = 1 - g'(\boldsymbol{I} + \boldsymbol{D}')^{-1} \boldsymbol{f}'^{\mathrm{T}} = \frac{\det\left(\boldsymbol{I} + \boldsymbol{D}' - \boldsymbol{f}'^{\mathrm{T}} g'\right)}{\det(\boldsymbol{I} + \boldsymbol{D}')} \equiv \frac{\det(\boldsymbol{I} + \boldsymbol{B}')}{\det(\boldsymbol{I} + \boldsymbol{D}')} \tag{78}$$

where use is made of Appendix A.1 and

$$B'_{nm} \equiv \left(\boldsymbol{D}' - \boldsymbol{f}'^{\mathrm{T}} \boldsymbol{g}'\right)_{nm} = \frac{z_m z_n \lambda_n}{\rho^2 \lambda_m} D'_{nm} = \frac{\lambda_n}{\lambda_m} B_{nm} \tag{79}$$

In the end, by substituting (67) and (78) into (54), we attain the N-soliton solution to the DNLS⁺ Eq. (3) under NVBC and reflectionless case (note that the time dependence of soliton solution naturally emerges in $c_n(t)$):

$$\overline{u}(x,t) = \frac{U_N}{V_N} = \rho \frac{\det(\boldsymbol{I} + \boldsymbol{A}) \det(\boldsymbol{I} + \boldsymbol{D}')}{\det(\boldsymbol{I} + \boldsymbol{B}) \det(\boldsymbol{I} + B')} \equiv \rho \frac{C_N D_N}{\overline{D}_N^2}, \tag{80}$$

here

$$C_N \equiv \det(\boldsymbol{I} + \boldsymbol{A}), \overline{D}_N \equiv \det(\boldsymbol{I} + \boldsymbol{B}) \tag{81}$$

The solution has a standard form as (80), that is

$$\det(\boldsymbol{I} + \boldsymbol{B}') = \det(\boldsymbol{I} + \boldsymbol{B}) = \overline{\det(\boldsymbol{I} + \boldsymbol{D}')} \equiv \overline{D}, \tag{82}$$

which can be proved by direct calculation for the $N = 1$ case and by some special algebra techniques for the $N > 1$ case.

3. Verification of standard form and the explicit breather-type multi-soliton solution

3.1 Verification of $\det(\boldsymbol{I}+\boldsymbol{B}')=\det(\boldsymbol{I}+\boldsymbol{B})$

In order to prove the first identity in (82), we firstly calculate $\overline{D}_N=\det(\boldsymbol{I}+\boldsymbol{B})$. By use of (60)–(62), Binet-Cauchy formula, (Appendix (A.2)) and an important determinant formula, (Appendix (A.3)), we have

$$\overline{\boldsymbol{D}}_{\boldsymbol{N}}\equiv\det(\boldsymbol{I}+\boldsymbol{B})=1+\sum_{r=1}^{2N}\sum_{1\le n_1<\cdots<n_r\le 2N}B(n_1,n_2,\cdots,n_r),\tag{83}$$

here $B(n_1,n_2,\cdots,n_r)$ is a rth-order principal minor of B consisting of elements belonging to not only rows $(n_1,n_2,\cdots,n_r)$ but also columns $(n_1,n_2,\cdots,n_r)$. Due to (62),

$$B(n_1,n_2,\cdots,n_r)=\prod_{n,m}f_n\left[\frac{\rho}{i\left(z_n^2-\rho^4z_m^{-2}\right)}\right]f_m\prod_{n<m}i\left(z_n^2-z_m^2\right)\cdot i\left[\left(-\rho^4z_n^{-2}\right)-\left(-\rho^4z_m^{-2}\right)\right]\tag{84}$$

in (84), $n,m\in(n_1,n_2,\cdots,n_r)$. The technique of calculating $B(n_1,n_2,\cdots,n_r)$ is to couple term $i\left(z_n^2-\rho^4z_m^{-2}\right)^{-1}$ with term $i\left(z_m^2-\rho^4z_n^{-2}\right)$ into pair, $(n\neq m)$, in the denominator of $\prod_{n,m}(\cdots)$, (with totally $r(r-1)/2$ pairs), and transplant them into the denominator of $\prod_{n<m}(\cdots)$, and combine with $i\left(z_n^2-z_m^2\right)i\left(\rho^4z_m^{-2}-\rho^4z_n^{-2}\right)$ in $\prod_{n<m}(\cdots)$ to form a typical factor as a whole, (with just totally $r(r-1)/2$ pairs). Note that if we define

$$z_n\equiv\rho e^{\delta_n+i\beta_n},\ \text{with}\ \delta_n>0,\beta_n\in(0,\pi/2),\tag{85}$$

and further define that

$$z_n^2/\rho^2=e^{2\delta_n+i2\beta_n}\equiv\tanh\Theta_n,\tag{86}$$

then the typical factor is

$$\frac{i\left(z_n^2-z_m^2\right)i\left(\rho^4z_m^{-2}-\rho^4z_n^{-2}\right)}{i\left(z_n^2-\rho^4z_m^{-2}\right)i\left(z_m^2-\rho^4z_n^{-2}\right)}=\left(\frac{z_n^2/\rho^2-z_m^2/\rho^2}{1-z_n^2z_m^2/\rho^4}\right)^2=\tanh^2(\Theta_n-\Theta_m)\tag{87}$$

and

$$\begin{aligned}B(n_1,n_2,\cdots,n_r)&=\prod_n f_n^2\left[\frac{\rho}{i\left(z_n^2-\rho^4z_n^{-2}\right)}\right]\prod_{n<m}\tanh^2(\Theta_n-\Theta_m)\\&=\prod_n F_n\prod_{n<m}\tanh^2(\Theta_n-\Theta_m)\end{aligned}\tag{88}$$

here $n,m\in(n_1,n_2,\cdots,n_r)$, and a typical function F_n is defined as

$$F_n\equiv B_{n\,n}=f_n^2\frac{\rho}{i\left(z_n^2-\rho^4z_n^{-2}\right)}=\frac{2\rho}{i\left(z_n^2-\rho^4z_n^{-2}\right)}c_n(t)e^{i2\Lambda_nx}\tag{89}$$

where use is made of formula (60) and (62), the time dependence of the solution naturally emerged in $c_n(t)$. Substituting Eq. (88) into (83) thus completes the computation of $\overline{D}_N$.

Secondly, let us calculate $\det(\boldsymbol{I}+\boldsymbol{D}')$. By use of (72) and (73), Binet-Cauchy formula, (Appendix A.2) and an important matrix formula, (Appendix A.3), we have

$$\det(\boldsymbol{I}+\boldsymbol{D}') = 1 + \sum_{r=1}^{2N} \sum_{1\le n_1<n_2<\cdots<n_r\le 2N} D'(n_1, n_2, \cdots, n_r), \tag{90}$$

where $D'(n_1, n_2, \cdots, n_r)$ is the principal minor of a $r'th$-order submatrix of $\boldsymbol{D}'$ consisting of elements belonging to not only rows $(n_1, n_2, \cdots, n_r)$ but also columns$(n_1, n_2, \cdots, n_r)$, and

$$D'(n_1, n_2, \cdots, n_r) = \prod_{n,m} f'_n \left[\frac{i\rho}{z_n^2 - \rho^4 z_m^{-2}}\right] f'_m \prod_{n<m} \left(z_n^2 - z_m^2\right)\left(\rho^4 z_m^{-2} - \rho^4 z_n^{-2}\right) \tag{91}$$

$n, m \in (n_1, n_2, \cdots, n_r)$. Using the same tricks as used in dealing with (84) leads to

$$D'(n_1, n_2, \cdots, n_r) = \prod_n (-1)^r f'^2_n \left[\frac{\rho}{i\left(z_n^2 - \rho^4 z_n^{-2}\right)}\right] \prod_{n<m} \frac{\left(z_n^2 - z_m^2\right)\left(\rho^4 z_m^{-2} - \rho^4 z_n^{-2}\right)}{\left(z_n^2 - \rho^4 z_m^{-2}\right)\left(z_m^2 - \rho^4 z_n^{-2}\right)}$$

$$= \prod_n \left(\frac{\rho}{z_n}\right)^2 F_n \prod_{n<m} \tanh^2(\Theta_n - \Theta_m) \tag{92}$$

Thirdly, let us calculate $\det\left(\boldsymbol{I}+\boldsymbol{D}'-\boldsymbol{f}'^{\mathrm{T}}g'\right) \equiv \det(\boldsymbol{I}+\boldsymbol{B}')$ with $\boldsymbol{B}' \equiv \boldsymbol{D}' - \boldsymbol{f}'^{\mathrm{T}}g'$.

According to (79) and Binet-Cauchy formula (Appendix (A.2)), similarly we have

$$\det\left(\boldsymbol{I}+\boldsymbol{D}'-\boldsymbol{f}'^{\mathrm{T}}\boldsymbol{g}'\right) = \det(\boldsymbol{I}+\boldsymbol{B}') = 1 + \sum_{r=1}^{2N} \sum_{1\le n_1<n_2<\cdots<n_r\le 2N} B'(n_1, n_2, \cdots, n_r) \tag{93}$$

$$B'(n_1, n_2, \cdots, n_r) = \prod_{n,m} \left(\frac{z_n z_m \lambda_n}{\rho^2 \lambda_m}\right) f'_n \left[\frac{-\rho}{i\left(z_n^2 - \rho^4 z_m^{-2}\right)}\right] f'_m \prod_{n<m} i\left(z_n^2 - z_m^2\right) i\left(\rho^4 z_m^{-2} - \rho^4 z_n^{-2}\right) \tag{94}$$

$n, m \in (n_1, n_2, \cdots, n_r)$. Using the same tricks as that used in treating (84) leads to

$$B'(n_1, n_2, \cdots, n_r) = \prod_n \left(\frac{z_n}{\rho}\right)^2 \left(\frac{i\rho}{z_n}\right)^2 \cdot f_n^2 \left(\frac{-\rho}{i\left(z_n^2 - \rho^4 z_n^{-2}\right)}\right) \prod_{n<m} \tanh^2(\Theta_n - \Theta_m)$$

$$= \prod_n f_n^2 \left(\frac{\rho}{i\left(z_n^2 - \rho^4 z_n^{-2}\right)}\right) \prod_{n<m} \tanh^2(\Theta_n - \Theta_m) = \prod_n F_n \prod_{n<m} \tanh^2(\Theta_n - \Theta_m)$$
$$\equiv B(n_1, n_2, \cdots, n_r) \tag{95}$$

Due to (95), comparing (83) and (93) results in the expected identity and completes the verification of the first identity in (82).

3.2 Verification of $\overline{\det(I+D')} = \det(I+B)$

Our most difficult and challenging task is to prove the second identity in (82). For convenience of discussion, we define that

$$z_{\hat{n}} \equiv \rho^2 \overline{z_n^{-1}} \tag{96}$$

then

$$z_{2n} = z_{\wedge 2n-1} = \rho^2 \overline{z_{2n-1}^{-1}}, z_{2n-1} = z_{\hat{2n}} = \rho^2 \overline{z_{2n}^{-1}} \tag{97}$$

or

$$\hat{2n} = 2n-1, 2n \hat{-} 1 = 2n, (n = 1, 2, \cdots, N) \tag{98}$$

Then the sequence of poles (25) is just in the same order as follows

$$z_{\hat{2}}, z_{\hat{1}}; z_{\hat{4}}, z_{\hat{3}}; \cdots; z_{\wedge 2N}, z_{\wedge 2N-1} \tag{99}$$

On the other hand, due to (28), (62), and (73), we have

$$D'_{nm} = \frac{\rho^2}{z_n z_m} \cdot f_n \frac{\rho}{i\left(z_n^2 - \rho^4 z_m^{-2}\right)} f_m \tag{100}$$

Then

$$\overline{D'_{nm}} = \frac{\rho^2}{\overline{z_n z_m}} \cdot \sqrt{4\overline{c}_n \overline{c}_m} \cdot \frac{\rho}{-i\left(\overline{z}_n^2 - \rho^4 \overline{z_m^{-2}}\right)} \cdot e^{-i\left(\overline{\eta_n \lambda_n} + \overline{\eta_m \lambda_m}\right)x} \tag{101}$$

Substituting $\overline{z}_n = \rho^2 z_{\hat{n}}^{-1}, \overline{z}_m = \rho^2 z_{\hat{m}}^{-1}$ into above formula and using following relation

$$\overline{\eta}_n \overline{\lambda}_n = -\overline{\eta}_{\hat{n}} \overline{\lambda}_{\hat{n}}, \overline{c}_n = \rho^2 z_{\hat{n}}^{-2} c_{\hat{n}} \tag{102}$$

We can get an important relation between D'_{nm} and $B_{\hat{m}\hat{n}}$

$$\overline{D}'_{nm} = f_{\hat{m}} \frac{\rho}{i\left(z_{\hat{m}}^2 - \rho^4 z_{\hat{n}}^{-2}\right)} f_{\hat{n}} = B_{\hat{m}\hat{n}} = B_{\hat{n}\hat{m}}^{\mathrm{T}} \tag{103}$$

On the other hand, an unobvious symmetry between matrices $(B_{nm})_{2N\times 2N}$ and $(B_{\hat{n}\hat{m}})_{2N\times 2N}$ is found

$$diag(\sigma_1, \cdots, \sigma_1)_{2N\times 2N} (B_{nm})_{2N\times 2N} diag(\sigma_1, \cdots, \sigma_1)_{2N\times 2N} = (B_{\hat{n}\hat{m}})_{2N\times 2N} \tag{104}$$

It can be rewritten in a more explicit form

$$
\begin{pmatrix} \boldsymbol{\sigma}_1 & 0 & \cdots & \cdots & 0 \\ 0 & \boldsymbol{\sigma}_1 & & & \vdots \\ \vdots & & \ddots & & \vdots \\ \vdots & & & \ddots & \vdots \\ 0 & \cdots & \cdots & \cdots & \boldsymbol{\sigma}_1 \end{pmatrix}
\begin{pmatrix} B_{11} & B_{12} & \cdots & \cdots & B_{1,2N} \\ B_{21} & B_{22} & \cdots & \cdots & B_{2,2N} \\ \vdots & \vdots & & & \vdots \\ \vdots & \vdots & & & \vdots \\ B_{2N,1} & B_{2N,2} & \cdots & \cdots & B_{2N,2N} \end{pmatrix}
\begin{pmatrix} \boldsymbol{\sigma}_1 & 0 & \cdots & \cdots & 0 \\ 0 & \boldsymbol{\sigma}_1 & & & \vdots \\ \vdots & & \ddots & & \vdots \\ \vdots & & & \ddots & \vdots \\ 0 & \cdots & \cdots & \cdots & \boldsymbol{\sigma}_1 \end{pmatrix}
$$

$$
= \begin{pmatrix} B_{22} & B_{21} & \cdots & B_{2,2N} & B_{2,2N-1} \\ B_{12} & B_{11} & \cdots & B_{1,2N} & B_{1,2N-1} \\ \vdots & \vdots & \ddots & \vdots & \vdots \\ B_{2N,2} & B_{2N,1} & \cdots & B_{2N,2N} & B_{2N,2N-1} \\ B_{2N-1,2} & B_{2N-1,1} & \cdots & B_{2N-1,2N} & B_{2N-1,2N-1} \end{pmatrix}
$$

$$
= \begin{pmatrix} B_{\hat{1}\hat{1}} & B_{\hat{1}\hat{2}} & \cdots & \cdots & B_{\hat{1}\widehat{2N}} \\ B_{\hat{2}\hat{1}} & B_{\hat{2}\hat{2}} & \cdots & \cdots & B_{\hat{2},\widehat{2N}} \\ \vdots & \vdots & & & \vdots \\ B_{\widehat{2N-1},\hat{1}} & B_{\widehat{2N-1},\hat{2}} & & & B_{\widehat{2N-1},\widehat{2N}} \\ B_{\widehat{2N},\hat{1}} & B_{\widehat{2N},\hat{2}} & \cdots & \cdots & B_{\widehat{2N},\widehat{2N}} \end{pmatrix}
\tag{105}
$$

The last equation in (105) is due to (97) and (99), thus from (103) and (104), we have

$$
\begin{pmatrix} \boldsymbol{\sigma}_1 & & 0 \\ & \ddots & \\ 0 & & \boldsymbol{\sigma}_1 \end{pmatrix} (\boldsymbol{I}+\boldsymbol{B})_{2N\times 2N} \begin{pmatrix} \boldsymbol{\sigma}_1 & & 0 \\ & \ddots & \\ 0 & & \boldsymbol{\sigma}_1 \end{pmatrix} = \left(\boldsymbol{I}+\overline{\boldsymbol{D}'}^{\mathrm{T}}\right) \tag{106}
$$

The determinants of matrices at both sides of (106) are equal to each other

$$
(\det\sigma_1)^N \det(\boldsymbol{I}+\boldsymbol{B})(\det\boldsymbol{\sigma_1})^N = \det\left(\boldsymbol{I}+\overline{\boldsymbol{D}'}^{\mathrm{T}}\right) = \overline{\det(\boldsymbol{I}+\boldsymbol{D}')} \tag{107}
$$

The left hand of (107) is just $\det(\boldsymbol{I}+\boldsymbol{B})$, and this completes verification of identity (82). From the verified (82), we know multi-soliton solution (80) is surely of a typical form as expected.

3.3 The explicit N-soliton solution to the DNLS⁺ equation with NVBC

In order to get an explicit N-soliton solution to the DNLS⁺ Eq. (1) with NVBC, firstly we need to make an inverse Galileo transformation of (2) by $(x,t) \to (x-\rho^2 t, t)$ in $F_n(x,t)$ in (89). Due to (51) and (85)–(87), the typical soliton kernel function F_n can be rewritten as

$$
F_n = \frac{2\rho}{i\left(z_n^2 - \rho^4 z_n^{-2}\right)} \frac{b_n(0)}{\dot{a}(z_n)} \exp i2\Lambda_n\left[x - \left(2\lambda_n^2 + \rho^2\right)t\right] \tag{108}
$$

$$
F_n \equiv \exp\left(-\theta_n + i\varphi_n\right) \tag{109}
$$

$$
\theta_n(x,t) = \left(\rho^2 \sin 2\beta_n ch2\delta_n\right)\left[(x-x_{n0}) - \rho^2\left(2 + \frac{cos2\beta_n ch4\delta_n}{ch2\delta_n}\right)t\right] \equiv \nu_n(x - \upsilon_n t - x_{n0}) \tag{110}
$$

$$\varphi_n(x,t) = \left(\rho^2 \cos 2\beta_n sh2\delta_n\right)\left[x - \rho^2\left(2 + \frac{\cos 4\beta_n ch2\delta_n}{\cos 2\beta_n}\right)t\right] + \varphi_{n0} \equiv \mu_n(x - \xi_n t) + \varphi_{n0} \tag{111}$$

$$\mu_n = \rho^2 \cos 2\beta_n sh2\delta_n, \nu_n = \rho^2 \sin 2\beta_n ch2\delta_n \tag{112}$$

$$\upsilon_n = \rho^2\left(2 + \frac{ch4\delta_n \cos 2\beta_n}{ch2\delta_n}\right), \xi_n = \rho^2\left(2 + \frac{ch2\delta_n \cos 4\beta_n}{\cos 2\beta_n}\right) \tag{113}$$

$$\frac{2\rho}{i\left(z_n^2 - \rho^4 z_n^{-2}\right)} \frac{b_n(0)}{\dot{a}(z_n)} \equiv \exp\left(\nu_n x_{n0}\right) \exp\left(i\varphi_{n0}\right) \tag{114}$$

$$\tanh\left(\Theta_n - \Theta_m\right) = -\frac{sh(\delta_n - \delta_m)\cos\left(\beta_n - \beta_m\right) + ich(\delta_n - \delta_m)\sin\left(\beta_n - \beta_m\right)}{sh(\delta_n + \delta_m)\cos\left(\beta_n + \beta_m\right) + ich(\delta_n + \delta_m)\sin\left(\beta_n + \beta_m\right)} \tag{115}$$

where in (114) the n'th pole-dependent constant factor has been absorbed by redefinition of the n'th soliton center and initial phase in (110)–(111).

Secondly, we need to calculate determinant $C_N = \det(\boldsymbol{I} + \boldsymbol{A}) \equiv \det\left(\boldsymbol{I} + \boldsymbol{B} - \boldsymbol{f}^{\mathrm{T}}\boldsymbol{g}\right)$. According to the definition of $\boldsymbol{A}$ in (68)–(69), using Binet-Cauchy formula, (Appendix (A.2)), leads to

$$C_N = \det(\boldsymbol{I} + \boldsymbol{A}) = \det\left(\boldsymbol{I} + \boldsymbol{B} - \boldsymbol{f}^{\mathrm{T}}\boldsymbol{g}\right) = 1 + \sum_{r=1}^{2N} \sum_{1 \le n_1 < n_2 < \cdots < n_r \le 2N} \boldsymbol{A}(n_1, n_2, \cdots, n_r) \tag{116}$$

where $\boldsymbol{A}(n_1, n_2, \cdots, n_r)$ is the determinant of a r'th-order minor of $\boldsymbol{A}$ consisting of elements belonging to not only rows $(n_1, n_2, \cdots, n_r)$ but also columns $(n_1, n_2, \cdots, n_r)$.

$$\boldsymbol{A}(n_1, n_2, \cdots, n_f) = \prod_{n,m}\left(\frac{z_n z_m \lambda_n}{\rho^2 \lambda_m}\right) f_n \left[\frac{\rho}{i\left(z_n^2 - \rho^4 z_m^{-2}\right)}\right] f_m \prod_{n<m} i\left(z_n^2 - z_m^2\right) i\left(\rho^4 z_m^{-2} - \rho^4 z_n^{-2}\right) \tag{117}$$

$n, m \in (n_1, n_2, \cdots, n_r)$. Using the same tricks as used in dealing with (84) leads to

$$\begin{aligned} \boldsymbol{A}(n_1, n_2, \cdots, n_r) &= \prod_n \frac{z_n^2}{\rho^2} f_n^2 \frac{\rho}{i\left(z_n^2 - \rho^4 z_n^{-2}\right)} \prod_{n<m} \tanh^2(\Theta_n - \Theta_m) \\ &= \prod_n F_n \tanh\Theta_n \prod_{n<m} \tanh^2(\Theta_n - \Theta_m) \end{aligned} \tag{118}$$

$n, m \in (n_1, n_2, \cdots, n_r)$. Substituting (108)–(115) into (88) and (83) gives the explicit values of $\overline{D_N} \equiv \det(\boldsymbol{I} + \boldsymbol{B})$ and D_N. Substituting (118) into (116) then completes calculation of C_N in (81). In the end, by substituting (83) and (116) into (80), we thus attain an explicit breather-type N-soliton solution of the DNLS⁺ Eq. (1) with NVBC under reflectionless case, based upon a revised and improved inverse scattering transform. Due to the limitation of space, the asymptotic behaviors of the N-soliton solution are just similar to that of the pure N-soliton solution in Ref. [7] and thus not discussed here, but it should be emphasized that in the limit of $t \to \pm\infty$, the N-soliton solution surely can be viewed as summation of N single solitons with a definite displacement and phase shift of each soliton in the whole process of elastic collisions.

4. The one and two-soliton solutions to DNLS⁺ equation with NVBC

We give two concrete examples – the one and two breather-type soliton solutions in illustration of the general explicit $N-$ soliton formula.

In the case of one-soliton solution, $N=1$, $z_1 \equiv \rho e^{\delta_1} e^{i\beta_1}, z_2 = \rho^2 \overline{z}_1^{-1} = \rho e^{-\delta_1} e^{i\beta_1}$, and $\delta_1 > 0,\ \beta_1 \in (0, \pi/2)$, using formula (82), (88), (116), (108)–(115), and

$$c_1(0) = \frac{b_{10}}{\dot{a}(z_1)} = b_{10} \frac{z_1^2 - \overline{z}_1^2}{2z_1} \frac{z_1^2 - \overline{z}_2^2}{z_1^2 - z_2^2} \frac{z_1 z_2}{\overline{z}_1 \overline{z}_2} \tag{119}$$

$$c_2(0) = \frac{b_{20}}{\dot{a}(z_2)} = b_{20} \frac{z_2^2 - \overline{z}_2^2}{2z_2} \frac{z_2^2 - \overline{z}_1^2}{z_2^2 - z_1^2} \frac{z_1 z_2}{\overline{z}_1 \overline{z}_2} \tag{120}$$

we have

$$\begin{aligned} \overline{D}_1 &= 1 + B(n_1 = 1) + B(n_1 = 2) + B(n_1 = 1, n_2 = 2) \\ &= 1 + F_1 + F_2 + F_1 F_2 \tanh^2(\Theta_1 - \Theta_2) \\ &= 1 + \frac{\sin 2\beta_1}{\sinh 2\delta_1} e^{i\beta_1} e^{-\theta_1} \left(e^{\delta_1} e^{i\varphi_1} + e^{-\delta_1} e^{-i\varphi_1} \right) - e^{i2\beta_1} e^{-2\theta_1} \end{aligned} \tag{121}$$

$$\begin{aligned} C_1 &= 1 + A(n_1 = 1) + A(n_1 = 2) + A(n_1 = 1, n_2 = 2) \\ &= 1 + F_1 \tanh\Theta_1 + F_2 \tanh\Theta_2 + F_1 F_2 \tanh\Theta_1 \tanh\Theta_2 \tanh^2(\Theta_1 - \Theta_2) \\ &= 1 + \frac{\sin 2\beta_1}{\sinh 2\delta_1} e^{i3\beta_1} e^{-\theta_1} \left(e^{3\delta_1} e^{i\varphi_1} + e^{-3\delta_1} e^{-i\varphi_1} \right) - e^{i6\beta_1} e^{-2\theta_1} \end{aligned} \tag{122}$$

where not as that in (114), we define

$$b_{10} e^{i2\lambda_1 \eta_1 (x - 2\lambda_1^2 t)} \equiv e^{-\theta_1} e^{i\varphi_1}, b_{10} = e^{\nu_1 x_{10}} e^{i\varphi_{10}} \tag{123}$$

$$-b_{20} e^{i2\lambda_2 \eta_2 (x - 2\lambda_2^2 t)} \equiv e^{-\theta_2} e^{i\varphi_2} \tag{124}$$

$$\theta_1(x,t) \equiv \nu_1 (x - \upsilon_1 t - x_{10}) \tag{125}$$

$$\varphi_1(x,t) \equiv \mu_1 (x - \xi_1 t) + \varphi_{10}, \tag{126}$$

$$\text{with } \mu_1 = \rho^2 \cos 2\beta_1 sh 2\delta_1, \nu_1 = \rho^2 \sin 2\beta_1 ch 2\delta_1, \text{ and} \tag{127}$$

$$\upsilon_1 = \rho^2 \left(2 + \frac{ch 4\delta_1 \cos 2\beta_1}{ch 2\delta_1} \right), \xi_1 = \rho^2 \left(2 + \frac{ch 2\delta_1 \cos 4\beta_1}{\cos 2\beta_1} \right) \tag{128}$$

$$\theta_2 = \theta_1, \varphi_2 = -\varphi_1 \tag{129}$$

It is different slightly from the definition in Eq. (114) for the reason that an additional minus sign "$-$" before b_{20} can support (131)–(133) due to $-b_{20} = \overline{b}_{10}$. Substituting (121)–(122) into the following formula gives the one-soliton solution of DNLS⁺ Eq. (1) with NVBC.

$$\overline{u}_1(x,t) = \rho C_1 D_1 / \overline{D}_1^2, \text{or } u_1(x,t) = \rho \overline{C}_1 \overline{D}_1 / D_1^2, \tag{130}$$

which is generally called a breather solution and shown as **Figure 2**.

Formula (130) includes the one-soliton solution of the DNLS equation with VBC as its limit case. In the limit of $\rho \to 0,\ \delta_1 \to \infty$ but an invariant ρe^{δ_1}, we have

$$\rho C_1 \to 4|\lambda_1| \sin 2\beta_1 e^{i3\beta_1} e^{-\theta_1} e^{i\varphi_1} \tag{131}$$

$$\overline{D}_1 \to 1 - e^{i2\beta_1} e^{-2\theta_1}, \text{and } D_1 \to 1 - e^{-i2\beta_1} e^{-2\theta_1} \tag{132}$$

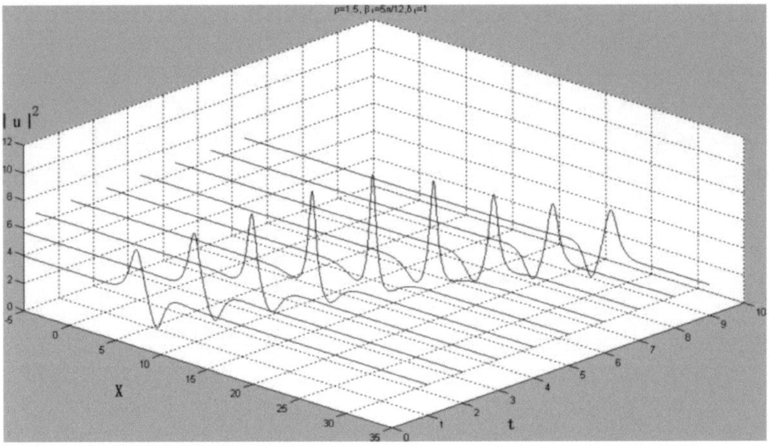

Figure 2.
The evolution of one-breather solution in time and space.

Substituting (131) and (132) into (130), we can attain

$$u_1(x,t) = 4|\lambda_1| \sin 2\beta_1 e^{-i3\beta_1} e^{-\theta_1} e^{-i\varphi_1} \left(1 - e^{i2\beta_1} e^{-2\theta_1}\right) / \left(1 - e^{-i2\beta_1} e^{-2\theta_1}\right)^2 \tag{133}$$

If we redefine $z_1 \equiv \rho e^{\delta_1} e^{i(\pi/2-\beta_1')}$, $z_2 \equiv \rho e^{-\delta_1} e^{i(\pi/2-\beta_1')}$, then $\bar{u}_1(x,t) = q_1(x,t)$, the complex conjugate of one-soliton solution (133), completely reproduce the one-soliton solution that gotten in [17–20, 23], under the VBC limit with $\rho \to 0$, $\delta_1 \to \infty$, but $\rho e^{\delta_1} = 2|\lambda_1'|$ invariant, up to a permitted global constant phase factor. This verifies the validity of our formula of N-Soliton solution and the reliability of the newly revised inverse scattering transform.

The degenerate case for $N = 1$, or the so-called pure one soliton solution, is also a typical illustration of the present improved IST. It can be dealt with by letting $\delta_1 \to 0$. The simple poles $z_1 (= \rho e^{i\beta_1})$ and $z_2 (= \rho^2 \bar{z}_1^{-1} = \rho e^{i\beta_1})$ are coincident, so do $z_3 (= -z_1)$ and $z_4 (= -\rho e^{i\beta_1})$. Meanwhile $\mu_1 \to 0$, $\varphi_1 \to 0$, $\nu_1 = \rho^2 \sin 2\beta_1$, $-ib_{10} \in \mathbb{R}$. Especially for the degenerate case, we have

$$a(z) = \frac{z^2 - z_1^2 \bar{z}_1}{z^2 - \bar{z}_1^2 z_1}, c_1(0) = \frac{b_{10}}{\dot{a}(z_1)} = b_{10} \frac{z_1^2 - \bar{z}_1^2}{2z_1} \frac{z_1}{\bar{z}_1} \tag{134}$$

$$-ib_{10} e^{i2\lambda_1\eta_1(x-2\lambda_1^2 t)} \equiv \varepsilon e^{-\theta_1}, \theta_1(x,t) \equiv \nu_1(x - v_1 t - x_{10}) \tag{135}$$

with $\nu_1 = \rho^2 \sin 2\beta_1$, $v_1 = \rho^2(1 + 2\cos^2\beta_1)$, $\varepsilon =$ sgn $(-ib_{10})$. Then we have

$$\overline{D} = 1 + \varepsilon e^{i\beta_1} e^{-\theta_1}, \text{or } D = 1 + \varepsilon e^{-i\beta_1} e^{-\theta_1}; C = 1 + z_1^2 F_1/\rho_1^2 = 1 + \varepsilon e^{i3\beta_1} e^{-\theta_1} \tag{136}$$

$$\bar{u}_1(x,t) = \rho \frac{C_1 D_1}{\overline{D}_1^2} = \rho \frac{\left(1 + \varepsilon e^{i3\beta_1} e^{-\theta_1}\right)\left(1 + \varepsilon e^{-i\beta_1} e^{-\theta_1}\right)}{\left(1 + \varepsilon e^{i\beta_1} e^{-\theta_1}\right)^2} = \rho \left[1 - \frac{4\varepsilon \sin^2\beta_1}{e^{\theta_1} e^{-i\beta_1} + e^{-\theta_1} e^{i\beta_1} + 2\varepsilon}\right] \tag{137}$$

where $\varepsilon = 1(-1)$ corresponds to dark (bright) soliton. Similarly if we redefine that $\beta_1 \equiv \pi/2 - \beta_1'$, then solution (137) is just the same as that gotten in [4, 5, 11, 12, 16] and called one-parameter pure soliton. This further convinces us of the validity and reliability of the newly revised IST for NVBC.

In the case of breather-type two-soliton solution, $N = 2$, we define that

$$z_1 \equiv \rho e^{\delta_1} e^{i\beta_1}, z_2 = \rho e^{-\delta_1} e^{i\beta_1}, z_3 \equiv \rho e^{\delta_3} e^{i\beta_3}, z_4 = \rho e^{-\delta_3} e^{i\beta_3} \tag{138}$$

$$F_j \equiv e^{-\theta_j} e^{i\varphi_j}, j = 1, 2, 3, 4 \tag{139}$$

which is just the same as that defined in (108)–(115), the pole z_j-related constant complex factor is absorbed into the j'th soliton center and the initial phase. Using formula (80)–(82), (88), (116), (108)–(115), we have

$$\overline{u}_2(x,t) = \rho C_2 D_2 / \overline{D}_2^2, \text{or } u_2(x,t) = \rho \overline{C}_2 \overline{D}_2 / D_2^2 \tag{140}$$

$$\begin{aligned}
\overline{D}_2 &= \det(\boldsymbol{I} + \boldsymbol{B}) = 1 + B(n_1 = 1) + B(n_1 = 2) + B(n_1 = 3) + B(n_1 = 4) \\
&\quad + B(n_1 = 1, n_2 = 2) + B(n_1 = 1, n_2 = 3) + B(n_1 = 1, n_2 = 4) + B(n_1 = 2, n_2 = 3) \\
&\quad + B(n_1 = 2, n_2 = 4) + B(n_1 = 3, n_2 = 4) + B(n_1 = 1, n_2 = 2, n_3 = 3) \\
&\quad + B(n_1 = 1, n_2 = 2, n_3 = 4) + B(n_1 = 1, n_2 = 3, n_3 = 4) \\
&\quad + B(n_1 = 2, n_2 = 3, n_4 = 4) + B(n_1 = 1, n_2 = 2, n_3 = 3, n_4 = 4) \\
&= 1 + F_1 + F_2 + F_3 + F_4 + F_1F_2 \tanh^2(\Theta_1 - \Theta_2) + F_1F_3 \tanh^2(\Theta_1 - \Theta_3) \\
&\quad + F_1F_4 \tanh^2(\Theta_1 - \Theta_4) + F_2F_3 \tanh^2(\Theta_2 - \Theta_3) \\
&\quad + F_2F_4 \tanh^2(\Theta_2 - \Theta_4) + F_3F_4 \tanh^2(\Theta_3 - \Theta_4) \\
&\quad + F_1F_2F_3 \tanh^2(\Theta_1 - \Theta_2) \tanh^2(\Theta_1 - \Theta_3) \tanh^2(\Theta_2 - \Theta_3) \\
&\quad + F_1F_2F_4 \tanh^2(\Theta_1 - \Theta_2) \tanh^2(\Theta_1 - \Theta_4) \tanh^2(\Theta_2 - \Theta_4) \\
&\quad + F_1F_3F_4 \tanh^2(\Theta_1 - \Theta_3) \tanh^2(\Theta_1 - \Theta_4) \tanh^2(\Theta_3 - \Theta_4) \\
&\quad + F_2F_3F_4 \tanh^2(\Theta_2 - \Theta_3) \tanh^2(\Theta_2 - \Theta_4) \tanh^2(\Theta_3 - \Theta_4) \\
&\quad + F_1F_2F_3F_4 \tanh^2(\Theta_1 - \Theta_2) \tanh^2(\Theta_1 - \Theta_3) \tanh^2(\Theta_1 - \Theta_4) \\
&\quad \cdot \tanh^2(\Theta_2 - \Theta_3) \tanh^2(\Theta_2 - \Theta_4) \tanh^2(\Theta_3 - \Theta_4)
\end{aligned} \tag{141}$$

Similarly we can attain C_2 from (116) and (118) as follows

$$\begin{aligned}
C_2 &= \det(\boldsymbol{I} + A) \\
&= 1 + F_1 \tan\Theta_1 + F_2 \tan\Theta_2 + F_3 \tan\Theta_3 + F_4 \tan\Theta_4 \\
&\quad + F_1F_2 \tanh\Theta_1 \tanh\Theta_2 \tanh^2(\Theta_1 - \Theta_2) + \tanh\Theta_1 \tanh\Theta_3 F_1F_3 \tanh^2(\Theta_1 - \Theta_3) \\
&\quad + F_1F_4 \tanh\Theta_1 \tanh\Theta_4 \tanh^2(\Theta_1 - \Theta_4) + F_2F_3 \tanh\Theta_2 \tanh\Theta_3 \tanh^2(\Theta_2 - \Theta_3) \\
&\quad + F_2F_4 \tanh\Theta_2 \tanh\Theta_4 \tanh^2(\Theta_2 - \Theta_4) + F_3F_4 \tanh\Theta_3 \tanh\Theta_4 \tanh^2(\Theta_3 - \Theta_4) \\
&\quad + F_1F_2F_3 \tanh\Theta_1 \tanh\Theta_2 \tanh\Theta_3 \tanh^2(\Theta_1 - \Theta_2) \tanh^2(\Theta_1 - \Theta_3) \tanh^2(\Theta_2 - \Theta_3) \\
&\quad + F_1F_2F_4 \tanh\Theta_1 \tanh\Theta_2 \tanh\Theta_4 \tanh^2(\Theta_1 - \Theta_2) \tanh^2(\Theta_1 - \Theta_4) \tanh^2(\Theta_2 - \Theta_4) \\
&\quad + F_1F_3F_4 \tanh\Theta_1 \tanh\Theta_3 \tanh\Theta_3 \tanh^2(\Theta_1 - \Theta_3) \tanh^2(\Theta_1 - \Theta_4) \tanh^2(\Theta_3 - \Theta_4) \\
&\quad + F_2F_3F_4 \tanh\Theta_2 \tanh\Theta_3 \tanh\Theta_4 \tanh^2(\Theta_2 - \Theta_3) \tanh^2(\Theta_2 - \Theta_4) \tanh^2(\Theta_3 - \Theta_4) \\
&\quad + F_1F_2F_3F_4 \tanh\Theta_1 \tanh\Theta_2 \tanh\Theta_3 \tanh\Theta_4 \tanh^2(\Theta_1 - \Theta_2) \tanh^2(\Theta_1 - \Theta_3) \tanh^2(\Theta_1 - \Theta_4) \\
&\quad \cdot \tanh^2(\Theta_2 - \Theta_3) \tanh^2(\Theta_2 - \Theta_4) \tanh^2(\Theta_3 - \Theta_4)
\end{aligned} \tag{142}$$

Substituting (141)–(142) into (140) completes the calculation of breather-type two-soliton solution. The evolution of breather-type two-soliton solution with respect to time and space is given in **Figure 3**. It clearly display the whole process of the elastic collision between two breather solitons, and in the limit of infinite time $t \to \pm\infty$, the breather-type two-soliton is asymptotically decomposed into two breather-type 1-solitons.

4.1 Explicit pure N-soliton solution to the DNLS⁺ equation with NVBC

When all the simple poles are on the circle (O, ρ) centered at the origin O, just as shown in **Figure 4**, our revised IST for DNLS⁺ equation with NVBC will give a typical pure N-soliton solution. The discrete part of $a(z)$ is of a slightly different form from that of the case for breather-type solution, and it can be expressed as

$$a(z) = \prod_{n=1}^{N} \frac{z^2 - z_n^2}{z^2 - \overline{z}_n^2} \frac{\overline{z}_n}{z_n}; \dot{a}(z_n) = \frac{2z_n}{z_n^2 - \overline{z}_n^2} \frac{\overline{z}_n}{z_n} \prod_{m=1, m \neq n}^{N} \frac{z_n^2 - z_m^2}{z_n^2 - \overline{z}_m^2} \frac{\overline{z}_m}{z_m} \tag{143}$$

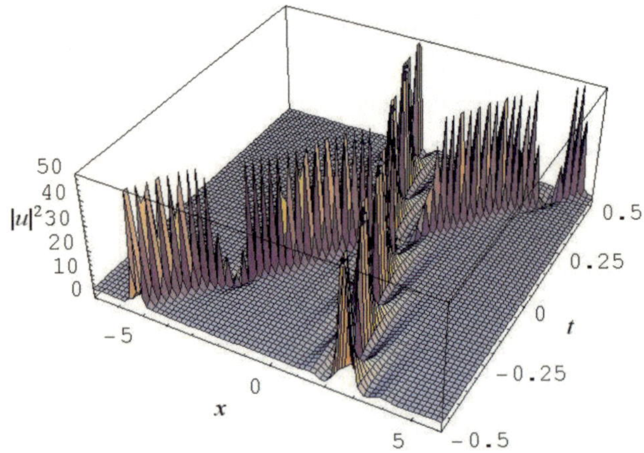

Figure 3.
Evolution of the square amplitude of a breather-type two-soliton with respect to time and space $\rho = 2; \delta_1 = 0.4; \delta_3 = 0.6; \beta_1 = \pi/5.0; \beta_3 = \pi/2.2; x_{10} = 0; x_{20} = 0 = 0; x_{30} = 0; x_{40} = 0; \varphi_{10} = 0; \varphi_{20} = 0; \varphi_{30} = 0; \varphi_{40} = 0.$

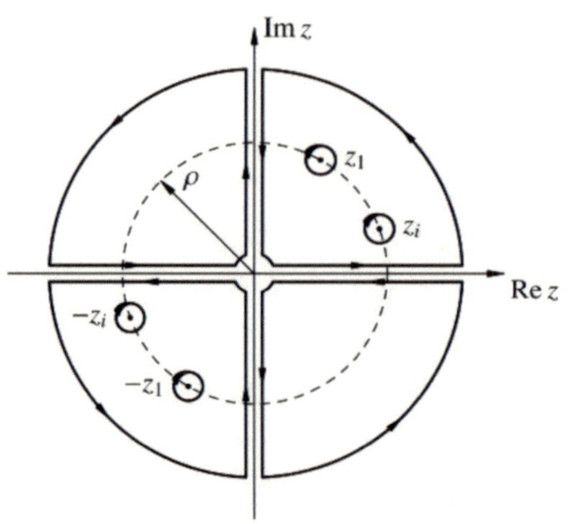

Figure 4.
Integral contour as all poles are on the circle of radiusρ.

At the zeros of $a(z)$, we have

$$\phi(x,z_n) = b_n\psi(x,z_n), \dot{a}(-z_n) = -\dot{a}(z_n), \overline{b}_n = -b_n \tag{144}$$

On the other hand, the zeros of $a(z)$ appear in pairs and can be designed by z_n, $(n = 1, 2, \cdots, N)$, in the I quadrant, and $z_{n+N} = -z_n$ in the III quadrant. The Zakharov-Shabat equation for pure soliton case of DNLS+ equation under reflectionless case can be derived immediately

$$\tilde{\psi}_1(x,z) = e^{-i\Lambda x} + \lambda\left[\sum_{n=1}^{N}\frac{2z}{\lambda_n}\frac{1}{z^2 - z_n^2}c_n\psi_1(x,z_n)e^{i\Lambda_n x}\right]e^{-i\Lambda x} \tag{145}$$

$$\tilde{\psi}_2(x,z) = i\rho z^{-1}e^{-i\Lambda x} + \lambda\left[\sum_{n=1}^{N}\frac{2z_n}{\lambda_n}\frac{1}{z^2 - z_n^2}c_n\psi_2(x,z_n)e^{i\Lambda_n x}\right]e^{-i\Lambda x} \tag{146}$$

Here $\Lambda = \kappa\lambda$, $\Lambda_n = \kappa_n\lambda_n$; Letting $z = \rho^2 z_m^{-1}$, $m = 1, 2, \cdots, N$, then

$$\psi_1(x,z_m) = -i\rho z_m^{-1}e^{i\Lambda_m x} + \sum_{n=1}^{N}\frac{\lambda_m c_n}{\lambda_n z_m^2}\cdot\frac{2\rho^3}{i\left(\rho^4 z_m^{-2} - z_n^2\right)}\psi_1(x,z_n)e^{i(\Lambda_n+\Lambda_m)x} \tag{147}$$

$$\psi_2(x,z_m) = e^{i\Lambda_m x} + \sum_{n=1}^{N}\frac{\lambda_m z_n c_n}{\lambda_n z_m}\cdot\frac{2\rho}{i\left(\rho^4 z_m^{-2} - z_n^2\right)}\psi_2(x,z_n)e^{i(\Lambda_n+\Lambda_m)x} \tag{148}$$

Different from that in breather-type case, we define $z_n \equiv \rho e^{\delta_n}e^{i\beta_n} = \rho e^{i\beta_n}$, with $\beta_n \in (0, \pi/2), \delta_n = 0$, $(i = 1, 2, \cdots, N)$, specially we have

$$c_{n0} = b_{n0}/\dot{a}(z_n) = ib_{n0}\rho\sin 2\beta_n e^{i\beta_n}\prod_{k=1;\,k\neq n}^{N}\frac{\sin\left(\beta_n + \beta_k\right)}{\sin\left(\beta_n - \beta_k\right)} \tag{149}$$

$$\tanh^2(\Theta_n - \Theta_m) = \sin^2(\beta_n - \beta_m)/\sin^2(\beta_n + \beta_m) \tag{150}$$

An inverse Galileo transformation $(x,t) \to (x - \rho^2 t, t)$ changes $F_n(x,t)$ into

$$\begin{aligned} F_n &\equiv f_n^2\frac{\rho}{i\left(z_n^2 - \rho^4 z_n^{-2}\right)} = \frac{2\rho}{i\left(z_n^2 - \rho^4 z_n^{-2}\right)}c_{no}e^{-i4\lambda_n^3\kappa_n t}e^{i2\Lambda_n x} \\ &= (-ib_{n0})\prod_{k=1;\,k\neq n}^{N}\frac{\sin\left(\beta_n + \beta_k\right)}{\sin\left(\beta_n - \beta_k\right)}e^{i2\Lambda_n\left[x - \left(2\lambda^2 + \rho^2\right)t\right] + i\beta_n} \end{aligned} \tag{151}$$

Due to $\overline{b}_{n0} = -b_{n0}$, $-ib_{n0} \in \mathbb{R}$, following equations hold:

$$F_n = \left(e^{i\beta_n}\prod_{k=1;\,k\neq n}^{N}\frac{\sin\left(\beta_n + \beta_k\right)}{\sin\left(\beta_n - \beta_k\right)}\right)\left(-ib_{n0}e^{i2\Lambda_n\left[x - \left(2\lambda_n^2 + \rho^2\right)t\right]}\right) \equiv \varepsilon_n E_n e^{i\beta_n}\exp\left(-\theta_n + i\varphi_n\right) \tag{152}$$

where

$$-ib_{n0}e^{i2\Lambda_n\left[x - \left(2\lambda_n^2 + \rho^2\right)t\right]} \equiv \varepsilon_n\exp\left(-\theta_n + i\varphi_n\right) \tag{153}$$

$$\theta_n(x,t) \equiv \nu_n(x - v_n t - x_{n0}), \varphi_n = 0, \tag{154}$$

$$\varepsilon_n \equiv \operatorname{sgn}\left(-ib_{n0}\right); |-ib_{n0}| \equiv e^{\nu_n x_{n0}} \tag{155}$$

$$\nu_n = \rho^2 \sin 2\beta_n, \upsilon_n = \rho^2 (1 + 2\cos^2\beta_n) \tag{156}$$

$$E_n \equiv \prod_{k=1;\, k\neq n}^{N} \frac{\sin(\beta_n + \beta_k)}{\sin(\beta_n - \beta_k)} \tag{157}$$

where E_n is also a real constant which is only dependent upon the order number n. The constant and positive real number $|-ib_{n0}|$ has been absorbed by redefinition of the n' th soliton center x_{n0} in (155). Thus the determinants in formula (83) for pure soliton solution can be calculated as follows

$$\overline{D}_N \equiv \det(I + B) = 1 + \sum_{r=1}^{N} \sum_{1\le n_1 < \cdots < n_r \le N} B(n_1, n_2, \cdots, n_r) \tag{158}$$

$$\begin{aligned} B(n_1, n_2, \cdots, n_r) &= \prod_n f_n^2 \left[\frac{\rho}{i(z_n^2 - \rho^4 z_n^{-2})}\right] \prod_{n<m} \frac{i(z_n^2 - z_m^2) i(\rho^4 z_m^{-2} - \rho^4 z_n^{-2})}{i(z_n^2 - \rho^4 z_m^{-2}) i(z_m^2 - \rho^4 z_n^{-2})} \\ &= \prod_n F_n \prod_{n<m} \tanh^2(\Theta_n - \Theta_m) \\ &= \prod_n \varepsilon_n E_n e^{i\beta_n} e^{-\theta_n} \prod_{n<m} \frac{\sin^2(\beta_n - \beta_m)}{\sin^2(\beta_n + \beta_m)}; n, m \in (n_1, n_2, \cdots, n_r) \end{aligned} \tag{159}$$

$$C_N = \det(I + A) = 1 + \sum_{r=1}^{N} \sum_{1\le n_1 < n_2 < \cdots < n_r \le N} \boldsymbol{A}(n_1, n_2, \cdots, n_r) \tag{160}$$

$$\begin{aligned} A(n_1, n_2, \cdots, n_f) &= \prod_n \frac{z_n^2}{\rho^2} F_n \prod_{n<m} \tanh^2(\Theta_n - \Theta_m) \\ &= \prod_n \varepsilon_n E_n e^{i3\beta_n} e^{-\theta_n} \prod_{n<m} \frac{\sin^2(\beta_n - \beta_m)}{\sin^2(\beta_n + \beta_m)}; n, m \in (n_1, n_2, \cdots, n_r) \end{aligned} \tag{161}$$

Substituting (149)–(157) into (158)–(161), and substituting (158)–(161) into the following formula, we attain the explicit pure N-soliton solution

$$\overline{u}_N \equiv \rho C_N D_N / \overline{D}_N^2 \text{ or } u_N \equiv \rho \overline{C}_N \overline{D}_N / D_N^2 \tag{162}$$

The $N = 2$ case, that is, the pure two-soliton is also a typical illustration of the general explicit N-soliton formula. According to (158)–(162), it can be calculated as follows

$$\begin{aligned} \overline{D}_2 &= 1 + B(1) + B(2) + B(1,2) \\ &= 1 + \varepsilon_1 E_1 e^{i\beta_1} e^{-\theta_1} + \varepsilon_2 E_2 e^{i\beta_2} e^{-\theta_2} + \varepsilon_1\varepsilon_2 E_1 E_2 e^{i(\beta_1+\beta_2)} e^{-(\theta_1+\theta_2)} \sin^2(\beta_1 - \beta_2)/\sin^2(\beta_1 + \beta_2) \\ &= 1 + \varepsilon_1 \frac{\sin(\beta_1 + \beta_2)}{\sin(\beta_1 - \beta_2)} e^{i\beta_1} e^{-\theta_1} - \varepsilon_2 \frac{\sin(\beta_1 + \beta_2)}{\sin(\beta_1 - \beta_2)} e^{i\beta_2} e^{-\theta_2} - \varepsilon_1\varepsilon_2 e^{i(\beta_1+\beta_2)} e^{-(\theta_1+\theta_2)} \end{aligned} \tag{163}$$

$$\begin{aligned} C_2 &= 1 + A(1) + A(2) + A(1,2) \\ &= 1 + \varepsilon_1 \frac{\sin(\beta_1 + \beta_2)}{\sin(\beta_1 - \beta_2)} e^{i3\beta_1} e^{-\theta_1} - \varepsilon_2 \frac{\sin(\beta_1 + \beta_2)}{\sin(\beta_1 - \beta_2)} e^{i3\beta_2} e^{-\theta_2} - \varepsilon_1\varepsilon_2 e^{i3(\beta_1+\beta_2)} e^{-(\theta_1+\theta_2)} \end{aligned} \tag{164}$$

$$\overline{u}_2(x,t) = \rho C_2 D_2 / \overline{D}_2^2 \tag{165}$$

The evolution of pure two-soliton solution with respect to time and space is given in **Figure 5**. It clearly demonstrates the whole process of the elastic collision between pure two solitons. If $0 < \beta_2 < \beta_1 < \pi/2$, then $\varepsilon_1 = 1$, $\operatorname{sgn} E_1 = 1$ and $\varepsilon_2 = -1$, $\operatorname{sgn} E_2 = -1$ correspond to double-dark pure 2-soliton solution as in **Figure 5a**; $\varepsilon_1 = -1$, $\operatorname{sgn} E_1 = 1$ and $\varepsilon_2 = 1$, $\operatorname{sgn} E_2 = -1$ correspond to a double-bright pure 2-soliton solution in **Figure 5c**; $\varepsilon_1 = 1$, $\operatorname{sgn} E_1 = 1$ and $\varepsilon_2 = 1$, $\operatorname{sgn} E_2 = -1$ correspond to a dark-bright-mixed pure 2-soliton solution in **Figure 5b**. In the limit of infinite time $t \to \pm\infty$, the pure 2-soliton solution is asymptotically decomposed into two pure 1-solitons.

By the way, it should be point out, although our method and solution have different forms from that of Refs. [7, 16], they are actually equivalent to each other. In fact if the constant E_n, $(n = 1, 2, \cdots, N)$, is also absorbed into the n'th soliton center x_{n0} just like $-ib_{n0}$ does in (152)–(154), and replace β_n with $\beta'_n = \pi/2 - \beta_n \in (0, \pi/2)$, the result for the pure soliton case in this section will reproduce the solution gotten in Refs. [7, 9, 16].

On the other hand, letting only part of the poles converge in pairs on the circle in **Figure 1** and rewriting the expression of $a_n(z)$ as in Ref. [7, 8, 12], our result can

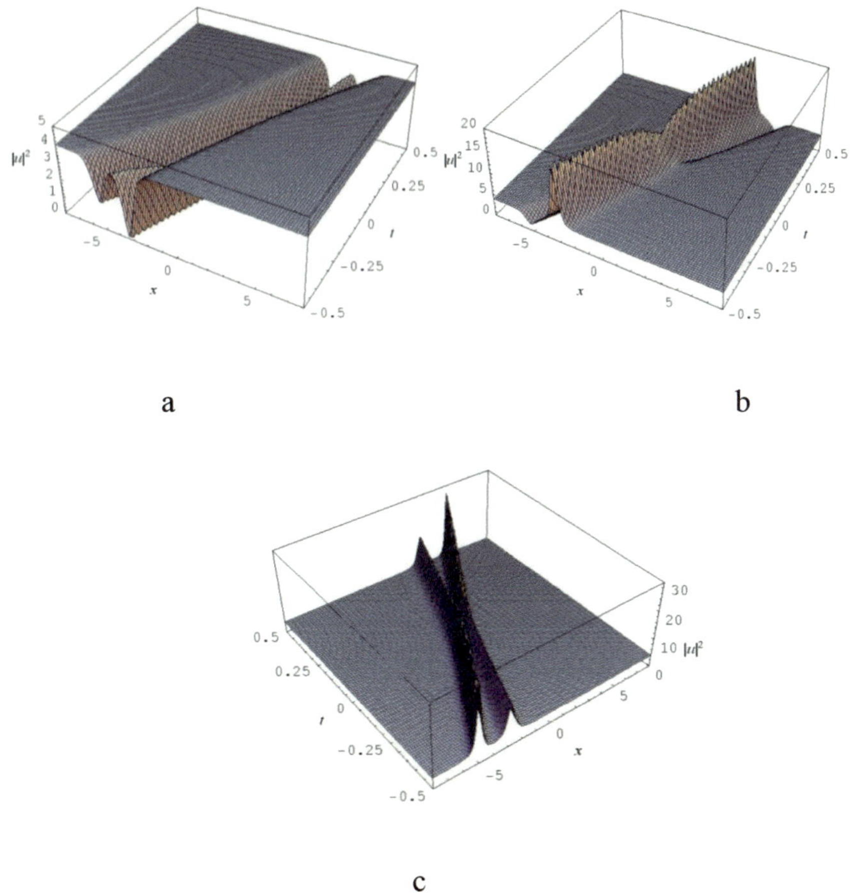

Figure 5.
Evolution of pure two soliton solution in time and space. (a) dark-dark pure 2-soilton, (b) dark-bright pure 2-soilton, and (c) bright-bright pure 2-soilton.

naturally generate the mixed case with both pure and breather-type multi-soliton solution.

4.2 The asymptotic behaviors of the N-soliton solution

Without loss of generality, we assume $\beta_1 > \beta_2 > \cdots > \beta_n > \cdots > \beta_N$; $v_1 < v_2 < \cdots < v_n < \cdots < v_N$ in (156), and define the n'th neighboring area as Υ_n : $x - x_{no} - v_n t \sim 0, (n = 1.2, \cdots, N)$. In the neighboring area of Υ_n,

$$\theta_j = \nu_j(x - x_{j0} - v_j t) \to \begin{cases} +\infty, & \text{for } j > n \\ -\infty, & \text{for } j < n \end{cases} \tag{166}$$

$$\overline{D} \approx B(1, 2, \ldots, n-1) + B(1, 2, \ldots, n-1, n) \tag{167}$$

$$C \approx A(1, 2, \ldots, n-1) + A(1, 2, \ldots, n-1, n) \tag{168}$$

where

$$B(1, 2, \ldots, n) = \varepsilon_n E_n e^{-\theta_n - i\beta_n} \prod_{j=1}^{n-1} \frac{\sin^2(\beta_j - \beta_n)}{\sin^2(\beta_j + \beta_n)} B(1, 2, \ldots, n-1) \tag{169}$$

$$A(1, 2, \ldots, n-1, n) = \varepsilon_n E_n e^{-\theta_n + i3\beta_n} \prod_{j=1}^{n-1} \frac{\sin^2(\beta_j - \beta_n)}{\sin^2(\beta_j + \beta_n)} A(1, 2, \ldots, n-1) \tag{170}$$

In the neighboring area of Υ_n, we have

$$u \simeq u_1\left(\theta_n + \Delta\theta_n^{(-)}\right) \tag{171}$$

With

$$\Delta\theta_n^{(-)} = 2\sum_{j=1}^{n-1} \ln\left|\frac{\sin(\beta_j - \beta_n)}{\sin(\beta_j + \beta_n)}\right| \tag{172}$$

As $t \to -\infty$, the N neighboring areas queue up in a descending series $\Upsilon_N, \Upsilon_{N-1}, \cdots, \Upsilon_1$, then

$$u_N \simeq \sum_{n=1}^{N} u_1\left(\theta_n + \Delta\theta_n^{(-)}\right) \tag{173}$$

the N-soliton solution can be viewed as N well-separated exact pure one solitons, each $u_1(\theta_n + \Delta\theta_n^{(-)})$, $(1, 2, \cdots, n)$ is a single pure soliton characterized by one parameter β_n, moving to the positive direction of the x-axis, queuing up in a series with descending order number n.

As $t \to \infty$, in the neighboring area of Υ_n we have

$$\theta_j = \nu_j(x - x_{j0} - v_j t) \to \begin{cases} -\infty, & \text{for } j > n \\ +\infty, & \text{for } j < n \end{cases} \tag{174}$$

$$\overline{D} \approx B(n, n+1, \ldots, N) + B(n+1, n+2, \ldots, N) \tag{175}$$

$$C \approx A(n, n+1, \ldots, N) + A(n+1, n+2, \ldots, N) \tag{176}$$

where

$$B(n, n+1, \dots, N) = \varepsilon_n E_n e^{-\theta_n - i\beta_n} \prod_{j=n+1}^{N} \frac{\sin^2\left(\beta_j - \beta_n\right)}{\sin^2\left(\beta_j + \beta_n\right)} B(n+1, n+2, \dots, N) \quad (177)$$

$$A(n, n+1, \dots, N) = \varepsilon_n E_n e^{-\theta_n + i3\beta_n} \prod_{j=n+1}^{N} \frac{\sin^2\left(\beta_j - \beta_n\right)}{\sin^2\left(\beta_j + \beta_n\right)} A(n+1, n+2, \dots, N) \quad (178)$$

$$u \simeq u_1\left(\theta_n + \Delta\theta_n^{(+)}\right) \quad (179)$$

$$\Delta\theta_n^{(+)} = 2\sum_{n+1}^{N} \ln\left|\frac{\sin\left(\beta_j + \beta_n\right)}{\sin\left(\beta_j - \beta_n\right)}\right| \quad (180)$$

$$u_N \simeq \sum_{n=1}^{N} u_1\left(\theta_n + \Delta\theta_n^{(+)}\right) \quad (181)$$

That is, the N-soliton solution can be viewed as N well-separated exact pure one solitons, queuing up in a series with ascending order number n such as $\Upsilon_1, \Upsilon_2, \cdots, \Upsilon_N$..

In the process of going from $t \to -\infty$ to $t \to \infty$, the n'th pure single soliton overtakes the solitons from the 1'th to $n-1$'th and is overtaken by the solitons from $n+1$'th to N'th. In the meantime, due to collisions, the n'th soliton got a total forward shift $\Delta\theta_n^{(-)}/\nu_n$ from exceeding those slower soliton from the 1'th to $n-1$'th, got a total backward shift $\Delta\theta_n^{(+)}/\nu_n$ from being exceeded by those faster solitons from $n+1$'th to N'th, and just equals to the summation of shifts due to each collision between two solitons, that is,

$$\Delta x_n = \left|\Delta\theta_n^{(+)} - \Delta\theta_n^{(-)}\right|/\nu_n \quad (182)$$

By introducing an suitable affine parameter in the IST and based upon a newly revised and improved inverse scattering transform and the Z-S equation for the DNLS$^+$ equation with NVBC and normal dispersion, the rigorously proved breather-type N-soliton solution to the DNLS$^+$ equation with NVBC has been derived by use of some special linear algebra techniques. The one- and two-soliton solutions have been given as two typical examples in illustration of the unified formula of the N-soliton solution and the general computation procedures. It can perfectly reproduce the well-established conclusions for the special limit case. On the other hand, letting part/all of the poles converge in pairs on the circle in **Figure 4** and rewriting the expression of $a_n(z)$ as in [7, 12, 13], can naturally generate the partly/wholly pure multi-soliton solution. Moreover, the exact breather-type multi-soliton solution to the DNLS$^+$equation can be converted to that of the MNLS equation by a gauge-like transformation [17].

Finally, the elastic collision among the breathers of the above multi-soliton solution has been demonstrated by the case of a breather-type 2-soliton solution. The newly revised IST for DNLS$^+$ equation with NVBC and normal dispersion makes corresponding Jost functions be of regular properties and asymptotic behaviors, and thus supplies substantial foundation for its direct perturbation theory.

5. Space periodic solutions and rogue wave solution of DNLS equation

DNLS equation is one of the most important nonlinear integrable equations in mathematical physics, which can describe many physical phenomena in different application fields, especially in space plasma physics and nonlinear optics [1, 2, 16, 24–29]. We have found that DNLS equation can generate not only some usual soliton solutions such as dark/bright solitons and pure/breather-type solitons, but also some special solutions – space periodic solutions and rogue wave solution [14].

There are two celebrated models of the DNLS equations. One equation is called Kaup-Newell (KN) equation [15]:

$$iu_t + u_{xx} + i\left(u^2\overline{u}\right)_x = 0 \tag{183}$$

and the other is called Chen-Lee-Liu (CLL) equation [30]:

$$iv_t + v_{xx} + iv\overline{v}v_x = 0 \tag{184}$$

Actually, there is a gauge transformation between these two Eqs. (183) and (184) [14, 30, 31]. Supposing u is one of the solutions of the KN Eq. (183), then

$$v = u \cdot \exp\left(\frac{i}{2}\int^x |u|^2 \mathrm{d}x\right) \tag{185}$$

will be the solution of the CLL equation.

This section focuses on the KN Eq. (183) with NVBC – periodic plane-wave background. The first soliton solution of (183) was derived by Kaup and Newell via inverse scattering transformation (IST) [3, 15, 32]. Whereafter, the multi-soliton solution was gotten by Nakamura and Chen by virtue of the Hirota method [30, 31]. The determinant expression of the N-soliton solution was found by Huang and Chen on the basis of the Darboux transformation (DT for brevity) [33], and by Zhou et al., by use of a newly revised IST [7, 11–13, 17].

Recently, rogue waves which seem to appear from nowhere and disappear without a trace have drawn much attention [34, 35]. The most significant feature of rogue wave is its extremely large wave amplitude and space-time locality [35]. The simplest way to derive the lowest order of rogue wave, that is, the Peregrine solution [35, 36], is to take the long-wave limit of an Akhmediev breather [37] or a Ma breather [38], both of which are special cases of the periodic solution. Thus, the key procedure of generating a rogue wave is to obtain an Akhmediev breather or a Ma breather. As far as we know, DT plays an irreplaceable role in deriving the rogue wave solution [39–41]. Because both Akhmediev breather and Ma breather can exist only on a plane-wave background; Darboux transformation has the special privilege that a specific background or, in other words, a specific boundary condition can be chosen as the seed solution used in DT. For instance, if we choose $q_0 = 0$ as the seed solution of the DT of the KN Eq. (183), then after 2-fold DT, a new solution will be gotten under VBC:

$$q^{[2]} = 4i\alpha\beta \frac{\left(-i\alpha_1 \cosh(2\Gamma) + \beta_1 \sinh(2\Gamma)\right)^3}{\left(\left(-\alpha_1^2 - \beta_1^2\right)\cosh(2\Gamma)^2 + \beta_1^2\right)^2} \tag{186}$$

(where all the parameters are defined in Ref. [38]). Similarly, setting a seed solution $q_0 = c\exp i[ax + (-c^2 + a)at]$, a plane-wave solution to Eq. (183), will generate a new solution after 2-fold DT under a plane-wave background. Therefore, there is no need to discuss the boundary conditions or background when applying DT to solve those nonlinear integrable equations. This makes DT the most effective and prevailing method in obtaining a rogue wave solution.

Compared with DT, the IST has its fatal flaw that the difficulty of dealing with the boundary condition is unavoidable, which limits the possible application of the IST. Although the KN equation has been solved theoretically by means of an improved IST for both VBC and the NVBC [7–9, 17–20], there is no report that the KN equation could be solved under a plane-wave background by means of IST. And consequently, it appears that rogue wave solutions cannot be obtained through the IST method. This major problem is caused by the difficulty of finding appropriate Jost solutions under the plane-wave background.

On the other hand, the Hirota's bilinear-derivative transform (HBDT for brevity) [42–46], though not as a prevalent method as DT, has its particular advantages. The core of this method is a bilinear operator D which is defined by:

$$D_t^n D_x^m A \bullet B \equiv \left(\frac{\partial}{\partial t} - \frac{\partial}{\partial t'}\right)^n \left(\frac{\partial}{\partial x} - \frac{\partial}{\partial x'}\right)^m A(x,t)B(x',t')|_{t'=t;x'=x} \tag{187}$$

where, at the left side of the above formula, a dot • between two functions $A(x,t)$ and $B(x',t')$ represents an ordered product. The HBDT method is very useful in dealing with periodic solutions for its convenience in computing the bilinear derivatives of an exponential function [44]:

$$\begin{aligned} &F(D_x, D_y, \cdots, D_t) \exp(kx + ly + \cdots + \omega t) \bullet \exp(k'x + l'y + \cdots + \omega' t) \\ &= F(k - k', l - l', \cdots, \omega - \omega') \exp\left[(k + k')x + (l + l')y + \cdots + (\omega + \omega')t\right] \end{aligned} \tag{188}$$

Here, F represents general function expressed by the finite or infinite power series expansion of the Hirota's bilinear differential operators. Formula (188) is the generalization of Appendix 5.5. Thus using HBDT method to find space periodic solutions of KN equation is practicable. The space periodic solutions possess the characters that they approach the plane-wave solution when $|t| \to \infty$ and are periodic in space. The first space periodic solution was found by Akhmediev with one parameter [45]. Actually, we can regard the space periodic solution as a special Akhmediev breather with a pure complex-valued wave number. Further, through a space periodic solution, a rogue wave solution can be constructed. This means besides DT, HBDT method is also an alternative and effective way to find rogue wave solution of KN equation.

5.1 Bilinear derivative transformation of DNLS equation

The Hirota bilinear transformation is an effective method which could help to solve the KN equation. Due to the similarity of the first equation of Lax pairs between that of DNLS equation and AKNS system, there is a direct inference and manifestation that $u(x,t)$ has a typical standard form [6, 7]:

$$u(x,t) = g\bar{f}/f^2 \tag{189}$$

where f and g are complex auxiliary functions needed to be determined. Applying the bilinear derivative transform to (189), we can rewrite the derivatives of $u(x,t)$ in the bilinear form [19, 20, 42–46]:

$$u_t = \left(f\bar{f}D_t g \cdot f - gfD_t f \cdot \bar{f}\right)/f^4 \tag{190}$$

$$u_{xx} = \left[f\bar{f}D_x^2 g\cdot f - 2(D_x g\cdot f)\left(D_x f\cdot \bar{f}\right) + gfD_x^2 f\cdot \bar{f} - 2g\bar{f}D_x^2 f\cdot f\right]/f^4 \quad (191)$$

$$\left(|u|^2 u\right)_x = (2g\bar{g}D_x g\cdot f + g^2 D_x \bar{g}\cdot f)/f^4 \quad (192)$$

Directly substituting the above Eqs. (190)–(192) into (183) gives:

$$f\bar{f}(iD_t + D_x^2)g\cdot f - gf(iD_t + D_x^2)f\cdot \bar{f} + f^{-2}D_x\left[f^3\cdot g\left(2D_x f\cdot \bar{f} - ig\bar{g}\right)\right] = 0 \quad (193)$$

Then the above transformed KN equation can be decomposed into the following bilinear equations:

$$\left(iD_t + D_x^2 - \lambda\right)g\cdot f = 0 \quad (194)$$

$$\left(iD_t + D_x^2 - \lambda\right)f\cdot \bar{f} = 0 \quad (195)$$

$$D_x f\cdot \bar{f} = ig\bar{g}/2 \quad (196)$$

where λ is a constant which needs to be determined. Notice that if $\lambda = 0$ then the above bilinear equations are overdetermined because we have only two variables but three equations. Actually, setting $\lambda = 0$ is the approach to search for the soliton solution of the DNLS equation under vanishing boundary condition [19, 20]. Here, we set λ as a nonzero constant to find solutions under a different boundary condition – a plane-wave background.

5.2 Solution of bilinear equations

5.2.1 First order space periodic solution and rogue wave solution

Let us assume that the series expansion of the complex functions f and g in (189) are cut off, up to the 2'th power order of ϵ, and have the following formal form:

$$f = f_0(1 + \epsilon f_1 + \epsilon^2 f_2); g = g_0(1 + \epsilon g_1 + \epsilon^2 g_2) \quad (197)$$

Substituting f and g into Eqs. (194)–(196) yields a system of equations at the ascending power orders of ϵ, which allows for determination of its coefficients [14, 19, 20]. We have 15 equations [14, 19, 20] corresponding to the different orders of ϵ. After solving all the equations, then we can obtain the solution of the DNLS equation:

$$u^{[1]}(x,t) = \bar{f}^{[1]} g^{[1]}/f^{[1]2} \quad (198)$$

with

$$g^{[1]} = \rho e^{i\omega t}\left(1 + a_1 e^{px+\Omega t+\phi_0} + a_2 e^{-px+\bar{\Omega}t+\bar{\phi}_0} + Ma_1 a_2 e^{(\Omega+\bar{\Omega})t+\phi_0+\bar{\phi}_0}\right) \quad (199)$$

$$f^{[1]} = e^{i\beta x}\left(1 + b_1 e^{px+\Omega t+\phi_0} + b_2 e^{-px+\bar{\Omega}t+\bar{\phi}_0} + Mb_1 b_2 e^{(\Omega+\bar{\Omega})t+\phi_0+\bar{\phi}_0}\right) \quad (200)$$

where

$$\omega = 3\rho^4/16; \beta = \rho^2/4 \quad (201)$$

$$a_1 = b_1 \frac{2\Omega + 2ip^2 - p\rho^2}{2\Omega - 2ip^2 - p\rho^2}; a_2 = b_2 \frac{2\overline{\Omega} + 2ip^2 + p\rho^2}{2\overline{\Omega} - 2ip^2 + p\rho^2} \tag{202}$$

$$b_2 = \frac{\overline{b}_1(\overline{\Omega} + ip^2 - p\rho^2)}{\overline{\Omega} - ip^2 - p\rho^2}; M = 1 + \frac{4p^4}{(\Omega + \overline{\Omega})^2} \tag{203}$$

Notice that ρ and M are real; b_1 and φ_0 are complex constants, so there are two restrictions for a valid calculation: (1) the wave number p must be a pure imaginary number; (2) the angular frequency Ω must not be purely imaginary number and must furthermore satisfy the quadratic dispersion relation:

$$4\Omega^2 + 4p\rho^2\Omega + 4p^4 + 3p^2\rho^4 = 0 \tag{204}$$

According to the test rule for a one-variable quadratic, there is a threshold condition under which Ω will not be a pure imaginary number:

$$2p^4 + p^2\rho^4 < 0 \tag{205}$$

The asymptotic behavior of this breather is apparent. Because the wave number p is a pure imaginary number, the breather is a periodic function of x. The quadratic dispersion relation (204) permits the angular frequency Ω to have two solutions:

$$\Omega_+ = \left[-p\rho^2 + \sqrt{-2(2p^4 + p^2\rho^4)}\right]/2 \tag{206}$$

$$\Omega_- = \left[-p\rho^2 - \sqrt{-2(2p^4 + p^2\rho^4)}\right]/2 \tag{207}$$

If we set $\Omega = \Omega_+$, because $\sqrt{-(2p^4 + p^2\rho^4)} > 0$, then $t \to -\infty$ will lead to:

$$g^{[1]} \to \rho \exp(i\omega t) \tag{208}$$

$$f^{[1]} \to \exp(i\beta x) \tag{209}$$

$$u^{[1]} \to \rho \exp i(-3\beta x + \omega t) \tag{210}$$

And $t \to \infty$ will lead to:

$$g^{[1]} \to \rho M a_1 a_2 \exp\left[\sqrt{-2(2p^4 + p^2\rho^4)} + \phi_0 + \overline{\phi}_0 + i\omega t\right] \tag{211}$$

$$f^{[1]} \to M b_1 b_2 \exp\left[\sqrt{-2(2p^4 + p^2\rho^4)} + \phi_0 + \overline{\phi}_0 + i\beta x\right] \tag{212}$$

$$u^{[1]} \to \rho \exp i(-3\beta x + \omega t + \varphi) \tag{213}$$

where φ is the phase shift across the breather:

$$\exp(i\varphi) = a_1 a_2 / b_1 b_2 \tag{214}$$

and due to $|a_1 a_2| = |b_1 b_2|$, thus the above phase shift φ is real and does not affect the module of the breather $u^{[1]}$ when $t \to \infty$. As for the other choice $\Omega = \Omega_-$, further algebra computation shows the antithetical asymptotic behavior of $g^{[1]}$, $f^{[1]}$, and $u^{[1]}$ when $|t| \to \infty$. In a nutshell, $u^{[1]}$ will degenerate into a plane wave.

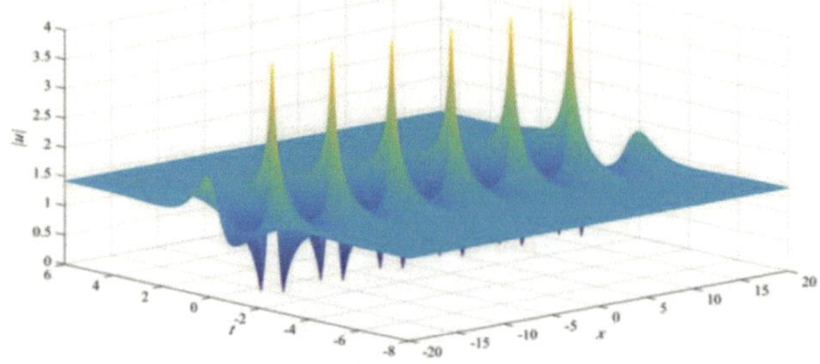

Figure 6.
The space-time evolution of the module of the 1st order space periodic solution in (198) with $p = i, \rho = \sqrt{2}, b_1 = i$ *and* $\Omega = \Omega_+$*, complex constant* φ_0 *is set to zero.*

Hereto, we have completed the computation of the 1st-order space periodic solution, the space-time evolution of its module is depicted in **Figure 6**. In what follows, we will take the long-wave limit, that is, $p \to 0$, to construct a rogue wave solution. Supposing $p = iq$, here q is a real value and $q \to 0$, then the asymptotic expansion of the angular frequency Ω is:

$$\Omega = q\rho^2(-i + \sigma)/2 + O(q^3) \tag{215}$$

where $\sigma = \pm\sqrt{2}$. For the sake of a valid form of the rogue wave solution, we need to set $b_1 = 1$ and $\varphi_0 = 0$ (of course, setting $b_1 = 1$ and $e^{\varphi_0} = -1$ is alright, all we need is to make sure that the coefficients of the q_0 and q_1 in the expansions of $f^{[1]}$ and $g^{[1]}$ are annihilated). Therefore, the expansions of $g^{[1]}$ and $f^{[1]}$ in terms of q are given by:

$$g^{[1]} = q^2 e^{i\omega t} \frac{-8(7i + 5\sigma) + 16x(1 - 2i\sigma)\rho^2 + 3(-i + \sigma)\rho^4(4x^2 - 4\rho^2 tx - 8it + 3\rho^4 t^2)}{12(-i + \sigma)\rho^3} + O(q^3) \tag{216}$$

$$f^{[1]} = q^2 e^{i\beta x} \frac{8(-i + \sigma) + 16x\rho^2 + (-i + \sigma)\rho^4(4x^2 - 4\rho^2 tx - 8it + 3\rho^4 t^2)}{4(-i + \sigma)\rho^4} + O(q^3) \tag{217}$$

Consequently, the rogue wave solution can be derived according to Eq. (198):

$$u_{RW} = \rho\, e^{i(-3\beta x + \omega t)} \left(g' \overline{f'}\right) / f'^2 \tag{218}$$

where

$$g' = -8(7i + 5\sigma) + 16x(1 - 2i\sigma)\rho^2 + 3(-i + \sigma)\rho^4\left(4x^2 - 4\rho^2 tx - 8it + 3\rho^4 t^2\right);$$

$$f' = 24(-i + \sigma) + 48x\rho^2 + 3(-i + \sigma)\rho^4\left(4x^2 - 4\rho^2 tx - 8it + 3\rho^4 t^2\right).$$

Here ω and β are given by Eq. (201), ρ is an arbitrary real constant. The module of rogue wave solution Eq. (218) is shown in **Figure 7**.

As we discussed in the Introduction section, there is a gauge transformation between KN Eq. (183) and CLL Eq. (184). Thus, it is instructive to use the integral transformation Eq. (185) to construct a solution of Eq. (184). Substituting the solution (198) into (185), further algebra computation will lead to a space periodic solution of the CLL equation:

$$v_c(x,t) = g^{[1]}/f^{[1]} \tag{219}$$

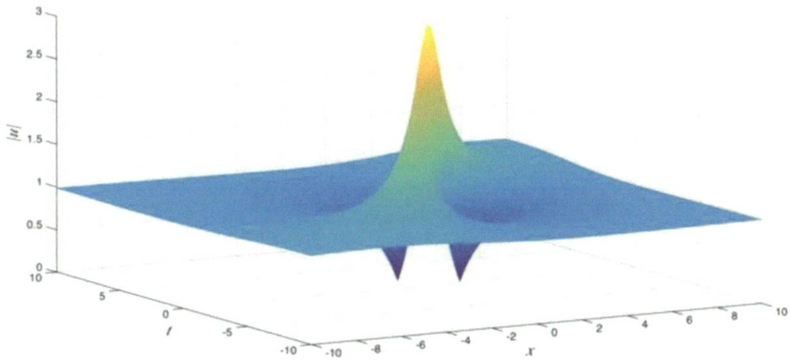

Figure 7.
The space-time evolution of the module of the rogue wave solution with $\rho = 1$ and $\sigma = \sqrt{2}$. The max amplitude is equal to 3 at the point $(x = -\sqrt{2}, t = -2\sqrt{2}/3)$.

where, $g^{[1]}$, $f^{[1]}$, and other auxiliary parameters are invariant and given by Eqs. (199)–(203). The same procedures which are used to derive the rogue wave solution of the KN equation can be used to turn v_c into a rogue wave solution of the CLL equation:

$$v_{c,RW} = \rho\, e^{i(-\beta x+\omega t)} g'/f' \tag{220}$$

which has the same parameters as u_{RW}. And this solution $v_{c,RW}$ has exactly the same form as the result given by ref. [46].

5.2.2 Second-order periodic solution

Taking the similar procedures described previously could help us to derive the 2nd-order space periodic solution. Assume the auxiliary functions f and g to have higher order expansions in terms of ϵ:

$$g = g_0\left(1 + \epsilon g_1 + \epsilon^2 g_2 + \epsilon^3 g_3 + \epsilon^4 g_4\right) \tag{221}$$

$$f = f_0\left(1 + \epsilon f_1 + \epsilon^2 f_2 + \epsilon^3 f_3 + \epsilon^4 f_4\right) \tag{222}$$

Similarly, substituting f and g into the bilinear Eqs. (194)–(196) leads to the 27 equations [14, 19, 20] corresponding to different orders of ϵ. Solving these equations is tedious and troublesome but worthy and fruitful. The results are expressed in the following form:

$$u^{[2]}(x,t) = \bar{f}^{[2]} g^{[2]} / f^{[2]2} \tag{223}$$

with

$$g^{[2]} = \rho e^{i\omega t}\left(1 + g_1 + g_2 + g_3 + g_4\right) \tag{224}$$

$$f^{[2]} = e^{i\beta x}\left(1 + f_1 + f_2 + f_3 + f_4\right) \tag{225}$$

$$\beta = \rho^2/4; \omega = 3\rho^4/16; \lambda = \rho^4/16 \tag{226}$$

$$g_1 = \sum_i a_i e^{\phi_i};\ f_1 = \sum_i b_i e^{\phi_i} \tag{227}$$

$$g_2 = \sum_{i<j} M_{ij} a_i a_j e^{\phi_i + \phi_j}; f_2 = \sum_{i<j} M_{ij} b_i b_j e^{\phi_i + \phi_j} \tag{228}$$

$$g_3 = \sum_{i<j<k} T_{ijk} a_i a_j a_k e^{\phi_i + \phi_j + \phi_k}; f_3 = \sum_{i<j<k} T_{ijk} b_i b_j b_k e^{\phi_i + \phi_j + \phi_k} \tag{229}$$

$$g_4 = A a_1 a_2 a_3 a_4 e^{\phi_1 + \phi_2 + \phi_3 + \phi_4}; f_4 = A b_1 b_2 b_3 b_4 e^{\phi_1 + \phi_2 + \phi_3 + \phi_4} \tag{230}$$

where $i, j, k = 1, 2, 3, 4,$ and the above parameters and coefficients are given respectively by:

$$p_2 = \overline{p}_1; p_4 = \overline{p}_3; \Omega_2 = \overline{\Omega}_1; \Omega_4 = \overline{\Omega}_3 \tag{231}$$

$$\phi_i = p_i x + \Omega_i t + \phi_{0i}; a_i = b_i 2\Omega_i + 2ip_i^2 - p_i \rho^2 / 2\Omega_i - 2ip_i^2 - p_i \rho^2 \tag{232}$$

$$b_2 = \overline{b}_1 \frac{\Omega_2 + ip_2^2 + p_2 \rho^2}{\Omega_2 - ip_2^2 + p_2 \rho^2}; b_4 = \overline{b}_3 \frac{\Omega_4 + ip_4^2 + p_4 \rho^2}{\Omega_4 - ip_4^2 + p_4 \rho^2} \tag{233}$$

$$M_{ij} = \frac{\left(\Omega_i p_j - \Omega_j p_i\right)^2 + p_i^2 p_j^2 \left(p_i - p_j\right)^2}{\left(\Omega_i p_j - \Omega_j p_i\right)^2 + p_i^2 p_j^2 \left(p_i + p_j\right)^2} \tag{234}$$

$$T_{ijk} = M_{ij} M_{jk} M_{ki}; A = \prod_{i<j} M_{ij} \tag{235}$$

Of course, for a valid and complete calculation, we are faced with the same situation as the 1st-order breather: ρ is real, b_1, b_3 and all φ_{0i} are complex constants. Certainly, each wave number p_i must be a pure imaginary number and each angular frequency Ω_i has to satisfy the quadratic dispersion relation:

$$4\Omega_i^2 + 4p_i \rho^2 \Omega_i + 4p_i^4 + 3p_i^2 \rho^4 = 0, (i = 1, 2, 3, 4) \tag{236}$$

And the threshold conditions for each complex-valued Ω_i share the same form as Eq. (205):

$$2p_i^4 + p_i^2 \rho^4 < 0 \tag{237}$$

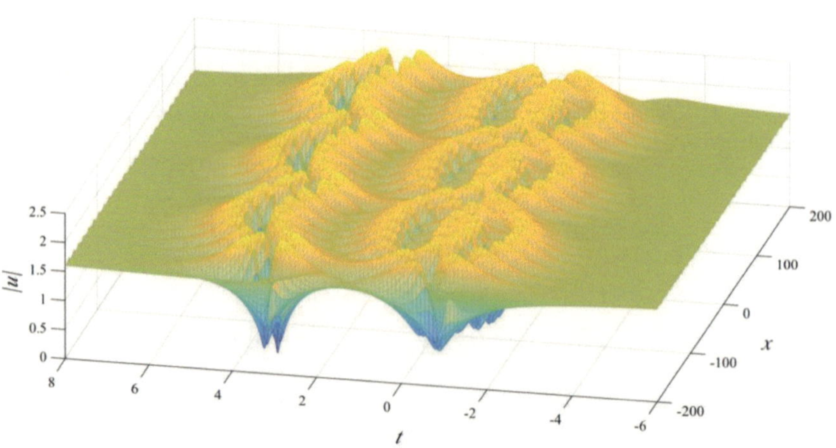

Figure 8.
The space-time evolution of the module of the 2nd order space periodic solution with $p_1 = 0.4i, p_3 = 0.75i, b_1 = i, b_3 = 1$ *and* $\rho = 1.6$. *Other phase factors* φ_1 *and* φ_3 *are set to zero.*

The space-time evolution of the module of the 2nd order space periodic solution (223) is shown in **Figure 8**. Paying attention to the form of this breather and the previous one, we will notice that this breather can exactly degenerate into the 1st-order breather if we take $p_3 = p_1$. Under this condition, $M_{13} = M_{24} = 0$, thus the higher order interaction coefficients T_{ijk} and A will vanish. Therefore, $g^{[2]}$ and $f^{[2]}$ will degenerate into the forms of $g^{[1]}$ and $f^{[1]}$, respectively:

$$g^{[2]}_{p_3=p_1} = g'^{[1]} = \rho e^{i\omega t}\left(1 + a_1' e^{\phi_1} + a_2' e^{\phi_2} + M_{12} a_1' a_2' e^{\phi_1+\phi_2}\right) \tag{238}$$

$$f^{[2]}_{p_3=p_1} = f'^{[1]} = e^{i\beta x}\left(1 + b_1' e^{\phi_1} + b_2' e^{\phi_2} + M_{12} b_1' b_2' e^{\phi_1+\phi_2}\right) \tag{239}$$

where $b_1' = \chi b_1, b_2' = \overline{\chi} b_2, a_1' = \chi a_1$ and $a_2' = \overline{\chi} a_2$ with $\chi = (b_1 + b_3)/b_1$. That is how $u^{[2]}$ can be reduced to $u^{[1]}$. Given to this reduction, a generalized form of these two breathers arises:

$$u^{[N]} = \overline{f}^{[N]} g^{[N]} / f^{[N]2}; (N = 1, 2) \tag{240}$$

$$g^{[N]} = \rho e^{i\omega t}\left(1 + \sum_{r=1}^{2N} \sum_{1\le n_1 < \cdots < n_r \le 2N} M(n_1, \cdots, n_r) \prod_{i=n_1}^{n_r} a_i e^{\phi_i}\right) \tag{241}$$

$$f^{[N]} = e^{i\beta x}\left(1 + \sum_{r=1}^{2N} \sum_{1\le n_1 < \cdots < n_r \le 2N} M(n_1, \cdots, n_r) \prod_{i=n_1}^{n_r} b_i e^{\phi_i}\right) \tag{242}$$

where the coefficient M is defined by:

$$M(i) = 1 \tag{243}$$

$$M(n_1, \cdots, n_r) = \prod_{i<j} M_{ij}; i, j \in (n_1, \cdots, n_r) \tag{244}$$

On the other hand, this breather possesses the same feature as the former one that it is periodic with respect to variable x due to the pure imaginary numbers p_1 and p_3. In addition, its asymptotic behaviors are analogical to the 1st-order space periodic solution. Each quadratic dispersion equation has two roots, respectively:

$$\Omega_{1\pm} = \left[-p_1\rho^2 \pm \sqrt{-2\left(2p_1^4 + p_1^2\rho^4\right)}\right]/2 \tag{245}$$

$$\Omega_{3\pm} = \left[-p_3\rho^2 \pm \sqrt{-2\left(2p_3^4 + p_3^2\rho^4\right)}\right]/2 \tag{246}$$

Thus, we will have four combinations of Ω_1 and Ω_2. Details are numerated in **Table 1**. The parameters φ_0, φ and φ' in **Table 1** are the phase shifts which are all real so that they will not change the module of $u^{[2]}$ when $t \to \infty$. And φ is given in Eq. (214), and others are determined by:

$$\exp\left(i\varphi_0\right) = a_1 a_2 a_3 a_4 / b_1 b_2 b_3 b_4 \tag{247}$$

$$\exp\left(i\varphi'\right) = a_3 a_4 / b_3 b_4 \tag{248}$$

From **Table 1**, we could draw the conclusion that this breather will also degenerate into the background plane wave as $|t| \to \infty$. Furthermore, there is a phase shift across the breather from $t = -\infty$ to $t = \infty$, which depended on the choice of Ω_1 and Ω_2.

Choice of Ω_i	Ω_{1+}, Ω_{2+}	Ω_{1+}, Ω_{2-}	Ω_{1-}, Ω_{2+}	Ω_{1-}, Ω_{2-}
$t \to -\infty$	$\rho e^{-3i\beta x+i\omega t}$	$\rho e^{-3i\beta x+i\omega t+i\varphi'}$	$\rho e^{-3i\beta x+i\omega t+i\varphi}$	$\rho e^{-3i\beta x+i\omega t+i\varphi_0}$
$t \to \infty$	$\rho e^{-3i\beta x+i\omega t+i\varphi_0}$	$\rho e^{-3i\beta x+i\omega t+i\varphi}$	$\rho e^{-3i\beta x+i\omega t+i\varphi'}$	$\rho e^{-3i\beta x+i\omega t}$

Table 1.
Asymptotic behaviors of $u^{[2]}$.

In this section, the 1st order and the 2nd order space periodic solutions of KN equation have been derived by means of HBDT. And after an integral transformation, these two breathers can be transferred into the solutions of CLL equation. Meanwhile, based on the long-wave limit, the simplest rogue wave model has been obtained according to the 1st order space periodic solution. Furthermore, the asymptotic behaviors of these breathers have been discussed in detail. As $|t| \to \infty$, both breathers will regress into the plane wave with a phase shift.

In addition, the generalized form of these two breathers is obtained, which gives us an instinctive speculation that higher order space periodic solutions may hold this generalized form, but a precise demonstration is needed. Moreover, higher order rogue wave models cannot be constructed directly by the long-wave limit of a higher order space periodic solution because the higher order space periodic solution has multiple wave numbers p_i, we are also interested in seeking an alternative method besides DT that could help us to determine the higher order rogue wave solutions.

6. Concluding remarks

In the end, as the author of the above two parts, part 1 and 2, I want to give some concluding remarks. As a whole, the two parts had taken the DNLS equation as a reference, systematically introduced several principal methods, such as IST, GLM (Marchenko) method, HBDT, to solve an integrable nonlinear equation under VBC and NVBC. We had gotten different kinds of soliton solutions, such as the light/dark soliton, the breather-type soliton, the pure soliton, the mixed breather-type and pure soliton, and especially the rogue-wave solution. We had also gotten soliton solutions in a different numbers, such as the one-soliton solution, the two-soliton solution, and the N-soliton solution. Nevertheless, I regret most that I had not introduced the Bäcklund transform or Darboux transform to search for a rogue wave solution or a soliton solution to the DNLS equation, just like professor Huang N.N., one of my guiders in my academic research career, had done in his paper [33]. Another regretful thing is that, limited to the size of this chapter, I had not introduced an important part of soliton studies, the perturbation theory for the nearly-integrable perturbed DNLS equation. Meanwhile, this chapter have not yet involved in the cutting-edge research of the higher-order soliton and rogue wave solution to the DNLS equation, which remain to be studied and concluded in the future.

A. Appendices

Some useful formulae.

A1, If A_1 and A_2 are $N \times 1$ matrices, A is a regular $N \times N$ matrix, then

$$A_1{}^T A^{-1} A_2 = \det\left(A + A_2 A_1{}^{\mathrm{T}}\right)/\det(A) - 1 \tag{A1}$$

A2, Binet-Cauchy formula: For a squared $N \times N$ matrix B

$$\det(I+B) = 1 + \sum_{r=1}^{N} \sum_{1 \le n_1 < n_2 < \cdots < n_r \le N} B(n_1, n_2, \cdots, n_r) \tag{A2}$$

where $B(n_1, n_2, \cdots, n_r)$ is a r'th-order principal minor of B.

A3, For a $N \times N$ matrix Q_1 and a $N \times N$ matrix Q_2,

$$\det(I + Q_1 Q_2) = 1 + \sum_{r=1}^{N} \sum_{1 \le n_1 < n_2 < \cdots < n_r \le N} \Omega_r(n_1, n_2, \cdots n_r)$$

$$= 1 + \sum_{r=1}^{N} \sum_{1 \le n_1 < \cdots < n_r \le N} \sum_{1 \le m_1 < \cdots < m_r \le N} Q_1(n_1, n_2, \cdots n_r; m_1, m_2, \cdots, m_r) Q_2(m_1, m_2, \cdots m_r; n_1, n_2 \cdots, n_r) \tag{A3}$$

where $Q_1(n_1, n_2, \cdots n_r; m_1, m_2, \cdots, m_r)$ denotes a minor, which is the determinant of a submatrix of Q_1 consisting of elements belonging to not only rows $(n_1, n_2, \cdots n_r)$ but also columns $(m_1, m_2, \cdots, m_r)$.

The above formula also holds for the case of $\det(I + \Omega_1 \Omega_2)$ With Ω_1 to be a $N \times (N+1)$ matrix and Ω_2 a $(N+1) \times N$ matrix.

A4, For a squared matrix C with elements $C_{jk-} = f_j g_k (x_j - y_k)^{-1}$,

$$\det(C) = \prod_j f_j g_j \prod_{j<j', k<k'} \left(x_j - x_{j'}\right)(y_{k'} - y_k) \prod_{j,k} (x_j - y_k)^{-1} \tag{A4}$$

A5, Some useful blinear derivative formulae.

$$\left(\frac{A}{B}\right)_x = \frac{D_x A \cdot B}{B^2} \tag{A5}$$

$$\left(\frac{A}{B}\right)_{xx} = \frac{D_x^2 A \cdot B}{B^2} - \frac{A}{B} \frac{D_x^2 B \cdot B}{B^2} \tag{A5.1}$$

$$D_x ab \cdot cd = bd D_x a \cdot c + ac D_x b \cdot d = bc D_x a \cdot d + ad D_x b \cdot c \tag{A5.2}$$

$$D_x^2 ab \cdot cd = bd D_x^2 a \cdot c + 2(D_x a \cdot c)(D_x b \cdot d) + ac D_x^2 b \cdot d \tag{A5.3}$$

$$D_t^n D_x^m \exp(\eta_1) \cdot \exp(\eta_2) = (\Omega_1 - \Omega_2)^n (\Lambda_1 - \Lambda_2)^m \exp(\eta_1 + \eta_2), \tag{A5.4}$$

where $\eta_i = \Omega_i t + \Lambda_i x + \eta_{0i}$ ，$i = 1, 2$; Ω_i, Λ_i, η_{0i} are complex constants.

Author details

Zhou Guo-Quan
Department of Physics, Wuhan University, Wuhan, P.R. China

*Address all correspondence to: zgq@whu.edu.cn

References

[1] Tzoar N, Jain M. Self-phase modulation in long-geometry optical waveguides. Physical Review A. 1981;**23**: 1266

[2] Anderson D, Lisak M. Nonlinear asymmetric self-phase modulation and self-steepening of pulses in long optical waveguides. Physical Review A. 1983;**27**: 1393

[3] Kawata T, Inoue H. Exact solution of derivative nonlinear Schrödinger equation under nonvanishing conditions. Journal of the Physical Society of Japan. 1978;**44**:1968

[4] Mjølhus E. Nonlinear Alfven waves and the DNLS equation: Oblique aspects. Physica Scripta. 1989;**40**:227

[5] Spangler, S.R. Nonlinear Waves and Chaos in Space Plasmas, Ed. T. Hada and T. Matsumoto. (Tokyo: Terrapub); 1997. p. 171

[6] Steudel H. The hierarchy of multi-soliton solutions of derivative nonlinear Schrödinger equation. Journal of Physics A. 1931;**2003**:36

[7] Chen XJ, Yang JK, Lam WK. N-soliton solution for the derivative nonlinear Schrödinger equation with nonvanishing boundary conditions. Journal of Physics A: Mathematical and General. 2006;**39**(13):3263

[8] Lashkin VM. N-soliton solutions and perturbation theory for DNLS with nonvanishing condition. Journal of Physics A. 2007;**40**:6119-6132

[9] Chen X-J et al. An inverse scattering transform for the derivative nonlinear Schrödinger equation. Physical Review E. 2004;**69**(6):066604

[10] Cai H. Research about MNLS Equation and DNLS Equation. Wuhan: Wuhan Univ; 2005

[11] Zhou G-Q. A newly revised inverse scattering transform for DNLS+ equation under nonvanishing boundary condition. Wuhan University. 2012; **17**(2):144-150

[12] Zhou G-Q. Explicit breather-type and pure *N*-Soliton solution of DNLS+ equation with nonvanishing boundary condition. Wuhan University. 2013; **18**(2):147-155

[13] Li X-J, Zhou G-Q. Mixed breather-type and pure Soliton solution of DNLS equation. Wuhan University. 2017; **22**(3):223-232

[14] Zhou G-Q, Li X-J. Space periodic solutions and rogue wave solution of the derivative nonlinear Schrödinger equation. Wuhan University. 2017; **22**(5):9-15

[15] Kaup DJ, Newell AC. An exact solution for a derivative nonlinear Schrödinger equation. Journal of Mathematical Physics. 1978;**19**:798

[16] Mjølhus E, Hada T. In: Hada T, Matsumoto T, editors. Nonlinear Waves and Chaos in Space Plasmas. Tokyo: Terrapub; 1997. p. 121

[17] Zhou G-Q, Huang N-N. An *N*-soliton solution to the DNLS equation based on revised inverse scattering transform. Journal of Physics A: Mathematical and Theoretical. 2007; **40**(45):13607

[18] Zhou G-Q. A multi-soliton solution of the DNLS equation based on pure Marchenko formalism. Wuhan University. 2010;**15**(1):36-42

[19] Zhou G-Q, Bi X-T. Soliton solution of the DNLS equation based on Hirota's bilinear derivative transform. Wuhan University. 2009;**14**(6):505-510

[20] Zhou G-Q. Hirota's bilinear derivative transform and soliton

solution of the DNLS equation. Physics Bulletin. Hebei, Baoding; 2014;**4**:93-97 (in Chinese)

[21] Huang N-N. Theory of Solitons and Method of Perturbations. Shanghai: Shanghai Scientific and Technological Education Publishing House; 1996

[22] Huang N-N et al. Hamilton Theory about Nonlinear Integrable Equations. Beijing: Science Press; 2005. pp. 93-95

[23] Huang N-N. Chinese Physics Letters. 2007;**24**(4):894-897

[24] Rogister A. Parallel propagation of nonlinear low-frequency waves in high-β plasma. Physics of Fluids. 1971;**14**:2733

[25] Ruderman MS. DNLS equation for large-amplitude solitons propagating in an arbitrary direction in a high- β hall plasma. Journal of Plasma Physics. 2002; **67**:271

[26] Govind PA. Nonlinear Fiber Optics. 3rd ed. New York: Academic Press; 2001

[27] Nakata I. Weak nonlinear electromagnetic waves in a ferromagnet propagating parallel to an external magnetic field. Journal of the Physical Society of Japan. 1991;**60**(11):3976

[28] Nakata I, Ono H, Yosida M. Solitons in a dielectric medium under an external magnetic field. Progress in Theoretical Physics. 1993;**90**(3):739

[29] Daniel M, Veerakumar V. Propagation of electromagnetic soliton in anti- ferromagnetic medium. Physics Letters A. 2002;**302**:77-86

[30] Nakamura A, Chen HH. Multi-soliton solutions of derivative nonlinear Schrödinger equation. Journal of the Physical Society of Japan. 1980;**49**:813

[31] Kakei S, Sasa N, Satsuma J. Bilinearization of a generalized derivative nonlinear Schrödinger equation. Journal of the Physical Society of Japan. 1995;**64**(5):1519-1523

[32] Yang J-K. Nonlinear Waves in Integrable and Nonintegrable Systems. Philadelphia: Society for Industrial and Applied Mathematics; 2010

[33] Huang NN, Chen ZY. Alfven solitons. Journal of Physics A: Mathematical and General. 1990;**23**:439

[34] Akhmediev N, Ankiewicz A, Taki M. Waves that appear from nowhere and disappear without a trace. Physics Letters A. 2009;**373**(6):675-678

[35] Akhmediev N, Soto-Crespo JM, Ankiewicz A. Extreme waves that appear from nowhere: On the nature of rogue waves. Physics Letters A. 2009; **373**(25):2137-2145

[36] Peregrine DH. Water waves, nonlinear Schrödinger equations and their solutions. Journal of the Australian Mathematical Society. Series B. Applied Mathematics. 1983;**25**(01):16-43

[37] Akhmediev NN, Korneev VI. Modulation instability and periodic solutions of the nonlinear Schrödinger equation. Theoretical and Mathematical Physics. 1986;**69**(2):1089-1093

[38] Ma YC. The perturbed plane wave solutions of the cubic Schrödinger equation. Studies in Applied Mathematics. 1979;**60**(1):43-58

[39] Xu S, He J, Wang L. The Darboux transformation of the derivative nonlinear Schrödinger equation. Journal of Physics A: Mathematical and Theoretical. 2011;**44**(30):305203

[40] Guo B, Ling L, Liu QP. High-order solutions and generalized Darboux transformations of derivative nonlinear Schrödinger equations. Studies in Applied Mathematics. 2013;**130**(4): 317-344

[41] Zhang Y, Guo L, Xu S, et al. The hierarchy of higher order solutions of the derivative nonlinear Schrödinger equation. Communications in Nonlinear Science and Numerical Simulation. 2014;**19**(6):1706-1722

[42] Hirota R. Direct method of finding exact solutions of nonlinear evolution equations. In: Miura RM, editor. Bäcklund Transformations, the Inverse Scattering Method, Solitons, and their Applications. Lecture Notes in Mathematics. Berlin, Heidelberg: Springer; 1976;**515**:40-68

[43] Hirota R. Exact solution of the Korteweg-de Vries equation for multiple collisions of solitons. Physical Review Letters. 1971;**27**(18):1192

[44] Hirota R. The Direct Method in Soliton Theory. Cambridge University Press; 2004

[45] Dysthe KB, Trulsen K. Note on breather type solutions of the NLS as models for freak-waves. Physica Scripta. 1999;**T82**:48

[46] Chan HN, Chow KW, Kedziora DJ, et al. Rogue wave modes for a derivative nonlinear Schrödinger model. Physical Review E. 2014;**89**(3):032

Chapter 4

Magnetic Solitons in Optical Lattice

Xing-Dong Zhao

Abstract

In this chapter, we discuss the magnetic solitons achieved in atomic spinor Bose-Einstein condensates (BECs) confined within optical lattice. Spinor BECs at each lattice site behave like spin magnets and can interact with each other through the static magnetic dipole-dipole interaction (MDDI), due to which the magnetic soliton may exist in blue-detuned optical lattice. By imposing an external laser field into the lattice or loading atoms in a red-detuned optical lattice, the light-induced dipole-dipole interaction (LDDI) can produce new magnetic solitons. The long-range couplings induced by the MDDI and ODDI play a dominant role in the spin dynamics in an optical lattice. Compared with spin chain in solid material, the nearest-neighbor approximation, next-nearest-neighbor approximation, and long-range case are discussed, respectively.

Keywords: spinor Bose-Einstein condensates, spin wave, optical lattice, magnetic soliton

1. Introduction

Soliton can be classified into different species, such as matter-wave soliton, magnetic soliton, optical soliton, and so on [1, 2]. The magnetic solitons, which describe the localized magnetization, are very important excitations in the Heisenberg spin chain in solid system in condensed matter [3]. Because of the defect and impurity in solid material, it is difficult to perform experimental observation and manipulation in these systems. In addition, magnetic solitons originate from the Heisenberg-like short-range exchange interaction between electrons in solid state systems, so the theoretical models and treatment are limited to only the approximation of nearest-neighbor interaction, it is disadvantageous to the study of long-range interaction-induced dynamics.

On the other hand, the spinor ultracold atoms in an optical lattice offer a pure and well-controlled platform to study spin dynamics. If the lattice trapping atoms is deep enough, the spinor atoms undergo a ferromagnetic-like phase transition that leads to a macroscopic magnetization of the condensates array; then, the individual sites become independent with each other. Spinor BECs at each lattice site behave like spin magnets and can interact with each other through the static MDDI, which can cause the ferromagnetic phase transition and the spin wave excitation. It is atomic spin chain in optical lattice [4–7]. By tuning the system parameters, magnetic soliton can be produced; however a key difference of the atomic spin chain from solid-state one is that we are facing a completely new spin system in which the long-range nonlinear interactions play a dominant role. To demonstrate the high controllability, an external laser can be imposed into the lattice and induce light-induced dipole-dipole interaction (LDDI) to produce new magnetic solitons. In fact, the LDDI can be induced by loading the atoms in a red-detuned optical lattice as used in [5]. However, the method used here has good feasibility, and some novel physical phenomena can be observed.

In order to make comparison with the Heisenberg spin chain, we used the nearest-neighbor approximation and the next-nearest-neighbor approximation in our discussion. In addition, the two approximations are effective and reasonable with some appropriate parameters chosen in experiments.

In this chapter, we are going to be seeing how magnetic soliton in atomic spin chain is, which will be theoretically described. We first discuss the formation of atomic spin chain in optical lattice in Section 2. Then we shift our focus to the condition for magnetic soliton in Section 3. Section 4 presents the conclusion and outlook.

2. Atomic spin chain

2.1 Atomic spin chain with magnetic dipole: dipole interaction

The spinor BECs can be trapped within one-, two-, and three-dimensional optical lattices with different experimental techniques; however, we only take one-dimensional case as an example. As depicted in **Figure 1**, the atomic gases with $F = 1$ are trapped in a one-dimensional optical lattice, which is formed by two π-polarized laser beams counter-propagating along the y-axis. The light-induced lattice potential takes the form:

$$V_L(\boldsymbol{r}) = U_L \exp\left(-r_\perp{}^2/W_L{}^2\right)\cos^2(k_L y) \tag{1}$$

where $r_\perp = \sqrt{x^2+z^2}$, $k_L = 2\pi/\lambda_L$ is the wave number, W_L the width of the lattice beams, and the potential depth $U_L = \hbar|\Omega|^2/6\Delta$ with Ω being the Rabi frequency and Δ being laser detuning.

It has been theoretically and experimentally demonstrated that the condensate trapped in optical lattice undergoes a superfluid-Mott-insulator transition as the depth of the lattice wells is increased. If the lattice potential is deep enough, the condensate is divided into a set of separated small condensates equally located at each lattice site, which forms the atomic chain. The condensate localized in each lattice site is approximately in its electronic ground state, while the two lattice laser beams are detuned far from atomic resonance frequency, and the laser detuning

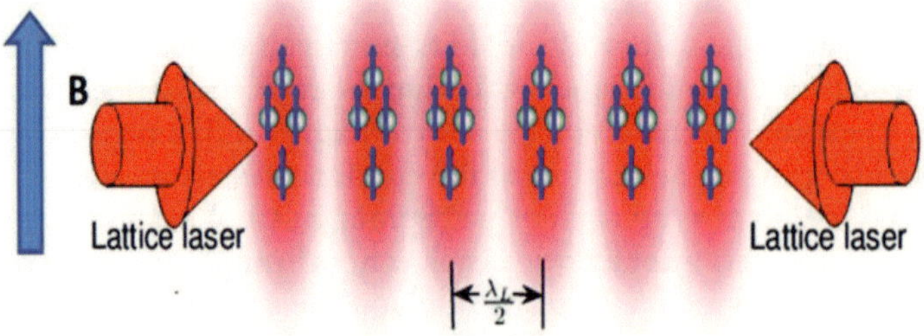

Figure 1.
Schematic diagram of spinor BECs confined in a one-dimensional optical lattice. A strong static magnetic field B is applied to polarize the ground-state spin orientations of the atomic chain along the quantized z-axis.

is defined as $\Delta = \omega_L - \omega_a$, ω_L being laser frequency and ω_a atomic resonant frequency. We classify the optical lattice into two categories: red-detuned optical lattice ($\Delta < 0$) and blue-detuned one $\Delta > 0$. For the former, the condensed atoms are trapped at the standing wave nodes where the laser intensity is approximately zero; the light-induced interactions are neglected in this case. In order to realize ferromagnetic phase transition, a strong static magnetic field $\boldsymbol{B}$ is applied to polarize the ground-state spin orientations of the atomic chain along the quantized z-axis. The Hamiltonian including the long-range MDDI between different lattice sites takes the following form [7–9]:

$$
\begin{aligned}
H_M = & \sum_m \int dr \left\{ \hat{\psi}_m^\dagger \left(-\frac{\hbar^2}{2M} \nabla^2 + V_L \right) \hat{\psi}_m \right\} \\
& + \sum_{m, m', n, n'} \left\{ \int dr \int dr' \hat{\psi}_m^\dagger \hat{\psi}_{m'}^\dagger \hat{\psi}_{n'} \hat{\psi}_n V_{mm'n'n}^{col}(\boldsymbol{r}, \boldsymbol{r}') \right. \\
& + \frac{1}{2} \int dr \int dr' \hat{\psi}_m^\dagger \hat{\psi}_{m'}^\dagger \hat{\psi}_{n'} \hat{\psi}_n V_1(\boldsymbol{r}, \boldsymbol{r}') \hat{\boldsymbol{F}}_{mn} \cdot \hat{\boldsymbol{F}}_{m'n'} \\
& \left. - \frac{1}{2} \int dr \int dr' \hat{\psi}_m^\dagger \hat{\psi}_{m'}^\dagger \hat{\psi}_{n'} \hat{\psi}_n V_2(\boldsymbol{r}, \boldsymbol{r}') \times \hat{\boldsymbol{F}}_{mn} \cdot (\boldsymbol{r} - \boldsymbol{r}') \hat{\boldsymbol{F}}_{m'n'} \cdot (\boldsymbol{r} - \boldsymbol{r}') \right\} \\
& + \sum_{m, n} \int dr \, \hat{\psi}_m^\dagger (-\boldsymbol{\mu} \cdot \boldsymbol{B})_{mn} \hat{\psi}_n
\end{aligned} \tag{2}
$$

where $\hat{\psi}_m(\boldsymbol{r}, t)(m = \pm 1, 0)$ are the hyperfine ground-state atomic field operators. The total angular momentum operator $\hat{\boldsymbol{F}}$ for the hyperfine spin of an atom has components represented by 3×3 matrices in the $|f = 1, m_f = m\rangle$ subspace. The final term in Eq. (2) represents the effect of polarizing magnetic field, and $\boldsymbol{\mu}$ is magnetic dipole moment. The interaction potential includes the two-body ground-state collisions and the magnetic dipole-dipole interactions, which can be described by the potentials:

$$V_{mm'n'n}^{col}(\boldsymbol{r}, \boldsymbol{r}') = \left[\lambda_s \delta_{m'n'} \delta_{mn} + \lambda_a \hat{\boldsymbol{F}}_{mn} \cdot \hat{\boldsymbol{F}}_{m'n'} \right] \delta(\boldsymbol{r} - \boldsymbol{r}'), \tag{3}$$

$$V_1(\boldsymbol{r}, \boldsymbol{r}') = \frac{\mu_0 {\gamma_B}^2}{4\pi} \frac{1}{|\boldsymbol{r} - \boldsymbol{r}'|^3}, \tag{4}$$

$$V_2(\boldsymbol{r}, \boldsymbol{r}') = \frac{3\mu_0 {\gamma_B}^2}{4\pi} \frac{1}{|\boldsymbol{r} - \boldsymbol{r}'|^5}. \tag{5}$$

In collision interaction potential in Eq. (3), λ_s and λ_a are related to the s-wave scattering length for the spin-dependent interatomic collisions. In the magnetic dipole-dipole interaction potential Eqs. (4) and (5), μ_0 is the vacuum permeability; the parameter $\gamma_B = -\mu_B g_F$ is the gyromagnetic ratio with μ_B being the Bohr magneton and g_F the Landé g factor.

Under tight-binding condition and single-mode approximation, the spatial wave function $\phi_m(\boldsymbol{r})$ of the m-th condensate is then determined by the Gross-Pitaevskii (GP) equation [5]:

$$\left[-\frac{\hbar^2}{2M} \nabla^2 + V_m(\boldsymbol{r}) + \lambda_s (N_m - 1) |\phi_m(\boldsymbol{r})|^2 \right] \phi_m(\boldsymbol{r}) = \mu_m \phi_m(\boldsymbol{r}), \tag{6}$$

with

$$V_m(r) = U_L \exp\left[-{r_L}^2 / W^2 \right] \cos^2(k_L y + m\pi), \; 0 < y < \lambda_L / 2, \tag{7}$$

and N_m is the number of condensed atoms at the m-th site; μ_m is the chemical potential. In this case, the individual condensate confined in each lattice behaves as spin magnets in the presence of external magnetic fields; such spin magnets form a 1D coherent spin chain with an effective Hamiltonian [5, 6]:

$$H_m = \sum_m \left(\lambda_a{}'\hat{\boldsymbol{S}}_m^2 - \gamma_B \hat{\boldsymbol{S}}_m \cdot \boldsymbol{B} - \sum_{n \neq m} J_{mn}^{mag} \hat{\boldsymbol{S}}_m \hat{\boldsymbol{S}}_n \right), \tag{8}$$

where we have defined collective spin operators $\hat{\boldsymbol{S}}_m$ with components $\hat{S}_m^{\{x,y,z\}}$ and

$$\lambda_a' = \frac{1}{2}\lambda_a \int d\boldsymbol{r} |\phi_m(\boldsymbol{r})|^4. \tag{9}$$

The parameter J_{mn}^{mag} describes the strength of the site-to-site spin coupling induced by the static MDDI. It takes the form:

$$J_{mn}^{mag} = \frac{\mu_0 \gamma_B{}^2}{16\pi\hbar^2} \int d\boldsymbol{r} \int d\boldsymbol{r}' \frac{R_1}{|\boldsymbol{r}'|^5} |\phi_m(\boldsymbol{r})|^2 |\phi_n(\boldsymbol{r} - \boldsymbol{r}')|^2, \tag{10}$$

where μ_0 is the vacuum permeability, $R_1 = |\boldsymbol{r}'|^2 - 3y'^2$. It will be seen from this that the MDDI is determined by many parameters; however, we need to choose a parameter which is easy controllable in the experiment. The realistic interaction strengths of the MDDI and LDDI coupling coefficients have been calculated in ref. [4], in which we find the dependency of the transverse width.

Here, we only consider the ferromagnetic condensates where the spins of atoms at each lattice site align up along the direction of the applied magnetic field (the quantized z-axis) in the ground state of the ferromagnetic spin chain. Because of the Bose-enhancement effect, the spins are very large in these systems. As a result, at low temperature the spins can be treated approximately as classical vectors, and we can replace $\hat{\boldsymbol{S}}_n$ with its average value S_n [1]. From the Hamiltonian (8), we can derive the Heisenberg equation of motion for operator $S_n^{\pm} = S_n^x \pm iS_n^y \, (i^2 = -1)$ (normalized by $\hbar$):

$$i\frac{dS_n^+}{dt} = \gamma_B B_z S_n^+ + \sum_{j \neq n} 2J_{nj}^{mag} \left(S_j^z S_n^+ - S_n^z S_j^+ \right). \tag{11}$$

Equation (11) determines the existence and the propagation of spin waves. From this equation, we can obtain the existence and dynamical evolving of the nonlinear magnetic soliton. In Sec.3, we attempt to show the long-range and controllable characteristics of MDDI and find how this long-range nonlinear dipole-dipole interaction affects the generation of the magnetic soliton. If the static magnetic field $\boldsymbol{B}$ is strong enough, we have:

$$S_n^z = \sqrt{S^2 - S_n^+ S_n^-} = \sqrt{S^2 - S_n^+ S_n^+}. \tag{12}$$

2.2 Atomic spin chain with light-induced dipole-dipole interaction

In fact, the site-to-site interaction can also be tuned by laser field, and the exchange interaction induced by the laser fields also plays an important role in atomic spin chain in optical lattice. We consider that an external modulational laser field is imposed into the system along the y-axis. This external laser field is properly chosen so that a new dipole-dipole interaction will be introduced into the system, i.e., the so-called LDDI. Its Hamiltonian takes the form [10]:

$$H_{opt} = \sum_{m} \sum_{n \neq m} J_{mn}^{opt} \left(\hat{S}_m^- \hat{S}_n^+ + S_m^+ S_n^- \right). \tag{13}$$

Here, the coefficient J_{mn}^{opt} describes the site-to-site spin coupling induced by the LDDI, which has the concrete form:

$$J_{mn}^{opt} = \frac{Q}{144\hbar k_L^3} \int dr \int dr' f_c(r') \exp\left(-\frac{R_2}{W_L^2}\right) \\ \times \cos(k_L y) \cos[k_L(y - y')] \boldsymbol{e}_{+1} \cdot \boldsymbol{W}(\boldsymbol{r}') \cdot \boldsymbol{e}_{-1} \times |\phi_m(\boldsymbol{r})|^2 |\phi_n(\boldsymbol{r} - \boldsymbol{r}')|^2. \tag{14}$$

In above integral we define $Q = \gamma |\Omega_1|^2 / \Delta_1^2$ to describe the magnitude of the LDDI; here γ is the spontaneous emission rate of the atoms, and Ω_1 is the Rabi frequency; Δ_1 is the optical detuning for the induced atomic transition. A cutoff function $f_c(\boldsymbol{r}) = \exp(-r/L_c)$ is introduced to describe the effective interaction range of the LDDI, L_c being the coherence length. $R_2 = r_\perp{}^2 + |\boldsymbol{r}_\perp - \boldsymbol{r}_\perp{}'|^2$ with the transverse coordinate $r_\perp = \sqrt{x^2 + z^2}$. The unit vectors have the form:

$$\boldsymbol{e}_0 = \boldsymbol{e}_z, \tag{15}$$

$$\boldsymbol{e}_{\pm 1} = \mp(\boldsymbol{e}_x \pm i\boldsymbol{e}_y), \tag{16}$$

and $\boldsymbol{e}_\alpha (\alpha = x, y, z)$ which are the usual geocentric Cartesian frame. The tensor $\boldsymbol{W}(\boldsymbol{r})$ describes the spatial profile of the LDDI and has the form:

$$\boldsymbol{W}(\boldsymbol{r}) = \frac{3}{4}\left[(1 - 3\cos^2\theta)\left(\frac{\sin\xi}{\xi^2} + \frac{\cos\xi}{\xi^3} \right) - \sin^2\theta \frac{\cos\xi}{\xi} \right], \tag{17}$$

where we define $\xi = k_L |\boldsymbol{r} - \boldsymbol{r}'|$, and θ is the angle between the dipole moment and the relative coordinate $\boldsymbol{r} - \boldsymbol{r}'$. Being different from MDDI in Eq. (8), it shows that the LDDI in this system leads to the spin coupling only in the transverse direction, i.e., the direction perpendicular to the external magnetic field, as shown in Eq. (13). The total Hamiltonian of the system can be written as [10].

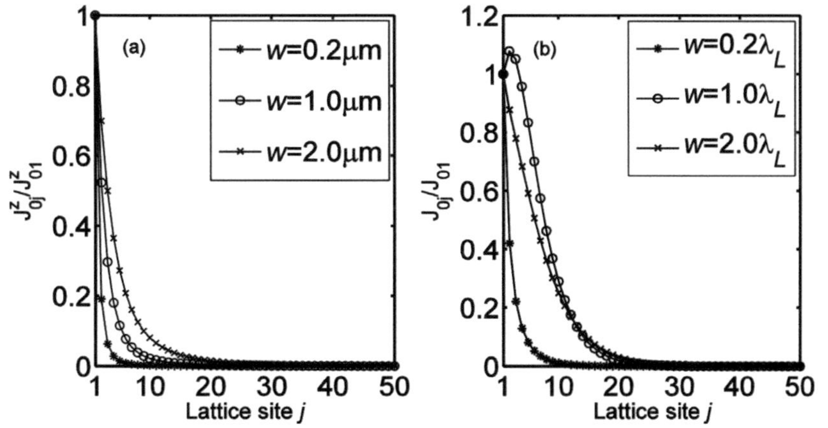

Figure 2.
The spin coupling coefficients $J_{mn}^z = J_{mn}^{mag}$ and J_{ij} as a function of the lattice site index for different transverse widths of the condensate in the MDDI and LDDI dominating optical lattice [4].

$$H = H_m + H_{opt}. \tag{18}$$

Function (11) now has the form:

$$i\frac{d\phi_n}{dt} = \gamma_B B_z \phi_n + \sum_{j=n\pm1} 2SJ_{nj}^{mag}\sqrt{1-\left|\phi_j\right|^2}\phi_n + \sum_{j=n\pm1} 4SJ_{nj}\sqrt{1-|\phi_n|^2}\phi_j. \tag{19}$$

The interaction coefficient J_{ij} satisfies $J_{ij} = J_{ij}^{opt} + \frac{1}{2}J_{ij}^{mag}$.

Figure 2 shows the two spin coupling coefficients varying with the lattice sites and the transverse width of the condensate. It is clear that the spin coupling coefficients are sensitive to the variation of the transverse width W of the condensate and exhibit the similar long-range behavior.

3. Magnetic soliton in atomic spin chain

3.1 Magnetic soliton under the MDDI

To confirm that the magnetic soliton can be formed in this system, we first consider the case that other interactions between spins are absent in this section. Namely if the solitons indeed exist, they can be generated by the MDDI alone $J_{ij} = J_{ij}^{mag}/2$. In the usual way, we set:

$$S_n^{\pm} = S_n^{\pm}/S, \tag{20}$$

$$S_n^{z} = S_n^{z}/S, \tag{21}$$

here S is the magnitude of spin for site n and only depends on the number of atoms trapped in each site. Under the external magnetic field $\boldsymbol{B}$ considered here, we make an assumption that the spin deviation from the quantized z-axis is considerable, which is reasonable when the number of atoms trapped in each site is large; it is achievable according to present experiments. We also assume that the magnitude of the spin deviation s_n will extend over a large number of lattice sites and vary slowly with site index n. Considering the spatial symmetry, we introduce a staggered variable:

$$\phi_n = (-1)^n s_n^{+}, \tag{22}$$

where ϕ_n is complex; to find the existence conditions of the magnetic soliton modes, we need firstly to get the nonlinear Schrödinger equation (NLSE) for ϕ_n in continuum form, which is expected to be qualitatively correct for solitons in discrete limit.

3.1.1 Nearest-neighbor approximation with the MDDI

Here we firstly consider only the nearest-neighbor (NN) interactions. In other words, we assume that the on-site spin couples only to the spins at its two neighboring sites, and the longer range couplings are artificially cut off. In above equation, for small retortion from the ground state, we have:

$$\sqrt{1-\left|\phi_j\right|^2} \approx 1 - \frac{1}{2}\left|\phi_j\right|^2. \tag{23}$$

Using the slow varying envelope approximation or long-wavelength limit, we have:

$$\left.\frac{d^2\phi(y,t)}{dy^2}\right|_{y=y_n} \approx \frac{\phi_{n+1}+\phi_{n-1}-2\phi_n}{a^2}. \tag{24}$$

where a is the lattice constant. Then we can obtain the continuum limit NLSE for $\phi(y,t)$, namely

$$i\frac{d\phi}{dt} = 2SJ_{01}^{mag}a^2\frac{d^2\phi}{dy^2} + \beta\phi - 4SJ_{01}^{mag}|\phi|^2\phi, \tag{25}$$

where $a = \lambda_L/2$ is the lattice spacing between two adjacent sites, with λ_L being the wave length of the lattice laser beams, and the coefficient $\beta = \gamma_B B_z + 8SJ_{01}^{mag}$. This NLSE is integrable, and it has bright soliton solution if the effects of dispersion and nonlinearity can cancel each other. In other words, the condition for Eq. (25) to have one bright soliton solution is that the coefficient of the second-order spatial derivative term and the coefficient of the cubic nonlinear term are of the same sign [11]. The criterion for the existence of solitons requires:

$$f(W) = -8S^2a^2J_{01}^{mag}2 > 0. \tag{26}$$

Since the spin coupling parameter J_{01}^{mag} is positive, this condition cannot be satisfied, that is to say, there is no one soliton solution in atomic spin chain under the NN approximation. It drives us to in-depth consideration; we should consider more complex interactions, i.e., more site-to-site interactions would be considered.

3.1.2 Next-nearest-neighbor approximation with the MDDI

Even though the long-range effect of the dipole-dipole interaction is obvious, in this section, we consider a case slightly beyond the NN approximation, i.e., the next-nearest-neighbor (NNN) approximation, which is used in some solid-state systems. Under this approximation, each on-site spin is coupled to the four near-neighbor spins, two on its right side and the other two on its left side, respectively. In this case, we set:

$$\alpha_j = J_{0j}^{mag}/J_{01}, \tag{27}$$

$$A_2 = J_{02}/J_{01}, \tag{28}$$

$$\eta = \gamma_B B_z + 4S\left(J_{01}^{mag} + 2J_{01} + J_{02}^{mag} - 2J_{02}\right). \tag{29}$$

Additionally, the second-order spatial derivative is introduced:

$$\left.\frac{d^2\phi(y,t)}{dy^2}\right|_{y=y_n} \approx \frac{\phi_{n+2}+\phi_{n-2}-2\phi_n}{(2a)^2}. \tag{30}$$

Then, we can easily get:

$$\begin{aligned} i\frac{d\phi(y,t)}{dt} &= 4SJ_{01}(1-4A_2)a^2\frac{d^2\phi(y,t)}{dy^2} + \eta\phi(y,t) \\ &- 2SJ_{01}(\alpha_1 + A_2\alpha_2 + 2 - 2A_2)|\phi(y,t)|^2\phi(y,t), \end{aligned} \tag{31}$$

where

$$\eta = \gamma_B B_z + 4S\left(J_{01}^{mag} + 2J_{01} + J_{02}^{mag} - 2J_{02}\right). \tag{32}$$

Because the LDDI is absent in this case, we have:

$$J_{01}^{opt} = 0, \tag{33}$$

$$J_{01} = \frac{1}{2}J_{01}^{mag}, \tag{34}$$

$$\alpha_1 = \alpha_2 = 2. \tag{35}$$

Then, Eq. (31) can be rewritten as [10]:

$$\begin{aligned} i\frac{d\phi(y,t)}{dt} &= 2SJ_{01}^{mag}(1-4A_2)a^2\frac{d^2\phi(y,t)}{dy^2} \\ &+ \eta\phi(y,t) - 4SJ_{01}^{mag}|\phi(y,t)|^2\phi(y,t), \end{aligned} \tag{36}$$

where $\eta = \gamma_B B_z + 8SJ_{01}^{mag}$. For two adjacent sites, $A_2 = J_{02}^{mag}/J_{01}^{mag}$ measures the relative strength of the NNN spin coupling to the NN spin coupling. The condition for Eq. (36) to have one soliton solution is

$$f(W) = -8S^2a^2J_{01}^{mag}2(1-4A_2) > 0; \tag{37}$$

we have $A_2 > 0.25$. From the definition of A_2, we can rewrite the condition as $0.25 < A_2 < 1$, which agrees with the instability condition of the extended nonlinear band edge spin wave modes in the linear stability analysis in Ref. [4]. It means that one soliton solution can exist in atomic spin chain after the NNN interactions between spins are considered. To a certain extent, this result is very useful because the NNN approximation is good enough to describe this system under appropriate parameters and it provides us an opportunity to make comparison between the atomic spin chain and Heisenberg spin chain.

3.1.3 Long-range case with the MDDI

To study the effects of the long-range spin coupling on the nonlinear spin wave dynamics, we need consider the spin coupling at all sites through the whole optical lattice, though the MDDI decreases rapidly in some cases. We firstly use Eq. (22) to rewrite the nonlinear motion equation of ϕ_n as

$$\begin{aligned} i\frac{d\phi_n}{dt} &= \left[\gamma_B B_z + \sum_{j\neq n} 2SJ_{nj}^{mag}\right]\phi_n - \sum_{j\neq n} SJ_{nj}^{z}\left|\phi_j\right|^2\phi_n \\ &- \sum_{j\neq n} 4SJ_{nj}\phi_j(-1)^{j-n} + \sum_{j\neq n} 2SJ_{nj}|\phi_n|^2\phi_j(-1)^{j-n}. \end{aligned} \tag{38}$$

Then we set:

$$\phi_n \to \phi(y,t), \tag{39}$$

$$J_{nj}^{mag} \to J^{mag}(y-y'), \tag{40}$$

$$J_{nj} \to J(y-y'), \tag{41}$$

$$\gamma_B B_z + \sum_{j\neq n} 2SJ_{nj}^{mag} \rightarrow \omega(y). \tag{42}$$

Considering the discreteness of the optical lattice, we can treat the symbolic terms as below:

$$(-1)^{j-n} = \cos[(j-n)\pi] = \cos\left[\frac{2\pi}{\lambda_L}(y-y')\right]. \tag{43}$$

After changing the sum to integration, Eq. (38) becomes:

$$\begin{aligned} i\frac{d\phi(y,t)}{dt} &= \omega(y)\phi(y,t) - \frac{2S}{\lambda_L}\phi(y,t)\int_{-\infty}^{\infty} J^{mag}(y-y')\phi^2(y',t)dy' \\ &- \frac{8S}{\lambda_L}\int_{-\infty}^{\infty} J(y-y')\phi(y',t)\cos\left[\frac{2\pi}{\lambda_L}(y-y')\right]dy' \\ &+ \frac{4S}{\lambda_L}\phi^2(y,t)\int_{-\infty}^{\infty} J(y-y')\phi(y',t)\cos\left[\frac{2\pi}{\lambda_L}(y-y')\right]dy'. \end{aligned} \tag{44}$$

Denoting $y - y' = \xi$, the $\phi(y',t)$ can be expanded as:

$$\phi(y',t) = \phi(y,t) + \frac{\partial\phi(y,t)}{\partial y}\xi + \frac{1}{2!}\frac{\partial^2\phi(y,t)}{\partial y^2}\xi^2 + \ldots. \tag{45}$$

Taking above series into Eq. (44), the integration variable is changed, i.e., $dy' \rightarrow -d\xi$, and using the assumption $\partial\phi(y,t)/\partial y \ll 1$, finally we get [10]:

$$\begin{aligned} i\frac{d\phi(y,t)}{dt} &= -2\beta_1\frac{\partial^2\phi(y,t)}{\partial y^2} + [\omega(y) - 4\beta_0]\phi(y,t) \\ &+ (2\beta_0 - \gamma)|\phi(y,t)|^2\phi(y,t), \end{aligned} \tag{46}$$

where we have used $J_{ij} = \frac{1}{2}J_{ij}^{mag}$, and the coefficients are defined as below:

$$\beta_n = \frac{S}{\lambda_L}\int_{-\infty}^{\infty} J^{mag}(\xi)\xi^{2n}\cos\left(\frac{2\pi}{\lambda_L}\xi\right)d\xi, n = 0, 1, \tag{47}$$

$$\gamma = \frac{2S}{\lambda_L}\int_{-\infty}^{\infty} J^{mag}(\xi)d\xi. \tag{48}$$

The condition for Eq. (46) to have envelope soliton solution is

$$f(W) = -2\beta_1(2\beta_0 - \gamma) > 0. \tag{49}$$

From the definitions of β_n and γ, we can easily get $2\beta_0 < \gamma$. Then Eq. (49) reduces to $\beta_1 > 0$, i.e.,

$$\int J^{mag}(\xi)\xi^2\cos\left(\frac{2\pi}{\lambda_L}\xi\right)d\xi > 0. \tag{50}$$

This result is consistent with the instability condition of modulational instability of nonlinear coherent spin wave modes near the Brillouin zone boundary under the long-wavelength modulation [4]. There, the modulational instability criteria take the corresponding discrete form:

$$1 - 4A_2 + 9A_3 - 16A_4 + \ldots < 0, \tag{51}$$

where $A_j = J_{0j}^{mag}/J_{01}^{mag}$ $(j = 2, 3\ldots)$ is the relative strength of the longitudinal spin coupling of site j to that of NN.

The function $f(W)$ versus the transverse width W of the condensate is plotted in **Figure 3** for three different approximations. $f(W) = 0$ marked in dashed line in **Figure 3** characterize critical value, above which the magnetic solitons can occur. It is clear that the NN approximation and the NNN approximation take effect for a small transverse width.

3.2 Magnetic soliton under the LDDI

When we choose red-detuned optical lattice or an external modulational laser field is imposed into the system, both the static MDDI and the laser-induced LDDI begin to take effect on the dynamics of the soliton excitation, and the criterion for the existence of solitons could be changed. Here we also consider three interaction distances as doing in Section 3.1.

3.2.1 Nearest-neighbor approximation with the LDDI

In this case, we can obtain the continuum limit NLSE for $\phi(y,t)$ with the form:

$$\begin{aligned} i\frac{d\phi(y,t)}{dt} &= 4SJ_{01}a^2\frac{d^2\phi(y,t)}{dy^2} + \left(\gamma_B B_z + 4SJ_{01}^{mag} + 8SJ_{01}\right)\phi(y,t) \\ &\quad - 2S\left(J_{01}^{mag} + 2J_{01}\right)|\phi(y,t)|^2\phi(y,t). \end{aligned} \tag{52}$$

The existence condition of magnetic soliton now has the form

$$f(W) = -8S^2a^2J_{01}\left(J_{01}^{mag} + 2J_{01}\right) > 0. \tag{53}$$

Obviously, magnetic soliton can be observed in this case, which is different from the case where only the MDDI plays role.

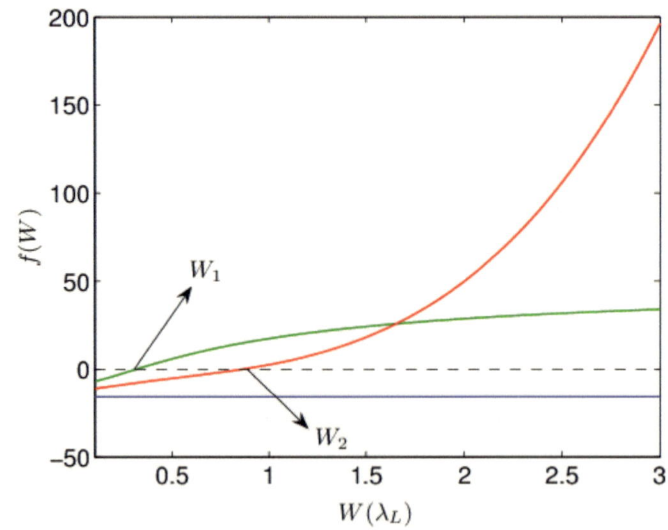

Figure 3.
The existence conditions for magnetic solitons in atomic spin chain dominated by the MDDI. The three lines correspond to the NN approximation (blue), the NNN approximation (green), and the long-range case, respectively. The critical value of the transverse width is W_1 and W_2 [10].

3.2.2 Next-nearest-neighbor approximation with the LDDI

In this case, the NLSE for $\phi(y,t)$ has the form:

$$i\frac{d\phi(y,t)}{dt} = 4SJ_{01}(1-4A_2)\frac{d^2\phi(y,t)}{dy^2}a^2 + \eta\phi(y,t) - 4SJ_{01}(1-4A_2)|\phi(y,t)|^2\phi(y,t). \quad (54)$$

We assume that the intensity of the external laser is strong enough which corresponds to strong LDDI; then the terms like J_{0j}^{mag}/J_{0j} have been ignored. So for two adjacent sites, $A_2 = J_{02}/J_{01}$ measures the relative strength of the NNN spin coupling to the NN spin coupling. The condition for Eq. (54) to have one soliton solution is

$$f(W) = -16S^2J_{01}^2(1-4A_2)(1-A_2) > 0. \quad (55)$$

3.2.3 Long-range case with the LDDI

In fact, once the external laser is imposed into the lattice, the LDDI dominates the system rapidly; the NLSE now takes the form [10]:

$$i\frac{d\phi(y,t)}{dt} = -2\beta_1\frac{\partial^2\phi(y,t)}{\partial y^2} + [\omega(y) - 4\beta_0]\phi(y,t) + (2\beta_0 - \gamma)|\phi(y,t)|^2\phi(y,t), \quad (56)$$

where the coefficients are defined as below:

$$\beta_n = \frac{2S}{\lambda_L}\int_{-\infty}^{\infty} J(\xi)\xi^{2n}\cos\left(\frac{2\pi}{\lambda_L}\xi\right)d\xi, n = 0, 1, \quad (57)$$

$$\gamma = \frac{2S}{\lambda_L}\int_{-\infty}^{\infty} J^{mag}(\xi)d\xi. \quad (58)$$

The value of β_n depends on both the MDDI and the LDDI. When a strong enough external modulated laser is imposed into the system, we have $J^{mag}(\xi) \ll J(\xi)$, which corresponds to $\gamma \ll 2\beta_0$. Thus the existence criteria of one soliton solution of Eq. (56) becomes $\beta_1\beta_0 < 0$, namely

$$f(W) = \int J(\xi)\xi^2\cos\left(\frac{2\pi}{\lambda_L}\xi\right)d\xi \int J(\xi)\cos\left(\frac{2\pi}{\lambda_L}\xi\right)d\xi < 0 \quad (59)$$

Similarly, Eq. (59) is consistent with the corresponding discrete form of MI condition of NCSW modes near the Brillouin zone boundary under the long-wavelength modulation [4]:

$$(1 - 4A_2 + 9A_3 - 16A_4 + ...)(1 - A_2 + A_3 - A_4 + ...) < 0, \quad (60)$$

where $A_j = J_{0j}/J_{01}$ ($j = 2, 3...$) is the relative strength of the transverse spin coupling of site j to that of NN.

We plot the existence regions of magnetic solitons in atomic spin chain dominated by the LDDI in **Figure 4**. The longitudinal coordinates stand for the intensity of the modulated laser. It is clear that the magnetic soliton can be achieved by

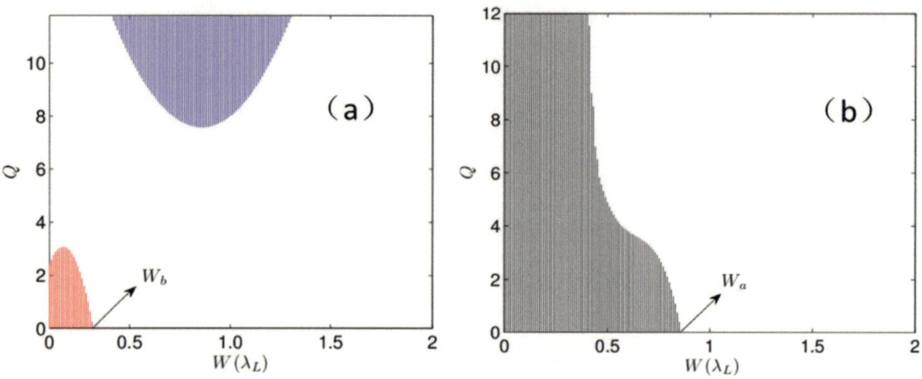

Figure 4.
The existence regions of magnetic solitons under the NNN approximation (a) and the long-range case (b) in atomic spin chain dominated by the LDDI, respectively. The critical value of the transverse width is W_a *and* W_b*. The blank regions correspond to the existence of solitons for both cases [10].*

tuning the transverse width and the modulated laser, which demonstrates the high controllability of optical lattice.

4. Conclusion

In this chapter, we have shown the existence conditions of magnetic solitons that can occur in optical lattice. Compared to the more conventional solid-state magnetic materials, we discuss how the NN, the NNN, and the long-range interactions (the MDDI and LDDI) dominate the system. We also show that they can be tuned by the laser parameters and the shape of the condensate including the laser detuning, the laser intensity, and the transverse width of the condensate.

Besides studying the magnetic solitons, the atomic spin chain in optical lattice provides us with a useful tool to study the fundamental static and dynamic aspects of magnetism and lattice dynamics [12–14]. In experimental applications, the atomic spin chain has become potential candidates for multi-bit quantum computation due to their long coherence and controllability [13, 15–17]. The theoretical study of magnetic solitons in optical lattices will play a guiding role when the optical lattice is used in cold atomic physics, condensed matter physics, and quantum information.

Acknowledgements

This work is supported by the National Natural Science Foundation of China under Grant No. 11604086. Thanks to Ying-Ying Zhang for her contributions to physics discussion and English writing of the chapter.

Author details

Xing-Dong Zhao
College of Physics and Materials Science, Henan Normal University, Xinxiang, Henan, China

*Address all correspondence to: phyzhxd@gmail.com

References

[1] Li Z-D, Li L, Liu WM, Liang J-Q, Ziman T. Exact soliton solution and inelastic two-soliton collision in a spin chain driven by a time-dependent magnetic field. Physical Review E. 2003; **68**:036102

[2] Xie Z-W, Zhang W, Chui ST, Liu WM. Magnetic solitons of spinor Bose-Einstein condensates in an optical lattice. Physical Review A. 2004;**69**: 053609

[3] Kjems JK, Steiner M. Evidence for soliton modes in the one-dimensional ferromagnet $CsNiF_3$. Physical Review Letters. 1978;**41**:1137

[4] Zhao X-D, Xie ZW, Zhang W. Modulational instability of nonlinear spin waves in an atomic chain of spinor Bose-Einstein condensates. Physical Review B. 2007;**76**:214408

[5] Zhang W, Pu H, Search C, Meystre P. Spin waves in a Bose-Einstei-condensed atomic spin chain. Physical Review Letters. 2002;**88**:6

[6] Pu H, Zhang W, Meystre P. Ferromagnetism in a lattice of Bose-Einstein condensates. Physical Review Letters. 2001;**87**:140405

[7] Zhao X-D, Zhao X, Jing H, Zhou L, Zhang W. Squeezed magnons in an optical lattice: Application to simulation of the dynamical Casimir effect at finite temperature. Physical Review A. 2013; **87**:053627

[8] Zhang W, Walls DF. Gravitational and collective effects in an output coupler for a Bose-Einstein condensate in an atomic trap. Physical Review A. 1998;**57**:1248

[9] Ho TL. Spinor Bose condensates in optical traps. Physical Review Letters. 1998;**81**:742

[10] Zhao X-D, Geng Z, Zhao X, Qian J, Zhou L, Li Y, et al. Controllable magnetic solitons excitations in an atomic chain of spinor Bose-Einstein condensates confined in an optical lattice. Applied Physics B: Lasers and Optics. 2014;**115**:451-460

[11] Zobay O, Pőting S, Meystre P, Wright EM, Wright EM. Creation of gap solitons in Bose-Einstein condensates. Physical Review Letters A. 1999;**59**:643

[12] Zhao X-D, Xie Z-W, Zhang W. Nonlinear spin wave s in a Bose condensed atomic chain. Acta Physica Sinica. 2007;**56**:6358

[13] Morsch O, Oberthaler M. Dynamics of Bose-Einstein condensates in optical lattices. Reviews of Modern Physics. 2006;**78**:179

[14] Zhao X-D, Zhang Y-Y, Liu W-M. Magnetic excitation of ultra-cold atoms trapped in optical lattice. Acta Physica Sinica. 2019;**68**:043703

[15] Sundar B, Mueller EJ. Universal quantum computation with Majorana fermion edge modes through microwave spectroscopy of quasi-one-dimensional cold gases in optical lattices. Physical Review A. 2013;**88**: 063632

[16] Brennen GK, Miyake A. Measurement-based quantum computer in the gapped ground state of a two-body Hamiltonian. Physical Review Letters. 2008;**101**:010502

[17] Wasak T, Chwedeńczuk J. Bell inequality, Einstein-Podolsky-Rosen steering, and quantum metrology with spinor Bose-Einstein condensates. Physical Review Letters. 2018;**120**: 140406

Chapter 5

Q-Switched 2 Micron Solid-State Lasers and Their Applications

Kejian Yang, Yaling Yang, Jingliang He and Shengzhi Zhao

Abstract

In this chapter, we overview the Q-switched 2 μm solid-state laser development achieved in recent years, including flash- and diode-pumped solid-state lasers based on active and passive modulators. In summary, active Q-switching is still the first choice for obtaining large pulse energy at 2 μm currently, while passive Q-switching based on saturable absorbers (SAs), especially the newly emerging broadband low-dimension nanomaterial, is becoming promising approach in generating Q-switched 2 μm lasers specially with high repetition rate, although the output power, pulse duration, and pulse energy needs further enhancement. Besides, some important applications of 2 μm lasers, such as medicine, laser radar, and infrared directional interference, have also been introduced in brief.

Keywords: 2 μm, solid-state laser, Q-switched, flash-pumped, diode-pumped

1. Introduction

In 1916, Einstein proposed the theory of stimulated radiation, which laid a theoretical foundation for the emerging of lasers. Till 1960, the world's first laser was invented by Mehman. Since then, flash-pumping has been widely employed for laser technology to generate beams with good monochromaticity, excellent coherence and promising directivity. Moreover, with the advent of laser diodes (LD), diode-pumped solid-state lasers have achieved rapid development for many advantages such as high efficiency, small size, and high beam quality compared with flash-pumped lasers. However, free-running lasers have relaxation oscillations, which deliver series of disordered small pulse spikes that do not possess high or stable peak powers. With the directional stored energy further modulated temporally, the laser operation regime would gradually evolve from continuous wave (CW) to Q-switching. It is well known that Q-switching is always an efficient method for generating pulsed lasers for any wavelength, since it provides an efficient way to obtain short laser pulses with high peak power, which is much beneficial for scientific or practical applications in terms of light-matter interactions. By using the Q-switching technology, the pulse peak power can reach KW or even MW level, and the pulse energy reaches mJ or even J level, which are very suitable for medical and industrial fields.

In recent years, 2 μm lasers located in the eye-safe spectral range have gained much attention because of their wide applications in the fields of environmental monitoring [1], laser medicine [2], laser radar [3, 4], micro-machining, material processing [5], and so on. In addition, the 2 μm lasers can also be used as pump sources for producing the mid- and far-infrared lasers. For example, they can be used as high-efficiency pumping sources for optical parametric oscillators (OPOs) and optical parametric amplifiers (OPAs) to achieve broadband tunable lasers in the mid-infrared spectral range of 3–5 and 8–12 μm [6–8]. However, it should be noticed that in practical applications, such as laser surgery, industrial processing, and nonlinear optics, laser pulses with high peak powers are required. Therefore, nanosecond 2 μm laser source with high peak power, large pulse energy and excellent stability is of great significance and has become an important topic of current research on 2 μm lasers.

As mentioned above, Q-switching technology is not an exceptional choice for obtaining nanosecond 2 μm solid-state lasers with high peak power. According to the different operation regimes, Q-switching technology can be divided into active and passive Q-switching (PQS). The basic principle of active Q-switching is regularly modulating the intracavity losses by a voltage controlled modulator, including electro-optic (EO) and acousto-optic (AO) modulators. Although actively Q-switched lasers are frequency adjustable, stable and reliable, they have the disadvantages of large size and expensive cost. Passive Q-switching technology is a good choice to overcome them. In passively Q-switched lasers, the saturable absorber (SA) is the key element and thus the output laser characteristics strongly depend on the nature of the SA materials. Up to now, semiconductor saturable absorber mirrors (SESAMs), chromium doped II-VI semiconductor materials (Cr^{2+}: ZnSe/S), and low-dimensional nanomaterials have been widely studied and regarded as reliable SAs for 2 μm lasers. However, the utilization of SESAMs is restricted by the complicated and expensive fabrication process as well as narrow absorption bandwidth. At present, the low dimensional nanomaterials with advantages of broadband absorption and low cost have been attractive candidates as SAs for PQS.

2. Flash-pumped 2 micron solid-state pulsed lasers

Up to now, flash-pumped solid-state lasers are still widely used in industry and medical science, because of the advantage of high energy output. In general, the flash-pumping way can support larger mode area than that of diode-pumping regime, which is critical for high energy laser output. To increase the absorption efficiency, Cr^{3+} ions are usually co-doped as sensitizer/activator for Tm or Tm-Ho host materials when flash-pumped. Under this situation, the energy level diagram for Cr,Tm,Ho: YAG crystal is illustrated in **Figure 1**. The flash pump light is absorbed by the broadband Cr^{3+} ions. After a nonradiative decay to and within the $^4T_2$ and ^2E states, the excitation is transferred from the Cr^{3+} ion to the $^3F_3$ and $^3H_4$ states of the Tm^{3+} ion, via dipole–dipole interactions. Nonradiative decay of the $^3F_3$ places virtually all the excited ions in the $^3H_4$ state. Each excited Tm^{3+} ion then interacts with a ground state of Tm^{3+} ion in a cross-relaxation process which gives rise to two Tm^{3+} ions in the $^3F_4$ state. Finally, these Tm^{3+} ions transfer their energy to two Ho^{3+} ions to populate the $^5I_7$ upper laser level, and lasing occurs between the $^5I_7$ and $^5I_8$ transition at 2.097 μm [9].

In 1962, Johnson et al. in Bell Labs, USA achieved the firstly 2.06 μm laser emission, but this laser was typically cooled by liquid-N_2 at 77 K and the output power was very low. Lately, it was found that some rare earth ions can sensitize Ho^{3+}, thus a Er,Tm,Ho:YAG laser was realized with a slope efficiency of 5% with aid of liquid-N_2 cooling. In 1985, Antipenko et al. increased the flash-pumping efficiency and achieved the first 2 μm laser output at room temperature by replacing Er^{3+} with Cr^{3+}. But CW operation cannot satisfy the requirements of high peak power and narrow

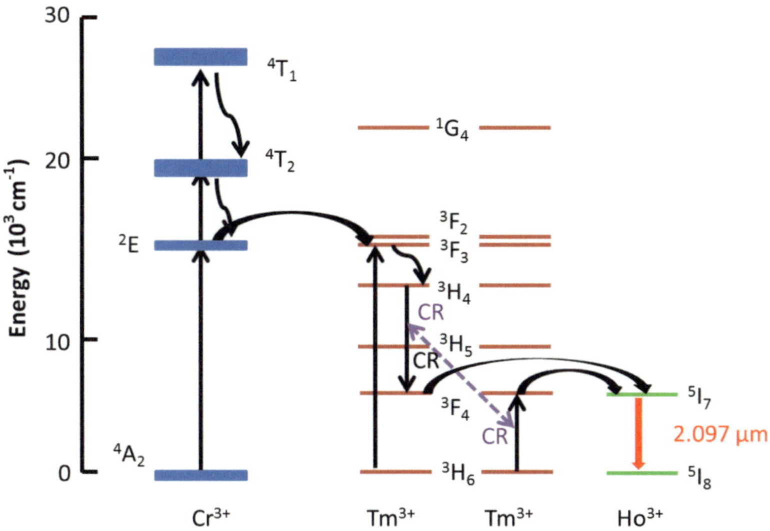

Figure 1.
Energy level diagram for Cr,Tm,Ho:YAG, CR: cross relaxation.

pulse width in many applications. Then Q-switching technology was utilized including passive and active Q-switching methods.

In 1996, Kuo et al. used $Ho:YVO_4$ and $Ho:CaF_2$ crystals as SAs for Tm,Cr:YAG lasers, and the corresponding pulse energies and pulse widths were 3.5 mJ, 45 ns and 5.1 mJ, 60 ns at 2.017 μm, respectively [10, 11]. In 2001, flash-lamp-pumped Ho:YAG (2090 nm) and Tm:YAG (2017 nm) lasers were passively Q-switched based on a Cr^{2+}:ZnSe SA for the first time, from which a Q-switched Ho laser with 1.3 mJ pulse energy and ~90 ns pulse duration and a Q-switched Tm laser with ~3.2 mJ pulse energy and 90 ns pulse duration were demonstrated [12]. In 2005, Gaponenko et al. realized passively Q-switched $Cr,Tm,Ho:Y_3Sc_2Al_3O_{12}$ and Cr,Tm,Ho:YAG lasers using PbS SAs, where the pulse energies and pulse widths were 2.4 mJ, 50 ns and 4.5 mJ, 70 ns at about 2.09 μm, respectively [13]. However, because the pulse energies were too low to satisfy the demands of practical application, active Q-switching technology was applied in flash-lamp-pumped 2 μm lasers.

In 1991, Bowman et al. designed an AO Q-switched Cr,Tm,Ho:YAG laser using a quartz crystal with Brewster angle placed in the cavity. At temperature of 20°C, the pulse energy of 110 mJ with the pulse width of 40 ns was obtained at the repetition rate of 1 Hz [14]. Two years late, they achieved AO Q-switched Cr,Tm,Ho:YAG laser with pulse energy of 24 mJ and duration of 90 ns at a repetition rate of 29 Hz [15]. In 2000, using flash-lamp-pumping, an AO Q-switched Cr,Tm:YAG laser was realized with a maximum pulse energy exceeding 0.8 J and a pulse width of 135 ns [16]. In 2005, Zheng et al. reported a AO Q-switched Cr,Tm,Ho:YAG laser delivering a pulse energy of 120 mJ and a pulse width of 90 ns at the repetition rate of 10 Hz [17].

Besides AO modulator, EO modulator is also a common method to modulate flash-lamp-pumped 2 μm lasers, which could further shorten the pulse duration. In 1981, Gettermy et al. reported a $LiNbO_3$ (LN) crystal based EO Q-switched Ho:YAG laser, which had to be cooled down to 77 K to obtain high energy and short pulses. A pulse energy of 80 mJ with a pulse width of 30 ns at the repetition rate of 5 Hz was obtained [18]. In 1990, Henderson et al. reported a EO Q-switched Cr,Tm,Ho:YAG laser using a LN crystal as modulator. At the repetition rate of 3 Hz, the pulse energy of 50 mJ with a corresponding pulse width of 150 ns was obtained

[19]. In 1993, Kim et al. also achieved a pulse energy of 50 mJ from a LN based EO Q-switched Cr,Tm,Ho:YAG laser at 170 K [20]. In 2008, Nieuwenhuis et al. utilized $RbTiOPO_4$ (RTP) crystal as EO modulator in a lamp-pumped Cr,Tm,Ho:YAG laser, and obtained a pulse width of 100 ns with a pulse energy of 42 mJ [21]. In 2012, a flash-lamp-pumped 2.09 μm Cr,Tm,Ho:YAG laser utilizing a $La_3Ga_5SiO_{14}$ (LGS) crystal as the EO modulator is proposed and demonstrated for the first time, which results are shown in **Figure 2**. Operated at a repetition rate of 3 Hz, a pulse energy as high as 520 mJ with a 35 ns pulse width was achieved by optimizing the delay time of EO modulator and compensating for the thermal depolarization with a quarter-wave plate. The corresponding pulse peak power was 14.86 MW, and the energy extraction efficiency was 66.3% [22].

In conclusion, as shown in **Table 1**, flash-pumped 2 μm solid-state lasers have been well developed for generating high pulse energy, especially in combination with active Q-switching methods. To now, the highest single pulse energy obtained at 2 μm was 800 mJ, with corresponding pulse duration of 135 ns by using an AO modulator. As for flash-pumped EO Q-switched 2 μm solid-state lasers, the highest single pulse energy of 520 mJ was achieved with a corresponding pulse width of 35 ns, which delivered the highest pulse peak power from the flash-pumped 2 μm

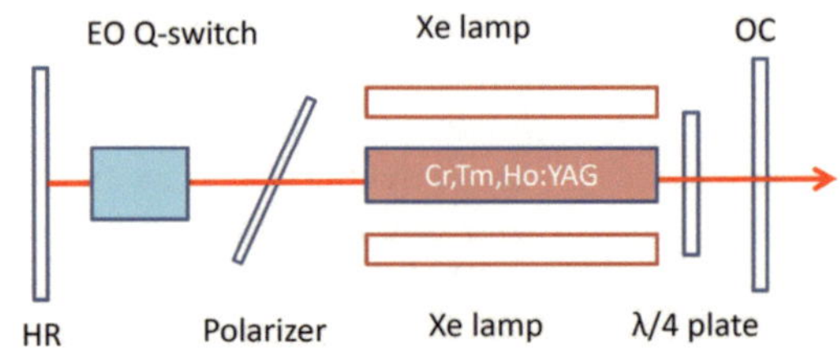

Figure 2.
Schematic setup of flash-lamp-pumped EO Q-switched Cr,Tm, Ho:YAG laser. [Reprinted/Adapted] With permission from Ref. [22] © The Optical Society.

Modulator	Gain media	Durations (ns)	Energy (mJ)	Time	Ref.
$Ho:YVO_4$	Tm,Cr:YAG	45	3.5	1966	[10]
$Ho:CaF_2$	Tm,Cr:YAG	30	5.1	1966	[11]
Cr:ZnSe	Ho:YAG	90	1.3	2001	[12]
Cr:ZnSe	Tm:YAG	90	3.2	2001	[12]
PbS	Cr,Tm,Ho:YSAG	50	2.4	2005	[13]
PbS	Cr,Tm,Ho:YAG	70	4.5	2005	[13]
AO	Cr,Tm,Ho:YAG	40	110	1991	[14]
AO	Cr,Tm,Ho:YAG	90	24	1993	[15]
AO	Cr,Tm:YAG	135	800	2000	[16]
AO	Cr,Tm,Ho:YAG	90	120	2005	[17]
LN	Ho:YAG	30	80	1981	[18]
LN	Cr,Tm,Ho:YAG	150	50	1990	[19]
LN	Cr,Tm,Ho:YAG	—	50	1993	[20]
RTP	Cr,Tm,Ho:YAG	100	42	2008	[21]
LGS	Cr,Tm,Ho:YAG	35	520	2012	[22]

Table 1.
Overview of flash-pumped Q-switched 2 micron solid-state lasers.

solid-state laser, to the best our knowledge. In the future, flash-pumped 2 μm solid-state lasers still have great potential for various applications demanding high energy.

3. Diode-pumped actively modulated 2 micron solid-state lasers

3.1 AO modulated 2 micron solid-state lasers

The AO modulator has the advantages of high damage threshold, easy operation, and low insertion loss. It is capable of generating peak power as high as several hundred kilowatts, high repetition frequency, and short pulse width with several tens of nanoseconds. Thus, AO Q-switching has attracted a lot in generating 2 μm laser pulses based on Tm^{3+} or Ho^{3+} ions doped crystals.

The actively Q-switched Tm^{3+} doped laser can be traced back to 1991, when Suni et al. realized a LD-pumped AO Q-switched Tm:YAG laser with the pulse width of 330 ns and the single pulse energy of 1 mJ at the repetition frequency of 100 Hz [23]. In 2004, Sullivan et al. realized a high-power Q-switched Tm:YAP laser with a maximum output power of 50 W at 1940 nm. At the repetition rate of 5 kHz, the maximum pulse energy was 7 mJ with a corresponding pulse width of 75 ns [24]. But it was low temperature of −10°C. In 2008, Cai et al. realized a diode-pumped AO Q-switched Tm:YAP laser with a maximum single pulse energy of 1.57 mJ and the minimum pulse width of 80 ns at a repetition rate of 1 kHz under room temperature [25]. In 2010, Li et al. showed another double-diode end-pumped AO Q-switched c-cut Tm:YAP laser, which delivered a maximum average output power of 12.5 W at a repetition rate of 10 kHz with a pulse width of 126 ns [26]. In 2015, Yumoto et al. realized AO Q-switched Tm:YAG laser at a center wavelength of 2.013 μm. The maximum pulse energy of 128 mJ was obtained at a repetition rate of 10 Hz, corresponding to a minimum pulse width of 160 ns [27], which was the largest pulse energy for AO Q-switched 2 μm laser. In 2015, Luan et al. realized a 790 nm diode-pumped doubly Q-switched Tm:LuAG laser simultaneously with AO modulator and multilayered graphene as Q-switches, from which a minimum pulse width of 170 ns with a corresponding pulse energy of 0.53 mJ was obtained at a repetition rate of 1 kHz [28]. The results well indicated that doubly Q-switching technique could efficiently shorten the pulse duration.

Efforts are not stopped being paid on exploring diode-pumped solid-state AO Q-switched lasers based on novel kinds of Tm^{3+} and Ho^{3+} doped crystals. In 2017, Liu et al. demonstrated an AO Q-switched Tm,Y:CaF_2 laser in a V-type cavity as shown in **Figure 3**, which could run at high repetition rates from 1 to 10 kHz [29]. Under the modulation frequency of 1 kHz, pulses with the shortest duration of 280 ns and the maximum pulse energy of 0.335 mJ were delivered, corresponding to a maximum peak power of 1.19 kW. In 2019, Zagumennyi et al. reported a novel AO Q-switched Tm:$Yb_3Al_5O_{12}$ (Tm:YbAG) laser pumped by the 1.678 μm laser. The pulse energy was 100 μJ with a pulse duration of 45 ns at the repetition rate of 6.7 kHz [30].

In addition to the Tm lasers emitting around 1.9–2.0 μm, Ho^{3+} ions can emit laser radiations at around 2.0–2.1 μm due to the transition $^5I_7$–$^5I_8$, which can keep away from the water vapor absorption and is helpful for the applications requiring free space transmission. Moreover, Ho^{3+} ion with long upper laser level lifetime are attractive for Q-switched operation. Since Tm^{3+} ions have absorption bands around 800 nm where commercial GaAs/AlGaAs diodes are available, a Tm sensitized Ho laser is an ideal way to achieve a ~ 2.1 μm laser. In 2005, Yao et al. realized the AO Q-switched Tm,Ho:$GdVO_4$ laser under liquid nitrogen refrigeration. And the maximum pulse energy of 1.1 mJ and minimum pulse duration of 23 ns were obtained at a repetition rate of 3 kHz [31]. In 2010, Yao et al. investigated the characteristics of AO Q-switched Tm,Ho:YVO_4 laser under low temperature conditions. A maximum output power of 19.4 W was achieved with a pulse width of 24.2 ns at a repetition rate of 15 kHz [32]. In 2017, Li et al. achieved a stable AO Q-switched Tm,Ho:YAP laser with a pulse width of 66.85 ns and a single pulse energy of 0.97 mJ at the repetition rate of 7.5 kHz and temperature of 77 K cooled by the liquid nitrogen in Dewar bottle [33]. However, the above Tm,Ho co-doped laser were operated under low temperature conditions, which limited their applications.

As for Ho laser emitting around 2 μm, another way is resonantly pumping Ho laser by an extracavity Tm laser, which has high conversion efficiency and low thermal load. Thus, Ho-doped laser is also a vital approach to obtain high output power and pulse energy laser at wavelength over 2 μm. In 2012, Lamrini et al. realized an actively Q-switched Ho:YAG laser based on an AO modulator at 2.09 μm by using a 1.9 μm LD as the pump source, which schematic setup is shown in **Figure 4**. At the repetition rate of 100 Hz, the obtained maximum pulse energy and minimum pulse width were 30 mJ and 100 ns, respectively [34]. In 2014, an AO Q-switched Ho:$GdVO_4$ was realized end-pumped by a 1942 nm Tm-fiber laser, delivering a pulse width of 4.7 ns and an output pulse energy of 0.9 mJ at the repetition frequency of 5 kHz, corresponding to a peak power of 187.2 kW [35]. In 2014, Wang et al. used a 1.908 μm Tm:YLF solid-state laser as the pump source to realize an actively Q-switched Ho:YAG ceramic laser at 2.097 μm based on an AO Q-switch. The maximum pulse energy and minimum pulse width were 10.2 mJ and 83 ns at the repetition rate of 100 Hz [36]. In 2016, Ji et al. used a 1.94 μm LD as the pump source to realize an actively Q-switched Ho:YLF solid-state laser based on an AO Q-switch at 2.06 μm. The maximum pulse energy of 1.1 mJ was achieved with a corresponding minimum pulse duration was 43 ns at the repetition rate of 100 Hz [37]. Up to present, many Ho lasers based on various hosts have been demonstrated. However, there is rare report on the Ho laser with more than 100 W output power.

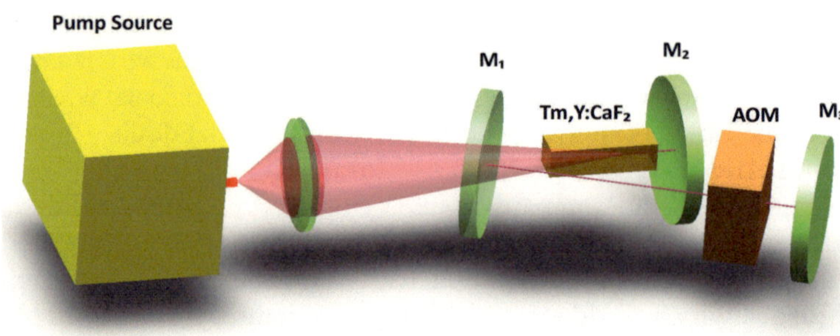

Figure 3.
Experimental setup of the actively Q-switched Tm,Y:CaF_2 laser. [Reprinted/Adapted] With permission from Ref. [29] © The Optical Society.

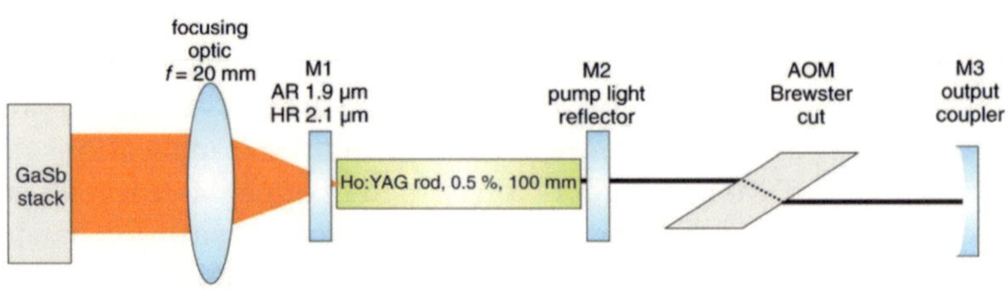

Figure 4.
Resonator of the AO Q-switched Ho(0.5%):YAG laser formed by mirror M1 and mirror M3 (output coupler). [Reprinted/Adapted] With permission from Ref. [34] © The Optical Society.

In 2018, Duan et al. reported a Ho:YAG laser with the output power of up to 108 W in CW mode and 106 W in Q-switching mode, respectively [38]. As far as we know, this is the highest output power of CW and Q-switched Ho lasers ever reported. A pulse energy of 5.3 mJ and a pulse duration of 21 ns were obtained, corresponding to a pulse peak power of approximately 252 kW.

In conclusion, AO Q-switching is an effective method to obtain pulsed 2 μm lasers. **Table 2** shows the summaries of diode-pumped AO Q-switched 2 micron solid-state lasers reported in recent years. From it, we can see that the maximum single pulse energy ever achieved based on AO Q-switching method is 128 mJ. However, for high-energy or high-power lasers, the turn-off capacity of AO switching is poor. In practical, most applications demand pulsed 2 μm lasers with large power, high pulse energy and repetition rate, which sets an important challenge for AO modulators to overcome.

3.2 EO modulated 2 micron solid-state lasers

EO Q-switching is an important way of obtaining high peak power and narrow pulse width which has the advantages of fast switching speed, high extinction ratio and small volume. The peak power of this kind of laser can easily reach hundreds of megawatts. The most important component in EO Q-switch is EO crystal, many EO crystals including $LiNbO_3$ (LN), $RbTiOPO_4$ (RTP) and $La_3Ga_5SiO_{14}$ (LGS) have been successfully applied in the 2 μm EO Q-switched lasers.

From 1970s on, LN crystals have been used as the EO crystal to generate laser pulses, which is exceptional for its no deliquescence, low half-wave voltage, and lateral modulation, etc. In 2015, based on LN crystal Liu et al. reported a diode-pumped EO Q-switched Tm:LuAG laser generating a pulse energy of 2.51 mJ with a pulse width of 88 ns at a repetition rate of 50 Hz [39]. In 2016, with Tm,Ho:YAP

Gain media	Durations (ns)	Frequency (kHz)	Energy (mJ)	Wavelength (μm)	Time	Ref.
Tm:YAG	330	0.1	1	2.015	1991	[23]
Tm:YAP	75	5	7	1.94	2004	[24]
Tm:YAP	80	1	1.57	1.99	2008	[25]
Tm:YAP	126	6	1.6	1.99	2010	[26]
Tm:YAG	160	0.01	128	2.015	2015	[27]
Tm:LuAG	170	1	0.53	2.023	2015	[28]
Tm,Y:CaF_2	280	1	0.335	1.912	2017	[29]
Tm:YbAG	45	6.7	0.1	2.02	2019	[30]
Tm,Ho:$GdVO_4$	23	3	1.1	2.05	2005	[31]
Tm,Ho:YVO_4	24.2	15	1.3	2.055	2010	[32]
Tm,Ho:YAP	66.85	7.5	0.97	2.119	2017	[33]
Ho:YAG	100	0.1	30	2.09	2012	[34]
Ho:$GdVO_4$	4.7	5	0.9	2.05	2014	[35]
Ho:YAG	83	0.1	10.2	2.097	2014	[36]
Ho:YLF	43	0.1	1.1	2.06	2016	[37]
Ho:YAG	21	20	5.3	2.09	2018	[38]

Table 2.
Overview of AO Q-switched 2 μm solid-state lasers.

crystal as gain material and EO Q-switch based on LN crystal, a maximum pulse energy of 1.65 mJ and a shortest pulse duration of 107.4 ns were obtained at a repetition rate of 200 Hz [40]. In 2018, to lower down the thermal effect of gain medium, Guo et al. incorporated diode-wing-pumping technique into a LN EO Q-switched Tm:LuAG laser. A maximum pulse energy of 10.8 mJ and a minimum pulse width of 52 ns at a repetition rate of 100 Hz was delivered, as shown in **Figure 5** [41]. However, further increase of pule energy is limited by the relatively low damage threshold of LN crystal.

RTP crystal belongs to orthorhombic crystal system and has large EO coefficient, high damage threshold and broad transmission range. In 2013, using a 1.94 μm Tm^{3+} fiber laser as the pump source, Fonnum et al. realized a RTP EO Q-switched Ho:YLF laser utilizing the setup shown in **Figure 6**. A maximum pulse energy of 550 mJ and a minimum pulse width of 14 ns at a repetition rate of 1 Hz were obtained [42]. In 2016, a diode-pumped RTP EO Q-switched Tm:YAG slab laser delivered a maximum pulsed energy of 7.5 mJ with a pulse width of 58 ns [43]. However, since RTP crystal is biaxial, two crystals with the same size and direction are required to combine with each other to offset natural birefringence, which greatly increases the difficulty of crystal processing. Moreover, under high-power operation regime, thermal induced birefringence can deteriorate switching performance of the RTP EO Q-switch, which subsequently limits the obtained pulse energy and output power.

In 2003, LGS crystal was firstly grown and studied in Institute of Crystal Materials of Shandong University. LGS crystal is uniaxial and thus free of birefringence induced problems. Compared with LN, the LGS crystal has a 9.5-times-higher anti-photo-damage threshold. Additionally, LGS crystal has a broadband transparency from 190 to 2400 nm, good physical and chemical stability, small thermal expansion coefficient and considerable electro-optic coefficient ($\gamma_{11} = 2.3 \times 10^{-12}$ m/V). Kong et al. firstly employed LGS crystal as EO Q-switch in a Nd:YAG laser and achieved pulsed operation at 1 μm [44]. Very recently, Ma et al. realized a LGS EO Q-switched Tm:YAP laser with a maximum repetition rate of 200 kHz, a maximum average output power of 2.79 W and a minimum pulse width of 5.5 ns [45].

In conclusion, EO Q-switch plays a crucial role in generating high-energy laser pulses and suitable choice of EO crystal can determine the performance of EO Q-switch. At present, LN, RTP, and LGS are the most successful EO crystals. For LD pumped 2 μm solid lasers, 550-mJ single pulse energy set a record, to the best of our knowledge. However, exploring excellent EO crystals with high damage threshold, low driven voltage, high repetition rate and the small volume are still on the way (**Table 3**).

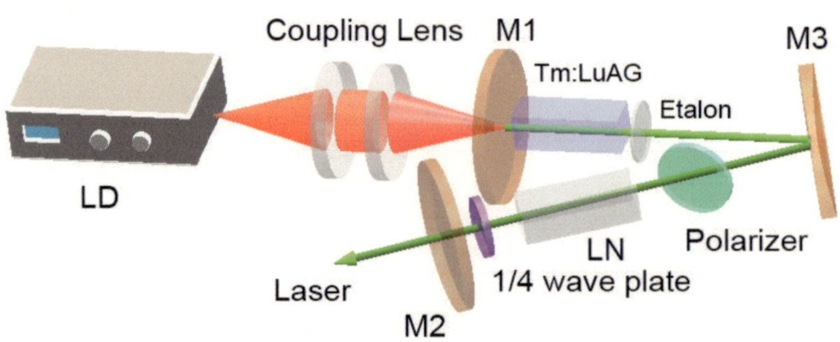

Figure 5.
Schematic of EO Q-switched Tm:LuAG laser. [Reprinted/Adapted] With permission from Ref. [41] © The Optical Society.

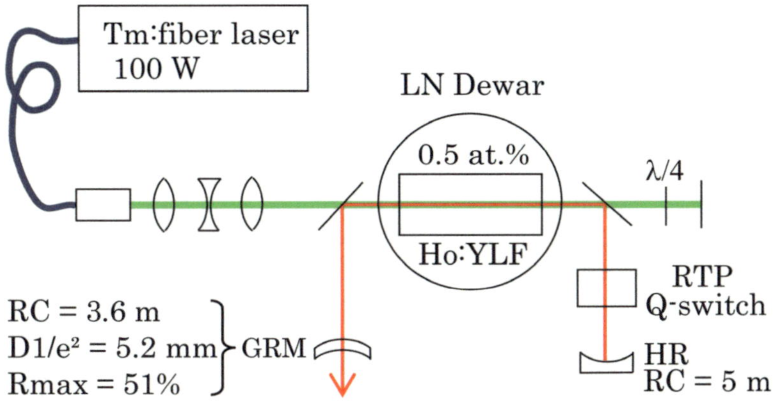

Figure 6.
Schematic setup of the Tm:fiber pumped RTP EO Q-Switched Ho:YLF laser. [Reprinted/Adapted] With permission from Ref. [42] © The Optical Society.

EO crystal	Gain media	Duration (ns)	Frequency (kHz)	Energy (mJ)	Wavelength (μm)	Time	Ref.
LN	Tm:LuAG	88	0.05	2.51	2.023	2015	[39]
LN	Tm,Ho:YAP	107	0.2	1.65	2.13	2016	[40]
LN	Tm:LuAG	52	0.1	10.8	2.023	2018	[41]
RTP	Tm:LuAG	3	100	0.0122	2.013	2016	[46]
RTP	Ho:YLF	14	0.001	550	2.051	2013	[42]
RTP	Tm:YAG	58	1	7.5	2.015	2016	[43]
LGS	Tm:YAP	5.5	200	0.0139	1.99	2019	[45]

Table 3.
Overview of EO Q-switched 2 μm solid-state lasers.

4. Diode-pumped SAs modulated 2 micron solid-state lasers

4.1 Two-dimensional nanomaterial modulated 2 micron solid-state lasers

In recent years, two-dimensional (2D) material based SAs have been widely used in generating laser pulses. The family of 2D materials includes graphene, black phosphorus (BP), transition metal dichalcogenides (TMDs) and topological insulators (TIs), and so on [47]. There are several methods available for fabricating 2D materials, including micromechanical exfoliation, chemical synthesis, pulsed laser deposition (PLD) and liquid phase exfoliation (LPE) [48, 49]. To fabricate 2D material, LPE is widely used because of its simplicity and effectiveness. Layered materials can be directly exfoliated from their bulk counterparts by this method.

Graphene, a single atomic layer of carbon atom, has attracted particular interest due to its broadband absorption, controllable modulation depth and low non-saturable loss. However, low absorption in graphene limits its applications. In 2012, graphene was firstly used as SAs at 2 μm region [50]. The obtained maximum average output power, pulse repetition rate, and single pulse energy were 38 mW, 27.9 kHz, and 1.74 μJ, respectively. Since then, graphene has been widely applied in many kinds of Tm⁻doped crystals based 2 μm lasers, such as Tm:LSO, Tm:YAP, Tm:LuAG, Tm,Y:CaF_2, and et al. [28, 51–53].

Another kind of 2D material, TMDs, has also been widely studied in recent years. Properties of few-layered TMDs depend on the number of layers. For instance, bulk MoS_2 has an indirect 1.29 eV (961 mm) bandgap, while monolayer MoS_2 has a direct 1.8 eV (689 nm) bandgap. However, many reports show the saturable absorption property of few-layered TMDs at near-infrared wavelength region, which corresponds to photon energies smaller than material bandgap for most of TMDs. In a perfect crystalline semiconductor, incident photons with energy lower than the bandgap cannot be absorbed. However, crystallographic defects, including edges and vacancies, enable such absorption [54]. In 2014, Wang et al. concluded that by introducing defects with a suitable quantity range, the bandgap of MoS_2 could be reduced from 1.08 (R = 1:2) to 0.08 eV (R = 1:2.09), corresponding to an absorption wavelength from 1.1 to 15.4 μm, which showed that MoS_2 with S defects could be used as a kind of broad SA. Passively Q-switched lasers based on MoS_2 SA in the range of 1–2 μm have been demonstrated [48].

BP, a layered allotrope of phosphorus, also exhibits a layer-dependent direct bandgap, which is tunable from 0.3 eV (bulk) to 2.0 eV (monolayer), corresponding to an absorption band from 620 nm to 4.13 μm. However, different from MoS_2, which owns indirect band-gap at multilayer format, BP always has the direct transition for all thickness. But it has an intrinsic disadvantage of easy oxidation. In 2015, Lu et al. reported the broadband and enhanced SA property of multi-layered BP (with a thickness of ~10 nm) by wide-band Z-scan method [49]. At the same year, Jiang et al. demonstrated passively Q-switched operation of Tm¯doped fiber laser based on BP SAs [55]. In 2016, Xie et al. achieved a BP Q-switched Tm:YAG ceramic laser generating pulses with a maximum average output power of 38.5 mW, pulse energy of 3.32 μJ, and pulse width of 3.12 μs [56]. Zhang et al. demonstrated a compact Q-switched Tm:YAP laser based on multi-layered BP nanoplatelets, which delivered an average output power of 3.1 W and a pulse duration of 181 ns at a repetition rate of 81 kHz [57].

Another group of 2D materials, TIs such as Bi_2Te_3 and Sb_2Te_3, have also been proposed and investigated in recent years. In general, bulk TIs materials have a small bandgap but a gapless metallic surface state is generated in layered 2D TIs, which is caused by strong spin-orbit coupling and time-reversal symmetry [58]. With such typical band structure, 2D TIs materials show broadband absorption like CNTs and graphene and have become one of promising optical modulator candidates in generating 2 μm laser pulses.

In conclusion, 2D material based SAs have attracted much attention in generating laser pulses due to its advantage of easy fabrication, compact setup, and broadband optical absorption. Related reports on passively Q-switched 2 μm solid-state lasers based on 2D material SAs are summarized in **Table 4**. However, the obtained output powers are in the order of several hundred mW and the pulse widths are in the order of several hundred ns, especially the pulse energies are only with tens of microjoules, which limits the pulse peak power and their applications furthermore. In the future, the damage thresholds of the 2D material based SAs should be greatly increased, which would benefit the increase of the output power and pulse energy as well as the long-term stabilities of the Q-switched lasers.

4.2 Cr^{2+}-doped crystal modulated 2 micron solid-state lasers

Since Podlipensky et al. demonstrated that Cr^{2+}:ZnSe crystals can be used as SAs for 1.54 μm Er:glass lasers [67], Cr^{2+} doped II-VI compounds like Cr^{2+}:ZnS and Cr^{2+}:ZnS have been widely used in the spectral range of 1.5–2.1 μm. **Figure 7** shows that Cr:ZnS and Cr:ZnSe have broadband absorption spectra form 1.5 to 2.2 μm [68].

SAs	Gain media	Output power (mw)	Duration (ns)	Frequency (kHz)	Energy (μJ)	Time	Ref.
Graphene	Tm:YAG	38	2250	27.9	3.8	2012	[50]
	Tm:LSO	106	7800	7.6	14	2013	[51]
	Tm:YAP	362	735	42.4	8.5	2014	[52]
	Ho:LuAG	370	752.2	48.8	7.5	2016	[28]
	Tm,Y:CaF_2	400	1316	20.22	20.4	2018	[53]
MoS_2	Tm,Ho:YGG	205.62	410	149	1.38	2014	[48]
	Tm:CLNGG	79.2	4840	110	0.72	2015	[59]
	Tm:$GaVO_4$	100	800	48.09	2.08	2015	[60]
	Tm,Ho:YAP	275	435	55	5	2017	[61]
	Tm:CYA	490	480	102.6	4.87	2017	[62]
BP	Tm:YAP	3199.5	181	81	39.5	2016	[57]
	Tm:YAP	150.92	1780	19.25	7.84	2016	[63]
	Tm:CaYA	12	3100	17.7	0.68	2016	[64]
	Tm:YAG	38.512	3120	11.6	3.32	2016	[56]
Bi_2Te_3	Tm:LuAG	2030	620	118	18.4	2017	[65]
Sb_2Te_3	Tm:$GdVO_4$	700	223	200	3.5	2018	[66]

Table 4.
Overview of 2D nanomaterials modulated 2 μm solid-state lasers.

Up to now, Cr:ZnSe/ZnS have been applied in many 2 μm pulsed lasers, as shown in **Table 5**. In 2015, Sebbag realized a Cr:ZnSe Q-switched Tm:YAP solid-state laser at 1935 nm, which delivered a pulse energy of 1.55 mJ and a pulse duration of 42.2 ns, corresponding to a peak power of 36.7 kW [71]. In 2017, Lan reported a Cr:ZnSe Q-switched Tm:CYA laser with a maximum repetition rate of 21.9 kHz, a minimum pulse width of 42.6 ns and a maximum pulse energy of 60.2 μJ [73]. In 2017, a Cr^{2+}:ZnSe passively Q-switched Nm–cut Tm:KLu$(WO_4)_2$ mini-slab laser was reported, where the obtained shortest pulse duration was 35 ns and maximum pulse energy was 0.3 mJ [90]. Due to the low absorption cross-section of Cr:ZnSe crystal around 2 μm, it was mainly used in Tm-doped crystal based lasers. In 2013, A Cr^{2+}:ZnS passively Q-switched Ho:YAG laser pumped by a Tm:YLF laser was demonstrated with a maximum pulse energy of 2.47 mJ, a minimum pulse duration of 36.6 ns at a repetition rate of 10.4 kHz [81]. In 2015, a diode-pumped Cr:ZnS passively Q-switched Tm:KLu$(WO_4)_2$ microchip laser generated sub-nanosecond (780 ps) pulses with a pulse repetition frequency of 5.6 kHz, which was the shortest pulse duration among the passively Q-switched 2 μm lasers ever achieved [85]. Besides, in 2017, another novel Cr^{2+} ions doped Cr:CdSe crystal was firstly demonstrated in a Ho: YAG laser at 2.09 μm. A maximum output pulse energy of 1.766 mJ at a repetition frequency of 685 Hz was obtained with a pulse duration of 15.4 ns, which indicated that a Cr:CdSe crystal could be a promising SA in passively Q-switched 2 μm lasers [89].

In conclusion, Cr^{2+} doped II-VI compounds have been successfully used as SAs in passively Q-switched 2 μm lasers. In this way, the obtained pulse energies and the pulse widths can reach hundreds of μJ and ns or even sub-ns level. The maximum pulse energy of 2.47 mJ, and the shortest pulses with sub-nanosecond (780 ps) duration were both realized with a kind of Cr:ZnS SA. Therefore, Cr^{2+} doped II-VI

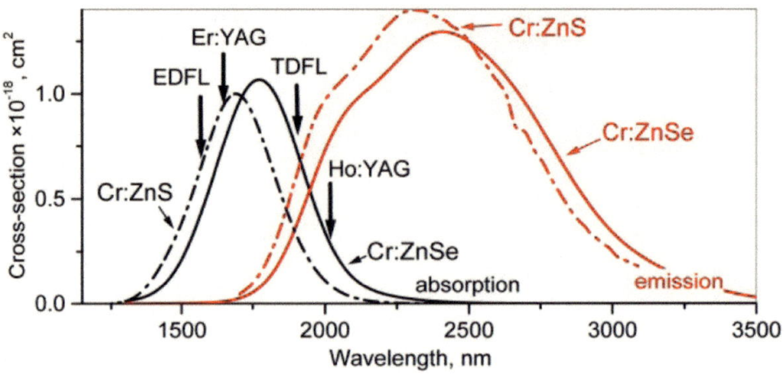

Figure 7.
Absorption (black) and emission (red) cross-sections of ZnS and ZnSe doped with Cr^{2+} ions. [Reprinted/Adapted] With permission from Ref. [68] © The Optical Society.

SAs	Gain media	Output power (W)	Duration (ns)	Frequency (kHz)	Energy (μJ)	Time	Ref.
Cr:ZnSe	Tm:YAG	1.6	300	4	400	2003	[69]
	Tm,Ho:YLF	0.1	40	4	25.6	2014	[70]
	Tm:YAP	0.87	42.2	0.561	1550	2015	[71]
	Tm:YLF	0.027	1200	2.1	13	2014	[72]
	Tm:CYA	1.32	42.6	21.9	60.2	2017	[73]
	Tm:CYA	0.25	107	5.85	48.2	2017	[74]
	Tm:CGA	0.64	44	13.9	46	2017	[75]
	Tm:KLuW	3.2	35	6.3	300	2017	[75]
	Ho:SSO	2.4	73.5	2.65	900	2017	[76]
Cr:ZnS	Tm:KLW	0.39	25	2.7	145	2012	[77]
	Tm:YLF	0.098	14	0.12	850	2012	[78]
	Tm,Ho:YLF	0.016	1250	1.3~2.6	4	2012	[79]
	Tm:LLF	0.2	7.6	0.161	1260	2012	[80]
	Ho:YAG	16.6	36.6	10.4	2470	2013	[81]
	Tm,Ho:$LuLiF_4$	0.074	1200	5.8	13	2013	[82]
	Tm,Ho:$GdVO_4$	3.2	354	52	70.5	2013	[83]
	Tm,Ho:YVO_4	0.3	500	65	3.5	2014	[84]
	Tm:KLW	0.146	0.78	5.6	25.6	2015	[85]
	Ho:YAG	14.8	29	24.4	600	2015	[86]
	Ho:LuAG	1.14	36	0.79	1540	2015	[87]
	Ho:YAP	6.1	93.6	7.5	830	2015	[88]
	Tm:YAP	0.89	35.8	0.481	1850	2015	[71]
	Tm,Ho:KLuW	0.131	14	8.2	10.4	2016	[77]
Cr:CdSe	Ho:YAG	1.2	15.4	0.685	1766	2017	[89]

Table 5.
Overview of Cr^{2+}-doped crystal modulated 2 μm solid-state lasers.

crystals are still irreplaceable SAs in generating passively Q-switched 2 μm lasers with pulse energies over mJ level and pulse width in order of several nanoseconds to sub-nanoseconds.

4.3 Gain crystal modulated 2 micron solid-state lasers

As mentioned above, the popular SAs used in the 2 μm Q-switched solid-state lasers are mainly low-dimensional nanomaterials and Cr-doped crystals. In addition, the absorption effect of Ho^{3+} doped gain crystals can also be used for SAs, which can be classified as a slow-relaxing solid-state SA due to possessing a long emission lifetime. However, the study of passively Q-switched solid-state lasers based on gain crystals is relatively rare. In 1994, Kuo et al. reported a flash-lamp pumped Tm,Cr:$Y_3Al_5O_{12}$ laser using a Ho:$YLiF_4$ as SA at room temperature [91]. A maximum single pulse energy of 11 mJ and a shortest pulse duration of 45 ns were obtained. It paved the way for Ho-doped crystals as SAs for passively Q-switched 2 μm lasers. Additionally, Ho:SSO has also been employed as an efficient SA. In 2018, a Ho:SSO crystal was used as SA for the Ho:YAG laser [92]. At a repetition rate of 42.1 kHz, a minimum pulsed duration of 48 ns and a maximum pulse energy of 2 mJ were obtained.

5. Applications of pulsed 2 micron solid-state lasers

2 μm laser is an ideal light source for the medical application, which has less damage to the rest of the tissue, and less bleeding during the operation. 2 μm lasers used in laser medical treatment are mainly YAG-based two pulsed lasers [93–96]. Ho:YAG has been successfully used in urology and orthopedics [96]. However, little is known about hard dental tissue ablation with Ho:YAG laser. For example, non-contact laser surgery offers several potential advantages in dental treatment, such as reduced pain and vibration, more precise control by electronic device and haemostatic effect. **Figure 8** shows cross-sectional images of ablation crater created on the urinary calculi surface by pulsed Ho:YAG laser (Wuhan National Laboratory for Optoelectronics) in air (dry condition) and underwater with various thickness [93]. The pulsed Ho:YAG laser emitted at 2.12 μm with the pulse energy of 2000 mJ and the pulse duration of 300/450 μs at the repetition rate of 20 Hz.

Lidar technology is an optical remote sensing technology that acquires the physical information of a target object by detecting the scattered light characteristics of a distant target object. Compared with the traditional radar technology, the laser radar technology realizes the information loading by modulating the amplitude and frequency of the laser beam, thereby having the advantages of high resolution and good anti-interference. In recent years, laser radar technology with 2 μm band laser as coherent light source has been proposed and made some progress. In 2006, the NICT agency in Japan reported on the airborne coherent wind lidar system. The system used a Tm:YAG laser with a center wavelength of 2.01 μm as the coherent light source, and the maximum single pulse energy was 7 mJ, and it was successfully applied to the test of atmospheric wind profile [97]. In 2007, NASA agencies in the United States reported on the vehicle's coherent wind Lidar system. The system utilized Tm,Ho:$LuLiF_4$ laser at 2.05 μm and a pulse energy of 100 mJ as the coherent light source, and successfully detected the wind field information of the boundary layer and the troposphere [98]. In 2013, the NICT facility in Japan used a differential absorption Lidar system to successfully measure CO_2 concentrations and mountain targets in the 7 km range. The system used a Ho:YLF laser as the light source with

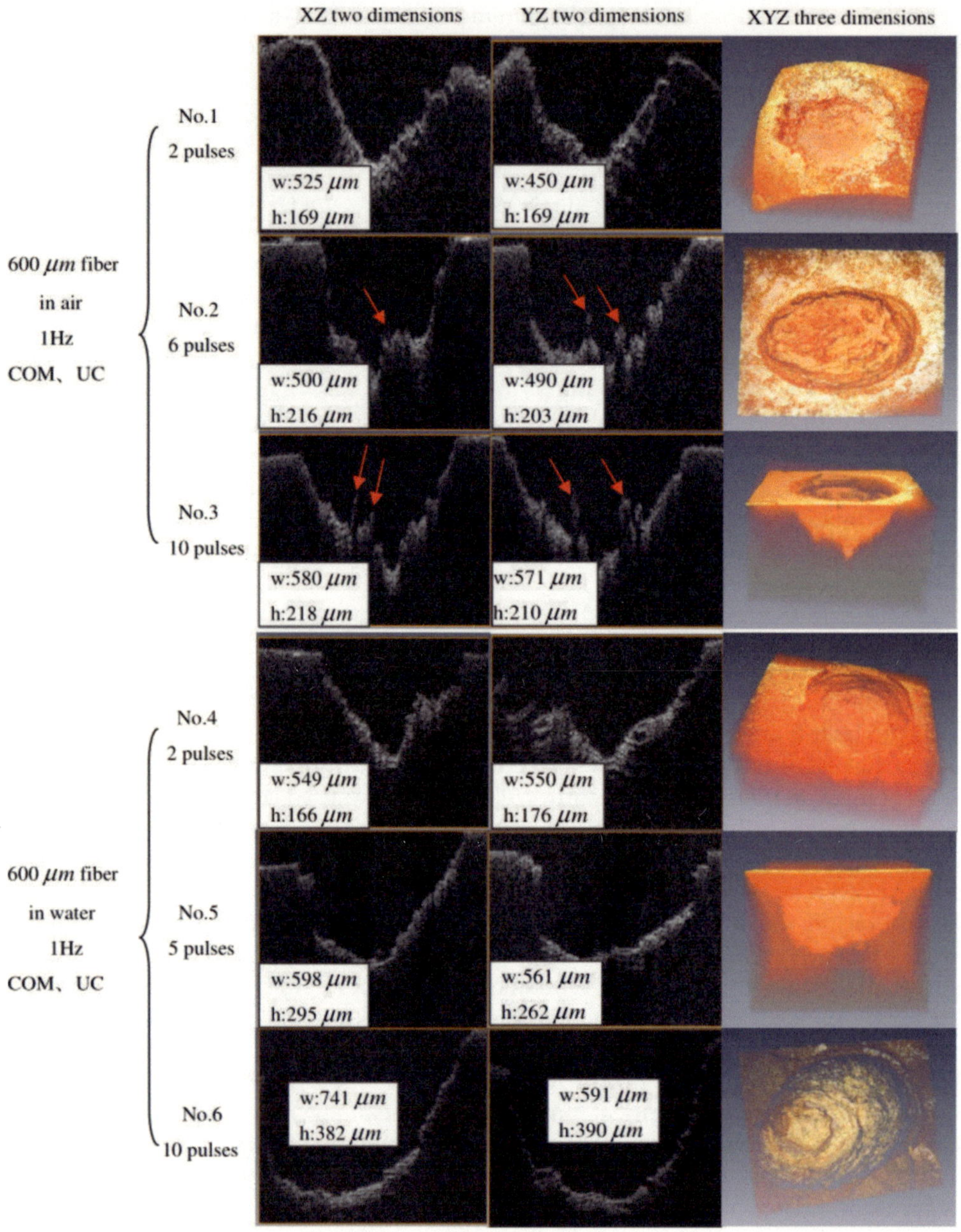

Figure 8.
(Color online) Cross-sectional topography of holmium laser-induced (300 μs, 300 mJ, 1 Hz) craters acquired with OCM system in air and in water. The 600 μm fiber is vertically in contact with the urinary calculi surface. w = width; h = height. [Reprinted/Adapted] With permission from Ref. [93] © The Optical Society.

a pulse energy of 50–80 mJ and a pulse width of 150 ns at the repetition frequency of 30 Hz [99]. In 2014, the French Polytechnic University reported a Ho:YLF multi-frequency single longitudinal mode laser with a center wavelength of 2.051 μm. The laser operated at a repetition rate of 2 kHz with a pulse energy of 13 mJ and a pulse width of 42 ns. The laser could be used as a source of differential absorption Lidar for the measurement of atmospheric CO_2 concentrations, as shown in **Figure 9** [100].

A pulsed 2 μm laser with high peak power is also a promising pump source for OPOs in the wavebands such as 3~5 and 8~12 μm. The laser in this waveband plays a key role in the directed infrared countermeasures. Compared with other mid-far

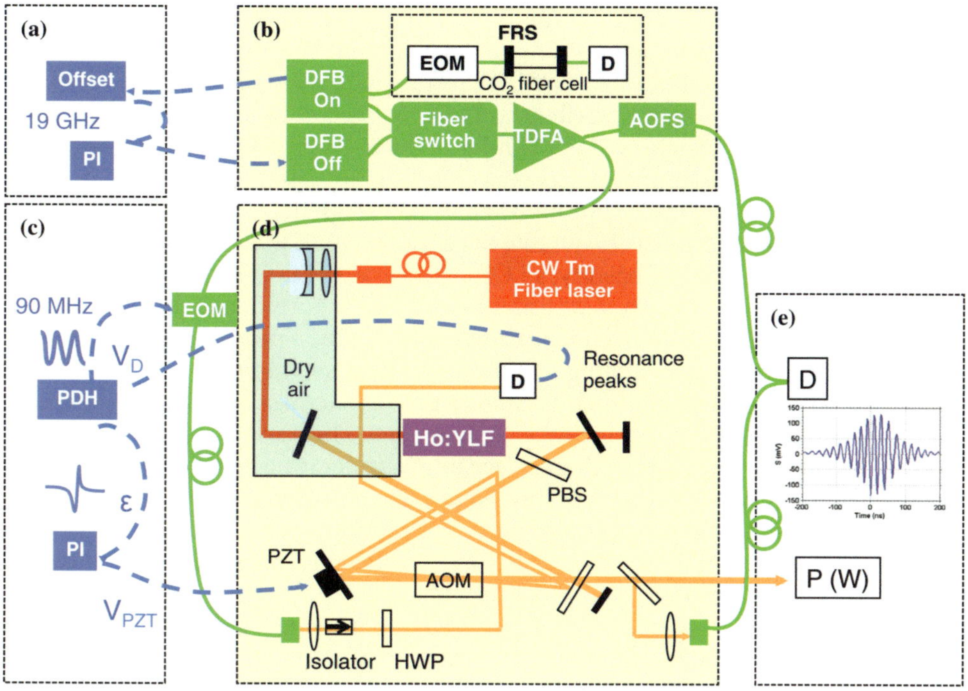

Figure 9.
Experimental setup diagram of differential absorption Lidar for the measurement of atmospheric CO_2 concentrations based on 2 μm laser. (a) Offset locking between On and Off distributed-feedback laser diode (DFB), (b) seeding architecture, (c) cavity length locking using Pound-Drever-Hall technique (PDH), (d) 2-μm pulsed oscillator, and (e) characterization: power meter, beam profiler, heterodyne detection [100]. Copyright © 2014, Springer Nature.

infrared lasers (such as CO lasers [101, 102], quantum cascade lasers [103, 104]), OPOs have the advantages of wideband tunability of output spectrum and narrow laser line width. The typical system is Directed Infrared Countermeasures (DIRCM) system, which can protect airborne platforms from infrared guided missile threats. The layout of the Ho:YAG master oscillator power amplifier (MOPA) in DIRCM system [105]. A Tm:YLF fiber laser pumped Ho:YAG laser system with pulse energies up to 90 mJ and the pulse width of 20 ns at 100 Hz. Then, the wavelength was converted into the 3 to 5-micron region using a zinc germanium phosphide (ZGP) crystal in a linear or ring resonator.

In conclusion, 2 μm wave is located in the weak absorption band of the atmosphere and the safe region of the human eye. Therefore, it is of great significance to study 2 μm Q-switched lasers in the areas such as laser medicine and laser lidar. Moreover, the 2 μm pulse laser with high peak power can also be used as an efficient pump source for OPOs and OPA to obtain the output of mid- and far- infrared laser. With the development of 2 μm pulsed laser, it will have more broad applying foreground in industrial processing, medicine, and military field.

6. Summary and outlook

2 μm pulsed laser with stable, compact and cost-effective characteristics has been a hot topic in recent years. To achieve pulsed laser sources with large pulse energy and high peak power, the traditional method is based on actively AO or EO Q-switching technology. For the flash-pumping operation regime, it has great advantages in generating high energy laser output with hundreds of mJ level due

to the larger mode area than that of diode-pumping case. However, diode-pumped solid-state 2 μm lasers have achieved rapid development because of the advantages such as high efficiency, small size, and high beam quality. Furthermore, in cased of the AO Q-switching, many excellent works have been done and the maximum single pulse energy ever achieved is 128 mJ. For diode-pumped EO Q-switched solid lasers, 550-mJ single pulse energy has set a record in the case of using a RTP crystal as EO modulator, to the best of our knowledge. Although the EO modulator has relatively good switching effect, suitable EO crystal with low driven voltage and high repetition rate is rare. Recently, a LGS crystal based EO Q-switched Tm:YAP laser with a maximum repetition rate of 200 kHz was reported, which indicated that the LGS crystal with high damage threshold could be a promising EO Q-switch. Compared with active Q-switching method, passive Q-switching technology has shown promising advantages, such as simplicity, compactness, and high repetition rate, although it usually works in low-energy regime. The most mature SAs are Cr^{2+} doped II-VI crystals which are still irreplaceable in generating passively Q-switched 2 μm laser with pulse energy over mJ level and pulse width in an order of several nanoseconds to sub-nanoseconds. Besides, the Ho^{3+} doped gain crystals can also be used as SAs for 2 μm solid-state lasers, however, this rarely employed method can only deliver tens of nanoseconds pulses with tens of kHz repetition rate. At present, the 2D nanomaterials based SAs have become hot spots in generating pulsed lasers due to its advantages of easy fabrication, compact setup, and broadband optical absorptions. However, the pulse energy was limited to μJ level determined by the low damage thresholds of the 2D material. At the same time, there is also a problem in fabricating high quality thin films with large area and uniform thickness. In the near future, active Q-switching technology would still be the main method in obtaining high pulse energy or short pulse width with excellent stability and controllability. Of course, more suitable EO crystals would be further explored. For SAs suitable for passively Q-switched lasers, novel kinds of 2D nanomaterials are emerging endlessly, and the preparation method for growing high quality 2D-SAs are coming to maturity.

As for the applications of Q-switched 2 μm lasers, the application fields are becoming broader and broader. When the 2 μm pulsed laser is applied on the tissue, it can generate the effects such as the vaporization, cutting, solidification, hemostasis and so on, so it plays an important role in medical surgery. Q-switched 2 μm lasers can also be effectively used in environmental monitoring and measurement of carbon dioxide, aerosol concentration, cloud layer and water vapor distribution in the atmosphere. Besides, it should be particularly stated that 2 μm pulsed laser with high peak power is an efficient pump source for OPOs or OPA to realize mid- and far- infrared laser sources, which have important applications in the fields such as Lidar, optoelectronic countermeasures, laser ranging and infrared guidance technology. With the continuous enhancement of Q-switched 2 μm laser performance, more complicated application fields would be further explored and developed in the future.

Acknowledgements

This work was partially supported by National Natural Science Foundation of China (NSFC) (61475088), Key research and development program of Shandong Province (2018GGX101006), Shenzhen Science and Technology Research and Development Funds (JCYJ20180305163932273), the Program of State Key Laboratory of Quantum Optics and Quantum Optics Devices (KF201908) and Young Scholars Program of Shandong University (2015WLJH38).

DOI: http://dx.doi.org/10.5772/intechopen.89548

Author details

Kejian Yang[1,3]*, Yaling Yang[2,3], Jingliang He[1] and Shengzhi Zhao[2]

1 Institute of Crystal Materials, Shandong University, Jinan, China

2 School of Information Science and Engineering, Shandong University, Qingdao, China

3 Shenzhen Research Institute of Shandong University, Shenzhen, China

*Address all correspondence to: k.j.yang@sdu.edu.cn

References

[1] Sigrist MW. Trace gas monitoring by laser photoacoustic spectroscopy and related techniques (plenary). The Review of Scientific Instruments. 2003;**74**(1):486-490. DOI: 10.1063/1.1512697

[2] Bilici T, Tabakoğlu Ö, Kalaycioğlu H, Kurt A, Sennaroglu A, Gülsoy M. Skin tissue ablation by thulium (Tm:YAP) laser at 1980 Nm. In: The European Conference on Lasers and Electro-Optics; 14-19 June 2009; Munich Germany; 2009

[3] Anderson SG. Confocal laser microscopes see a wider field of application. Laser Focus World. 1994;**30**(2):83-86. DOI: 10.1364/JOSAB.11.000394

[4] Koch GJ, Barnes BW, Petros M, et al. Coherent differential absorption lidar measurements of CO_2. Applied Optics. 2004;**43**(26):5092-5099. DOI: 10.1364/AO.43.005092

[5] Gower MC. Industrial applications of laser micromachining. Optics Express. 2000;**7**(2):56-67. DOI: 10.1364/OE.7.000056

[6] Lippert E, Fonnum H, Arisholm G, et al. A 22-watt mid-infrared optical parametric oscillator with V-shaped 3-mirror ring resonator. Optics Express. 2010;**18**(25):26475-26483. DOI: 10.1364/OE.18.026475

[7] Lippert E, Fonnum H, Stenersen K. High power multi-wavelength infrared source. Proceedings of SPIE, Technologies for Optical Countermeasures VII. Toulouse, France: International Society for Optics and Photonics. 2010;**7836**:78360D

[8] Dergachev A, Moulton PF. High-power, high-energy diode-pumped Tm:YLF-Ho:YLF-ZGP laser system. Advanced Solid-State Photonics. San Antonio, Texas, USA: Optical Society of America. 2003:137

[9] Quarles GJ, Rosenbaum A, Marquardt CL, et al. Flash-pumped, room-temperature 2-μm laser with 5% slope efficiency. Solid State Lasers. International Society for Optics and Photonics. 1990;**1223**:221-231

[10] Kuo YK, Milton M, Stultz RD. Ho:YVO_4 solid-state saturable-absorber Q switch for a 2 μm Tm,Cr:$Y_3Al_5O_{12}$. Applied Optics. 1996;**35**(6):881-884. DOI: 10.1109/LEOS.1995.484664

[11] Kuo YK, Birnbaum M, Unlu F, et al. Ho:CaF_2 solid-state saturable-absorber Q switch for a 2-μm Tm,Cr:$Y_3Al_5O_{12}$. Applied Optics. 1996;**35**(15):2576-2579. DOI: 10.1364/AO.35.002576

[12] Tsai TY, Birnbaum M. Q-switched 2-μm lasers by use of a Cr^{2+}:ZnSe saturable absorber. Applied Optics. 2001;**40**(36):6633-6637. DOI: 10.1364/AO.40.006633

[13] Gaponenko MS, Malyarevich AM, Yumashev KV, et al. Passive Q-switching of 2-μm holmium lasers with PbS-quantum dot-doped glass. International Conference on Lasers, Applications, and Technologies 2005: Advanced Lasers and Systems. International Society for Optics and Photonics. 2006;**6054**:605403. DOI: 10.1117/12.660492

[14] Bowman SR, Winings MJ, Searles S, et al. Short-pulsed 2.1 μm laser performance of Cr,Tm,Ho:YAG. IEEE Journal of Quantum Electronics. 1991;**27**(5):1129-1131. DOI: 10.1109/3.83365

[15] Lynn JG, Bowman SR, Searles SK, et al. High-average-power operation of a Q-switched diode-pumped holmium laser. Optics Letters. 1993;**18**(20): 1724-1726. DOI: 10.1364/OL.18.001724.

[16] Li C, Song J, Shen D, et al. Flash-lamp-pumped acousto-optic Q-switched Cr-Tm:YAG laser. Optical Review. 2000;7(1):58-61. DOI: 10.1007/s10043-000 -0058-1

[17] Zheng J, Liu J, Tang Y, et al. AO Q-switching of 2.1-μm laser. High-power lasers and applications III. International Society for Optics and Photonics. 2005;**5627**:147-154. DOI: 10.1117/12.574541

[18] Barnes N, Gettemy D. Pulsed Ho:YAG oscillator and amplifier. IEEE Journal of Quantum Electronics. 1981;**17**(7):1303-1308. DOI: 10.1109/JQE.1981.107 1266

[19] Henderson SW, Hale CP, Magee JR. Injection-seeded operation of a Q-switched Cr,Tm,Ho:YAG laser. Advanced Solid State Lasers. Optical Society of America, 1990: ML1. DOI:10.1364/ASSL.1990.ML1

[20] Kim KH, Choi YS, Norman P, et al. Investigation of 2.1 μm lasing properties of Ho:Tm:Cr:YAG crystals under flash-lamp pumping at various operating conditions. Applied Optics. 1993;**32**(12):2066-2074. DOI: 10.1364/AO.32.002066

[21] Nieuwenhuis AF, Lee CJ, van der Slot PJM, et al. High-efficiency mid-infrared $ZnGeP_2$ optical parametric oscillator directly pumped by a lamp-pumped, Q-switched Cr,Tm,Ho:YAG laser. Optics Letters. 2008;**33**(1):52-54. DOI: 10.1364/ol.33.000052

[22] Wang L, Cai X, Yang J, et al. 520 mJ langasite electro-optically Q-switched Cr, Tm,Ho:YAG laser. Optics Letters. 2012;**37**(11):1986-1988. DOI: 10.1364/ol.37.001986

[23] Suni PJM, Henderson SW. 1-mJ/pulse Tm:YAG laser pumped by a 3-W diode laser. Optics Letters. 1991;**16**(11):817-819. DOI: 10.1364/OL.16.000817

[24] Sullivan AC, Zakel A, Wagner GJ, et al. High power Q-switched Tm:YALO lasers. Advanced Solid-State Photonics. Optical Society of America. 2004:329. DOI: 10.1364/ASSP.2004.329

[25] Cai SS, Kong J, Wu B, et al. Room-temperature cw and pulsed operation of a diode-end-pumped Tm:YAP laser. Applied Physics B. 2008;**90**(1):133-136. DOI: 10.1007/s00340-007-2840-2

[26] Li G, Yao BQ, Meng PB, et al. Efficient continuous wave and Q-switched operation of a dual-end-pumped c-cut Tm:YAP laser. Laser Physics. 2010;**20**(10):1871-1876. DOI: 10.1134/s1054660x10190102

[27] Yumoto M, Saito N, Urata Y, et al. 128 mJ/pulse, laser-diode-pumped, Q-switched Tm:YAG laser. IEEE Journal of Selected Topics in Quantum Electronics. 2015;**21**(1):364-368. DOI: 10.1109/JSTQE.2014.2338872

[28] Luan C, Yang KJ, Zhao J, et al. Dual-loss-modulated Q-switched Tm:LuAG laser with AOM and monolayer graphene. Applied Optics. 2015;**54**(27):8024-8029. DOI: 10.1364/AO.54.008024

[29] Liu X, Yang K, Zhao S, et al. Kilo-hertz-level Q-switched laser characteristics of a Tm,Y:CaF_2 crystal. Optical Materials Express. 2017;**7**(12):4352-4357. DOI: 10.1364/OME.7.004352

[30] Zagumennyi AI, Zavartsev YD, Kalachev YL, et al. Acousto-optic Q-switched lasing in Tm:YbAG crystal. Physics of Wave Phenomena. 2019;**27**(1):20-23. DOI: 10.3103/s1541308x19010047

[31] Yao BQ, Wang YZ, Ju YL, et al. Performance of AO Q-switched Tm,Ho: $GdVO_4$ laser pumped by a 794 nm laser diode. Optics Express. 2005;**13**(13):5157-5162. DOI: 10.1364/OPEX.13.005157

[32] Yao BQ, Li G, Meng PB, et al. High power diode-pumped continuous wave and Q-switch operation of Tm,Ho:YVO_4 laser. Laser Physics Letters. 2010;7(12):857-861. DOI: 10.1002/lapl.201010074

[33] Li G, Xie W, Yang X, et al. Performance of continuous wave and acousto-optically Q-switched Tm,Ho:YAP laser pumped by diode laser. IOP Conference Series: Earth and Environmental Science. IOP Publishing. 2018;**113**(1):012173. DOI: 10.1088/1755-1315/113/1/012173

[34] Lamrini S, Koopmann P, Schäfer M, et al. Directly diode-pumped high-energy Ho:YAG oscillator. Optics Letters. 2012;**37**(4):515-517. DOI: 10.1364/OL.37.000515

[35] Yao BQ, Ding Y, Duan XM, et al. Efficient Q-switched Ho:$GdVO_4$ laser resonantly pumped at 1942 nm. Optics Letters. 2014;**39**(16):4755-4757. DOI: 10.1364/OL.39. 004755

[36] Wang L, Gao C, Gao M, et al. A resonantly-pumped tunable Q-switched Ho: YAG ceramic laser with diffraction-limit beam quality. Optics Express. 2014;**22**(1):254-261. DOI: 10.1364/OE.22.000254

[37] Ji E, Liu Q, Nie M, et al. High-slope-efficiency 2.06 μm Ho:YLF laser in-band pumped by a fiber-coupled broadband diode. Optics Letters. 2016;**41**(6):1237-1240. DOI: 10.1364/OL.41.001237

[38] Duan XM, Shen YJ, Yao BQ, et al. A 106 W Q-switched Ho: YAG laser with single crystal. Optik. 2018;**169**:224-227. DOI: 10.1016/j.ijleo.2018.05.094

[39] Liu C, Yang K, Zhao S, et al. 88 ns multi-millijoule $LiNbO_3$ electro-optically Q-switched Tm:LuAG laser. Optics Communication. 2015;**355**:167-171. DOI: 10.1016/j.optcom.2015.06.036

[40] Liu C, Zhao S, Li Y, et al. Stable kilo-hertz electro-optically Q-switched Tm, Ho:YAP laser at room temperature. Optics and Laser Technology. 2016;**81**:189-193. DOI: 10.1016/j.optlastec.2016.01.037

[41] Guo L, Zhao SZ, Li T, et al. Diode-wing-pumped electro-optically Q-switched 2 μm laser with pulse energy scaling over ten millijoules. Optics Express. 2018;**26**(13):17731-17738. DOI: 10.1364/OE.26.017731

[42] Fonnum H, Lippert E, Haakestad MW. 550 mJ Q-switched cryogenic Ho: YLF oscillator pumped with a 100 W Tm:Fiber laser. Optics Letters. 2013;**38**(11):1884-1886. DOI: 10.1364/OL.38.001884

[43] Jin L, Liu P, Huang H, et al. Short pulse diode-pumped Tm:YAG slab laser electro-optically Q-switched by $RbTiOPO_4$ crystal. Optical Materials. 2016;**60**:350-354. DOI: 10.1016/j.optmat.2016.08.011

[44] Kong H, Wang J, Zhang H, et al. Growth, properties and application as an electro-optic Q-switch of langasite crystal. Journal of Crystal Growth. 2003;**254**(3-4):360-367. DOI: 10.1016/S0022-0248(03)01106-0

[45] Ma S, Lu D, Yu H, et al. Langasite electro-optic Q-switched 2μm laser with high repetition rates and reduced driven voltages. Optics Communication. 2019;**447**:13-17. DOI: 10.1016/j.optcom.2019.04.014

[46] Zhang Y, Yao B, Dai T, et al. Electro-optically cavity-dumped 3 ns Tm:LuAG laser. Applied Optics. 2016;**55**(11):2848-2851. DOI: 10.1364/AO.55.002848

[47] Zhang H. Ultrathin two-dimensional nanomaterials. ACS Nano. 2015;**9**(10):9451-9469. DOI: 10.1021/acsnano.5b05040

[48] Wang S, Yu H, Zhang H, et al. Broadband few-layer MoS_2 saturable absorbers. Advanced Materials. 2014;**26**(21):3538-3544. DOI: 10.1002/adma.20130 63 22

[49] Lu SB, Miao LL, Guo ZN, et al. Broadband nonlinear optical response in multi-layer black phosphorus: An emerging infrared and mid-infrared optical material. Optics Express. 2015;**23**(9):11183-11194. DOI: 10.1364/OE.23.0111 83

[50] Wang Q, Teng H, Zou Y, et al. Graphene on SiC as a Q-switcher for a 2 μm laser. Optics Letters. 2012;**37**(3):395-397. DOI: 10.1364/OL.37.000395

[51] Feng TL, Zhao SZ, Yang KJ, et al. Diode-pumped continuous wave tunable and graphene Q-switched Tm:LSO lasers. Optics Express. 2013;**21**(21):2466-24673. DOI: 10.1364/OE.21.024665

[52] Hou J, Zhang B, He J, et al. Passively Q-switched 2 μm Tm:YAP laser based on graphene saturable absorber mirror. Applied Optics. 2014;**53**(22):4968-4971. DOI: 10.1364/AO.53.004968

[53] Zhang C, Liu J, Fan X, et al. Compact passive Q-switching of a diode-pumped Tm,Y:CaF_2 laser near 2 μm. Optics and Laser Technology. 2018;**103**:89-92. DOI: 10.1016/j.optlastec.2018.01.029

[54] Woodward R, Kelleher E. 2D saturable absorbers for fibre lasers. Applied Sciences. 2015;**5**(4):1440-1456. DOI: 10.3390/app5041440

[55] Jiang T, Yin K, Zheng X, et al. Black phosphorus as a new broadband saturable absorber for infrared passively Q-switched fiber lasers. arXiv preprint. 2015;arXiv:1504.07341

[56] Xie Y, Kong L, Qin Z, et al. Black phosphorus-based saturable absorber for Q-switched Tm:YAG ceramic laser. Optical Engineering. 2016;**55**(8):081307. DOI: 10.1117/1.OE.55.8.081307

[57] Zhang H, He J, Wang Z, et al. Dual-wavelength, passively Q-switched Tm: YAP laser with black phosphorus saturable absorber. Optical Materials Express. 2016;**6**(7):2328-2335. DOI: 10.1364/OME.6.002328

[58] Lee J, Jung M, Koo J, et al. Passively Q-switched 1.89-μm fiber laser using a bulk-structured Bi_2Te_3 topological insulator. IEEE Journal of Selected Topics in Quantum Electronics. 2014;**21**(1):31-36. DOI: 10.1109/JSTQE.2014.2329 934

[59] Kong LC, Xie GQ, Yuan P, et al. Passive Q-switching and Q-switched mode-locking operations of 2 μm Tm:CLNGG laser with MoS_2 saturable absorber mirror. Photonics Research. 2015;**3**(2):A47-A50. DOI: 10.1364/PRJ.3.000A47

[60] Ge P, Liu J, Jiang S, et al. Compact Q-switched 2 μm Tm:$GdVO_4$ laser with MoS_2 absorber. Photonics Research. 2015;**3**(5):256-259. DOI: 10.1364/PRJ.3.000256

[61] Luan C, Zhang X, Yang K, et al. High-peak power passively Q-switched 2-μm laser with MoS_2 saturable absorber. IEEE Journal of Selected Topics in Quantum Electronics. 2016;**23**(1):66-70. DOI: 10.1109/JSTQE.2016.2537980

[62] Lan J, Zhang X, Zhou Z, et al. Passively Q-switched Tm:$CaYAlO_4$ laser using a MoS_2 saturable absorber. IEEE Photonics Technology Letters. 2017;**29**(6):515-518. DOI: 10.1109/LPT.2017.2662699

[63] Chu Z, Liu J, Guo Z, et al. 2 μm passively Q-switched laser based on black phosphorus. Optical Materials Express. 2016;**6**(7):2374-2379. DOI: 10.1364/OME.6.002374

[64] Kong L, Qin Z, Xie G, et al. Black phosphorus as broadband saturable absorber for pulsed lasers from 1 μm to 2.7 μm wavelength. Laser Physics Letters. 2016;**13**(4):045801. DOI: 10.1088/1612-2011/13/4/045801

[65] Liu X, Yang K, Zhao S, et al. High-power passively Q-switched 2 μm all-solid-state laser based on a Bi_2Te_3 saturable absorber. Photonics Research. 2017;**5**(5):461-466. DOI: 10.1364/PRJ.5.000461

[66] Loiko P, Bogusławski J, Serres JM, et al. Sb_2Te_3 thin film for the passive Q-switching of a Tm:$GdVO_4$ laser. Optical Materials Express. 2018;**8**(7):1723-1732. DOI: 10.1364/OME.8.001723

[67] Podlipensky AV, Shcherbitsky VG, Kuleshov NV, et al. Cr^{2+}:ZnSe and Co^{2+}:ZnSe saturable absorber Q switches for 1.54-μm Er:Glass lasers. Optics Letters. 1999;**24**(14):960-962. DOI: 10.1364/OL.24.000960

[68] Sergey V, Igor M, Mike M, et al. Ultrafast middle-IR lasers and amplifiers based on polycrystalline Cr:ZnS and Cr:ZnSe. Optical Materials Express. 2017;7(7):2636-2650. DOI: 10.1364/OME.7.002636

[69] Mond M, Heumann E, Huber G, et al. Passive Q-switching of a diode-pumped Tm:YAG laser by Cr^{2+}:ZnSe. In: Conference on Lasers and Electro-Optics Europe (CLEO/Europe 2003). IEEE; 2003:38. DOI: 10.1109/CLEOE.2003.1312100

[70] Oreshkov B, Gianfrate A, Veronesi S, et al. Generation of 40 ns laser pulses by a diode-pumped passively Q-switched Tm,Ho:YLF laser. Laser Physics Letters. 2014;**11**(11):115801. DOI: 10.1088/1612-2011/11/11/115801

[71] Sebbag D, Korenfeld A, Ben-Ami U, et al. Diode end-pumped passively Q-switched Tm:YAP laser with 1.85-mJ pulse energy. Optics Letters. 2015;**40**(7):1250-1253. DOI: 10.1364/OL.40.001250

[72] Canbaz F, Yorulmaz I, Sennaroglu A. 2.3-μm Tm^{3+}:YLF laser passively Q-switched with a Cr^{2+}:ZnSe saturable absorber. Optics Letters. 2017;**42**(9):1656-1659. DOI: 10.1364/OL.42.001656

[73] Lan J, Xu B, Zhou Z, et al. High-power CW and Q-switched Tm:$CaYAlO_4$ lasers at 1.94 μm for shallow water absorption. IEEE Photonics Technology Letters. 2017;**29**(23):2127-2130. DOI: 10.1109/LPT.2017.2767200

[74] Lan J, Guan X, Xu B, et al. A diode-pumped Tm:$CaYAlO_4$ laser at 1851 nm. Laser Physics Letters. 2017;**14**(7):075801. DOI: 10.1088/1622-202x/aa7609

[75] Lan J, Zhou Z, Guan X, et al. Passively Q-switched Tm:$CaGdAlO_4$ laser using a Cr^{2+}:ZnSe saturable absorber. Optical Materials Express. 2017;**7**(6):1725-1731. DOI: 10.1364/OME.7.001725

[76] Yang XT, Liu L, Xie WQ. Passively Q-switched Ho:SSO laser by use of a Cr^{2+}:ZnSe Saturable absorber. Chinese Physics Letters. 2017;**2**:9. DOI: 10.1088/0256-307X/34/2/024201

[77] Serres JM, Loiko P, Mateos X, et al. Passive Q-switching of a Tm,Ho:KLu$(WO_4)_2$ microchip laser by a Cr:ZnS saturable absorber. Applied Optics. 2016;**55**(14):3757-3763. DOI: 10.1364/AO.55.003757

[78] Faoro R, Kadankov M, Parisi D, et al. Passively Q-switched Tm:YLF laser. Optics Letters. 2012;**37**(9):1517-1519. DOI: 10.1364/OL.37.001517

[79] Zhang X, Bao X, Li L, et al. Laser diode end-pumped passively Q-switched Tm,Ho:YLF laser with

Cr:ZnS as a saturable absorber. Optics Communications. 2012;**285**(8):2122-2127. DOI: 10.1016/j.optcom.2011.12.081

[80] Yu H, Petrov V, Griebner U, et al. Compact passively Q-switched diode-pumped Tm: $LiLuF_4$ laser with 1.26 mJ output energy. Optics Letters. 2012;**37**(13):2544-2546. DOI: 10.1364/OL.37.002544

[81] Chen ZY, Yao BQ, Du YQ, et al. A Cr:ZnS saturable absorber for a Tm:YLF pumped passively Q-switched Ho:YAG laser. Laser Physics Letters. 2013;**10**(10):105001. DOI: 10.1088/1612-2011/10/10/105001

[82] Zhang X, Yu L, Zhang S, et al. Diode-pumped continuous wave and passively Q-switched Tm, Ho:LLF laser at 2 μm. Optics Express. 2013;**21**(10):12629-12634. DOI: 10.1364/OE.21.012629

[83] Du Y, Yao B, Duan X, et al. Cr:ZnS saturable absorber passively Q-switched Tm,Ho:$GdVO_4$ laser. Optics Express. 2013;**21**(22):26506-26512. DOI: 10.1364/OE.21.026506

[84] Yan Qiu D, Bao-Quan Y, Zheng C, et al. Passively Q-switched Tm,Ho:YVO_4 laser with Cr:ZnS saturable absorber at 2 μm. Chinese Physics Letters. 2014;**31**(6):064209. DOI: 10.1088/0256-307X/31/6/064209

[85] Loiko P, Serres JM, Mateos X, et al. Subnanosecond Tm:KLuW microchip laser Q-switched by a Cr:ZnS saturable absorber. Optics Letters. 2015;**40**(22):5220-5223. DOI: 10.1364/OL.40.005220

[86] Yao B, Yuan J, Li J, et al. High-power Cr^{2+}:ZnS saturable absorber passively Q-switched Ho:YAG ceramic laser and its application to pumping of a mid-IR OPO. Optics Letters. 2015;**40**(3):348-351. DOI: 10.1364/OL.40.000348

[87] Yao BQ, Cui Z, Duan XM, et al. Passively Q-switched Ho:LuAG laser with near-diffraction-limited beam quality. Applied Physics B. 2015;**118**(2):235-239. DOI: 10.1007/s00340-014-5976-x

[88] Cui Z, Duan XM, Yao BQ, et al. Resonantly pumped 2.118 μm Ho:YAP laser Q-switched by a Cr^{2+}:ZnS as a saturable absorber. Laser Physics Letters. 2015;**12**(10):105002. DOI: 10.1088/1612-2011/12/10/105002

[89] Ji EC, Liu Q, Yao Y, et al. 2-μm Cr^{2+}:CdSe passively Q-switched laser. Solid State Lasers XXVII: Technology and Devices. International Society for Optics and Photonics. 2018;**10511**:105110Z

[90] Vatnik SM, Vedin IA, Kurbatov PF, et al. Spectral and power characteristics of a 5% Tm:KLu $(WO_4)_2$ nm-cut minislab laser passively Q-switched by a Cr^{2+}:ZnSe crystal. Quantum Electronics. 2017;**47**(11):981. DOI: 10.1070/QEL16494

[91] Kuo YK, Birnbaum M, Chen W. Ho:$YLiF_4$ saturable absorber Q-switch for the 2 μm Tm,Cr:$Y_3Al_5O_{12}$ laser. Applied Physics Letters. 1994;**65**(24):3060-3062. DOI: 10.1063/1.112506

[92] Yang X, Mu Y, Zhao N. Ho:SSO solid-state saturable-absorber Q switch for pulsed Ho:YAG laser resonantly pumped by a Tm:YLF laser. Optics and Laser Technology. 2018;**107**:398-401. DOI: 10.1016/j.optlastec.2018.06.021

[93] Lü T, Xiao Q, Li ZJ. Influence of water environment on holmium laser ablation performance for hard tissues. Applied Optics. 2012;**51**(13):2505-2514. DOI: 10.1364/AO.51.002505

[94] Elsheemy MS, Maher A, Mursi K, et al. Holmium: YAG laser ureteroscopic lithotripsy for ureteric calculi in children: Predictive factors

for complications and success. World Journal of Urology. 2014;**32**(4):985-990. DOI: 10.1007/s00345-013-1152-x

[95] Kraaij G, Malan DF, van der Heide HJL, et al. Comparison of Ho:YAG laser and coblation for interface tissue removal in minimally invasive hip refixation procedures. Medical Engineering and Physics. 2012;**34**(3):370-377. DOI: 10.1016/j.medengphy.2011.07.029

[96] Chu EA, Li M, Lazarow FB, Wong BJ. Mid-infrared laser orbital septal tightening: Ex vivo dosimetry study and pilot clinical study. JAMA Facial Plastic Surgery. 2014;**16**(6):425-431. DOI: 10.1001/jamafacial.2014.671

[97] Ishii S, Mizutaani K, Itabe T. Development of 2 μm airborne coherent Doppler lidar at NICT. Proceedings of SPIE 6409, Lidar Remote Sensing for Environmental Monitoring VII. Goa, India: SPIE. 2006;**6409**:60490J. DOI: 10.1117/12.693526

[98] Koch GJ, Beyon JY, Barnes BW, et al. High-energy 2 μm Doppler lidar for wind measurements. Optical Engineering. 2007;**46**(11): 116201-116214. DOI: 10.1117/1.2802584

[99] Ishii S, Koyama M, Baron P, et al. Ground-based integrated path coherent differential absorption lidar measurement of CO_2: Foothill target return. Atmospheric Measurement Techniques. 2013;**6**(5):1359-1369. DOI: 10.5194/amt-6-1359-2013

[100] Gibert F, Edouart D, Cénac C, et al. 2-μm high-power multiple-frequency single-mode Q-switched Ho:YLF laser for DIAL application. Applied Physics B. 2014;**116**(4):967-976. DOI: 10.1007/s00340-014-5784-3

[101] Sato S, Shimizu K, Shimamoto K. Efficient room-temperature CO laser with high specific output. Optics Letters. 1994;**19**(10):719-721. DOI: 10.1364/OL.19.000719

[102] Ma Y, Liang D. Tunable and frequency-stabilized CO_2 waveguide laser. Optical Engineering. 2002;**41**(12):3319-3323

[103] Rao GN, Karpf A. External cavity tunable quantum cascade lasers and their applications to trace gas monitoring. Applied Optics. 2011;**50**(4):A100-A115. DOI: 10.1364/AO.50.00A100

[104] Xie F, Caneau C, Le Blanc HP, et al. 5 μm strain balanced GaInAs/AlInAs quantum cascade lasers operating at room temperature. Proceedings of SPIE, Novel In-Plane Semiconductor Lasers X. San Francisco, California, USA. International Society for Optics and Photonics. 2011;**7953**:79531A. DOI: 10.1117/12.882762

[105] Renz G, Bohn W. Two-micron thulium-pumped-holmium laser source for DIRCM applications. Proceedings of SPIE, Laser Source Technology for Defense and Security III. Orlando, Florida, USA: International Society for Optics and Photonics. 2007;**6552**:655202. DOI: 10.1117/12.722248

Chapter 6

Thermal Fields in Laser Cladding Processing: A "Fire Ball" Model. A Theoretical Computational Comparison, Laser Cladding versus Electron Beam Cladding

Mihai Oane, Ion N. Mihăilescu and Carmen-Georgeta Ristoscu

Abstract

Laser cladding processing can be found in many industrial applications. A lot of different materials processing were studied in the last years. To improve the process, one may evaluate the phenomena behaviour from a theoretical and computational point of view. In our model, we consider that the phase transition to the melted pool is treated using an absorption coefficient which can underline liquid formation. In the present study, we propose a semi-analytical model. It supposes that melted pool is in first approximation a "sphere", and in consequence, the heat equation is solved in spherical coordinates. Using the Laplace transform, we will solve the heat equation without the assumption that "time" parameter should be interpolated linearly. 3D thermal graphics of the Cu substrate are presented. Our model could be applied also for electron cladding of metals. We make as well a comparison of the cladding method using laser or electron beams. We study the process for different input powers and various beam velocities. The results proved to be in good agreement with data from literature.

Keywords: laser cladding, electron beam cladding, heat equation, computer simulations

1. Introduction

The laser processing of materials is a continuous subject of study from a practical and theoretical point of view [1–3]. Laser cladding is a very important application in laser processing [4]. Laser cladding started in the 1980s and was widely implemented in industry. Meanwhile, the application of the laser cladding has exploded especially in 3D additive manufacturing at a relatively low production cost. From theoretical point of view, the mentioned application was studied in Refs. [5, 6]. In the present study, we want to generalize the existent theory to laser beams with different transverse multimode intensities. We will use the heat diffusion equation for the melted pool [5, 6]. We note the depth of the melted pool with H. It is reasonable to consider that the depth of the melted pool varies between 0.2 and 2.5 mm, for a CO_2 incident laser beam of 1 KW. The speed of laser is considered to vary from 0 to 100 mm/s. The main purpose of engineering technology science is to achieve the best quality of the product with the maximum use of facilities and resources. In these terms, laser cladding is a very delicate process. In consequence, all kinds of modelling are welcome. In general, we find in literature a lot of experiments, but for a laboratory which wants to start to build up a laser cladding set-up for the first time, theoretical and computer modelling are essential for the experimental success.

The powder attenuation is defined as the following ratio:

$$X_p = \frac{P_L - P'_L}{P_L} \tag{1}$$

where P_L is the laser power and P'_L is the transmitted laser power, which is in interaction with the work piece surface. We will focus to determine the temperature in the melted pool. For this we will choose a more realistic model (regarding the "time" parameter), writing the heat carried into the melted pool by the expression:

$$Q_P = I(X,Y,Z) \cdot (\alpha_P X_P + \alpha_P X_P (1-\alpha_P)(1-\alpha_W)(h(t) - h(t-t_0)) = Q_P\,(r,\theta,\varphi) \tag{2}$$

where t_0 is the exposure time, α_P is powder absorption coefficient and α_W is the workpiece absorption coefficient.

2. The analytical model

The novelty of the proposed model is that we consider the melted pool like a sphere with diameter H. Using the Laplace transform, we will solve the heat equation avoiding making the assumption that "time" parameter should be interpolated linearly. The heat equation in spherical coordinates is:

$$\frac{\partial^2 T}{\partial r^2} + \frac{2}{r}\frac{\partial T}{\partial r} + \frac{1}{r^2}\left[\frac{\partial}{\partial \mu}\left(1-\mu^2\right)\frac{\partial T}{\partial \mu}\right] + \frac{1}{r^2}\cdot\frac{1}{(1-\mu^2)}\frac{\partial^2 T}{\partial \varphi^2} = \frac{1}{\gamma}\frac{\partial T}{\partial t} - \frac{Q_P(r,\theta,\varphi,t)}{k} \tag{3}$$

In Eq. (3), T is temperature variation; r, θ and φ are the spherical coordinates; γ is thermal diffusivity; and k represents the thermal conductivity. For simplicity, we will note for the rest of the present study $Q = Q_P$.

We have the following relationships:

$$T|_{t=0} = 0 \text{ and } \mu = \cos\theta \tag{4}$$

The boundary conditions are:

$$k\left.\frac{\partial T}{\partial r}\right|_{r=a} + h \cdot T = 0 \tag{5}$$

where $a = H/2$ is the irradiated sphere radius and h is the thermal transfer coefficient.

We have the following relationships that are necessary to eliminate the variable φ:

$$\frac{\partial^2 F_1}{\partial \varphi^2} + m^2 F_1 = 0 \text{ and } F_1|_{\varphi=0} = F_1|_{\varphi=2\pi} \tag{6}$$

We obtain:

$$F_1(\varphi) = \cos m\varphi \text{ for } p = 2m \tag{7}$$

and

$$F_2(\varphi) = \sin m\varphi \text{ for } p = 2m - 1 \tag{8}$$

Such conditions lead to:

$$\frac{\partial^2 \overline{T}}{\partial r^2} + \frac{2}{r}\frac{\partial \overline{T}}{\partial r} + \frac{1}{r^2}\left\{\frac{\partial}{\partial \mu}\left(1-\mu^2\right)\frac{\partial \overline{T}}{\partial \mu} - \frac{m^2}{1-\mu^2}\overline{T}\right\} = \frac{1}{\gamma}\frac{\partial \overline{T}}{\partial t} - \frac{\overline{Q(r,\theta,p,t)}}{k} \tag{9}$$

where

$$\overline{T}(r,\theta,p,t) = \overline{T}(r,\theta,m,t) = \int_0^{2\pi} T\,(r,\theta,\varphi,t)F_{1p}(\varphi)\,d\varphi; p = \begin{cases} 2m \\ 2m-1 \end{cases} \tag{10}$$

Now, in order to eliminate the variable θ, we assume that:

$$F_3\overline{T} = \frac{\partial}{\partial \mu}\left[\left(1-\mu^2\right)\frac{\partial \overline{T}}{\partial \mu}\right] - \frac{m^2}{1-\mu^2}\overline{T} \tag{11}$$

We have:

$$\frac{\partial}{\partial \mu}\left[\left(1-\mu^2\right)\frac{\partial F_3}{\partial \mu}\right] + \left(\lambda - \frac{m^2}{1-\mu^2}\right)F_3 = 0 \tag{12}$$

where

$$\lambda = n(n+1)(n = 0,1,2,3....) \text{ and } F_3 = P_{nm}(\cos(\theta)) \tag{13}$$

where P_{nm} are the associated Legendre polynomials.
To eliminate the variable r, we have to consider that:

$$\frac{\partial^2 \widetilde{T}}{\partial r^2} + \frac{2}{r}\frac{\partial \widetilde{T}}{\partial r} - \frac{n(n+1)}{r^2}\widetilde{T} = \frac{1}{k}\frac{\partial \widetilde{T}}{\partial t} - \widetilde{Q}\,(r,\gamma,p,t) \tag{14}$$

where $\gamma = \begin{cases} 2m \\ 2m-1 \end{cases}$ and the heat equation is the following:

$$\widetilde{T} = \frac{\widetilde{v}}{r^{1/2}}\frac{\partial^2 \widetilde{v}}{\partial r^2} + \frac{1}{r}\frac{\partial \widetilde{v}}{\partial r} - \frac{\left(n+\frac{1}{2}\right)^2}{r^2}\widetilde{v} = \frac{1}{\gamma}\frac{\partial \widetilde{v}}{\partial t} - \widetilde{Q}(r,\gamma,p,t) \tag{15}$$

We have:

$$\frac{\partial^2 F_3}{\partial r^2} + \frac{1}{r}\frac{\partial F_3}{\partial r} + \left[\lambda^2 - \frac{\left(n+\frac{1}{2}\right)^2}{r}\right]F_3 = 0 \tag{16}$$

The theory says that:

$$|F_3||_{r=0} < \infty \; si \; k\frac{\partial F_3}{\partial r}\bigg|_{r=a} + hF_3 = 0 \tag{17}$$

The obtained result is:

$$F_3 = J_{n+\frac{1}{2}}(\lambda a) \tag{18}$$

and

$$k\left(J_{n-\frac{1}{2}}(\lambda_{ns}a) - J_{n+\frac{1}{2}}(\lambda_{ns}a)\right) + hJ_{n+\frac{1}{2}}(\lambda_{ns}a) = 0 \tag{19}$$

To eliminate the temporal variable, we use the direct and reverse Laplace transform. Thus we obtain [1]:

$$\begin{aligned} T(r,\theta,\varphi,t) = {} & \frac{1}{r^{\frac{1}{2}}}\sum_{m=0}^{\infty}\sum_{n,\,s=1}^{\infty}\frac{1}{C_{mn}\cdot C_{ns}}\cdot\frac{1}{\lambda_{ns}^2}\left[1-e^{-\lambda_{ns}^2 t}-\left(1-e^{-\lambda_{ns}^2\cdot(t-t_0)}\right)\cdot H(t-t_0)\right] \\ & \cdot J_{n+\frac{1}{2}}(\lambda_{ns}r)\left[P_{n\,m}(\cos\theta)\cos(m\varphi)\cdot\left(\int_0^a\int_0^{2\pi}\int_0^{\theta_{max}}\cdot\frac{E(E_0,r,\cos\theta)}{C}\cdot r^{\frac{3}{2}}\cdot J_{n+\frac{1}{2}}(\lambda_{ns}r)\right.\right. \\ & \left.\cdot P_{n\,m}(\cos\theta)\cdot\cos(m\varphi)drd\theta d\varphi\right) + P_{n\,m}(\cos\theta)\cdot\sin(m\varphi) \\ & \left.\cdot\left(\int_0^a\int_0^{2\pi}\int_0^{\theta_{max}}\cdot\frac{E(E_0,r,\cos\theta)}{C}\cdot r^{\frac{3}{2}}\cdot J_{n+\frac{1}{2}}(\lambda_{ns}r)\cdot P_{n\,m}(\cos\theta)\cdot\sin(m\varphi)drd\theta d\varphi\right)\right] \end{aligned} \tag{20}$$

In the above relationship:

$$C_{m\,n} = \int_{-1}^{+1}\left[P_{m\,n}(\mu)\right]^2 d\mu = \frac{2\delta}{2\,n+1}\frac{(n+m)!}{(n-m)!} \tag{21}$$

and

$$\delta = \begin{cases} 2\,for\;m=0 \\ 1\,for\;m\neq 0 \end{cases} \tag{22}$$

but also

$$C_{ns} = \frac{1}{2}\,a\left[J'_{n+\frac{1}{2}}(\lambda_{ns}a)\right] \tag{23}$$

The laser beam as compared to electron beam may be considered to be a sum of decoupled transverse modes, and one can write using a superposition of different transverse modes:

$$I = \sum_{i,\,m,\,n} p_i I_{mn} \tag{24}$$

where p_i are real numbers chosen in such a way to obtain the wanted laser intensity (from spatial distribution and intensity values' point of view).

3. Simulations and comments

Let us consider the cladding processing on a Cu substrate. The input parameters corresponding to **Figures 1–7** are collected in **Table 1**. We have chosen various situations, for example, different transverse modes (for laser beam), various velocities, incident powers and values of H.

For electron beam processing [7], one may consult the Katz and Penfolds absorption law [8] and also Tabata-Ito-Okabe absorption law [9].

In **Figure 1**, the thermal field for Gaussian laser beam is presented, when $V = 0$ mm/s, $P = 1$ KW, $H = 2$ mm and the substrate is Cu. In **Figure 2** the thermal field for Gaussian laser beam is given when $V = 10$ mm/s, $P = 1$ KW and $H = 2$ mm. In **Figure 3** the thermal field for TEM_{03} laser beam is presented, when $V = 0$ mm/s, $P = 2$ KW and $H = 3$ mm. **Figure 4** shows the thermal field for TEM_{03} laser beam, when $V = 10$ mm/s, $P = 4$ KW and $H = 4$ mm. **Figure 5** represents the thermal field for TEM_{03} laser beam, when $V = 100$ mm/s, $P = 10$ KW and $H = 3$ mm.

If one compares **Figures 1** and **2**, the differences in the spatial distribution of thermal field for the two cases can be seen. On the other hand, the comparison of **Figures 3–5** shows that for TEM_{03} we do not have significant changes in thermal profile but a proportional increase of the incident power with H.

In **Figures 6** and 7, we have as scanning source an electron beam of power $P = 1$ KW. If in **Figure 6** $V = 0$ mm/s, while in **Figure** 7 the speed is $V = 10$ mm/s. Our simulations show a decrease of thermal field with the increase of scanning velocity.

As observed from **Table 2**, Cu behaves very similarly with Au, Ag and Al from a thermal point of view [10]. Accordingly, we may consider that **Figures 1–7** are also meaningful if we use substrates from Au, Ag or Al.

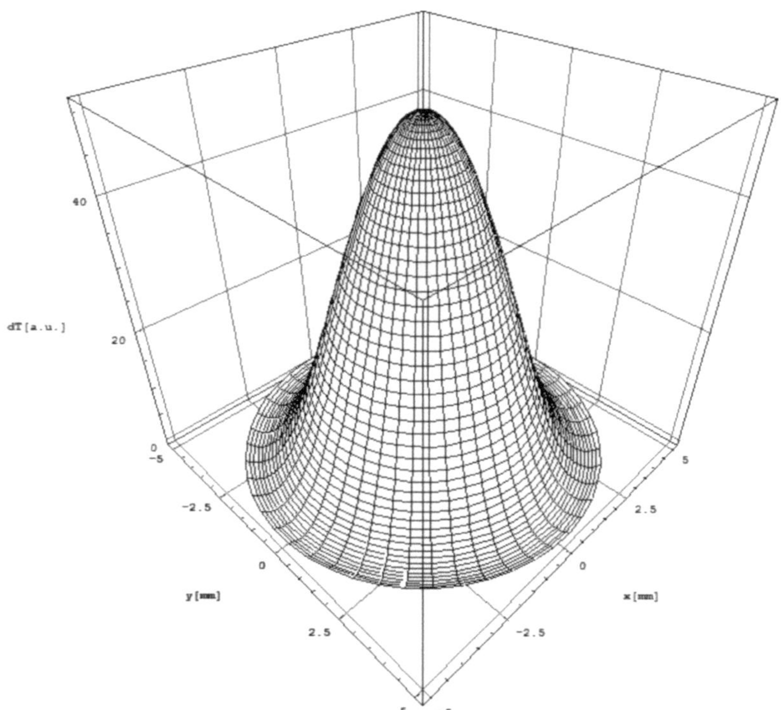

Figure 1.
The thermal field for a Gaussian laser beam when V = 0 mm/s, P = 1 KW and H = 2 mm. The substrate is supposed to be from Cu.

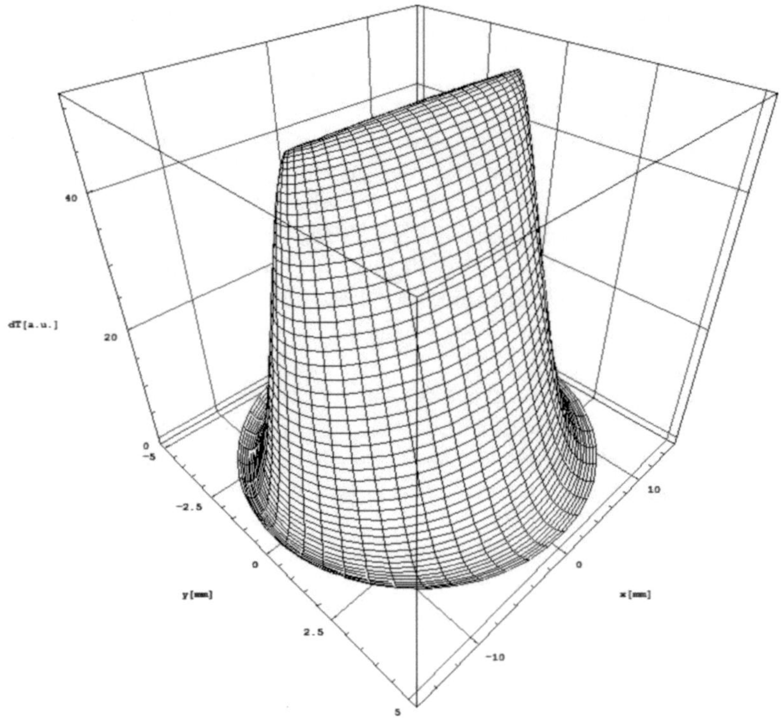

Figure 2.
The thermal field for a Gaussian laser beam when V = 10 mm/s, P = 1 KW and H = 2 mm. The substrate is supposed to be from Cu.

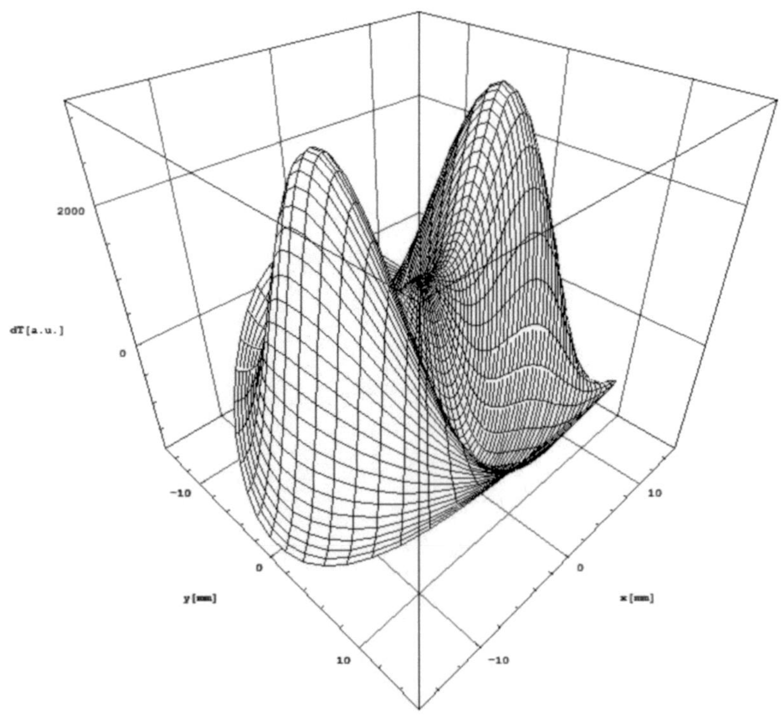

Figure 3.
The thermal field for a TEM_{03} laser beam when V = 0 mm/s, P = 2 KW and H = 3 mm. The substrate is supposed to be from Cu.

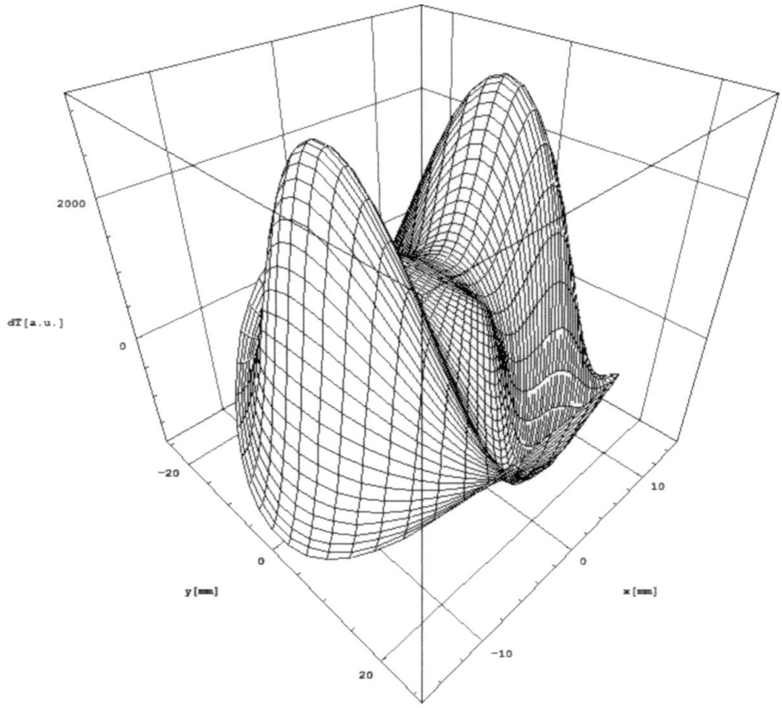

Figure 4.
The thermal field for a TEM_{03} laser beam when V = 10 mm/s, P = 4 KW and H = 4 mm. The substrate is supposed to be from Cu.

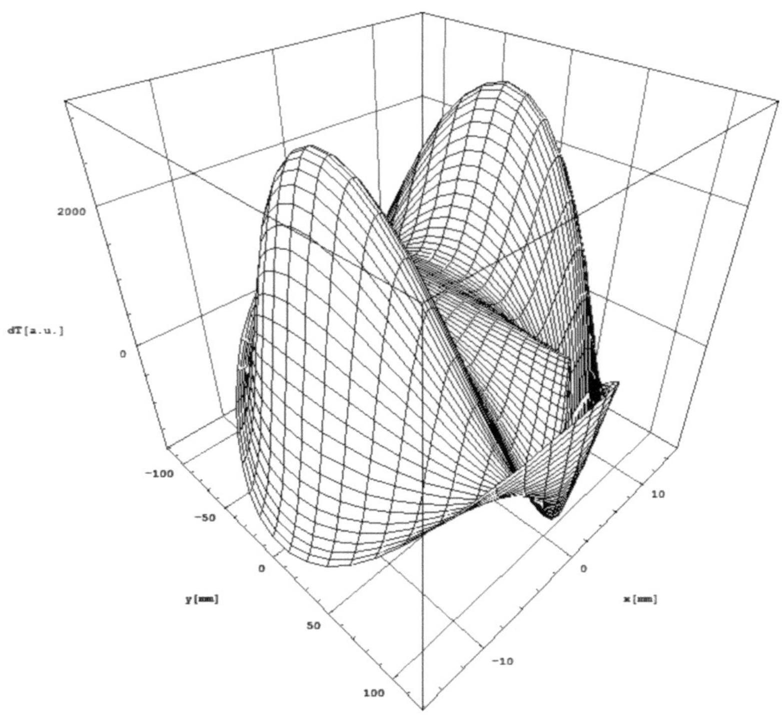

Figure 5.
The thermal field for a TEM_{03} laser beam, when V = 100 mm/s, P = 10 KW and H = 3 mm. The substrate is supposed to be from Cu.

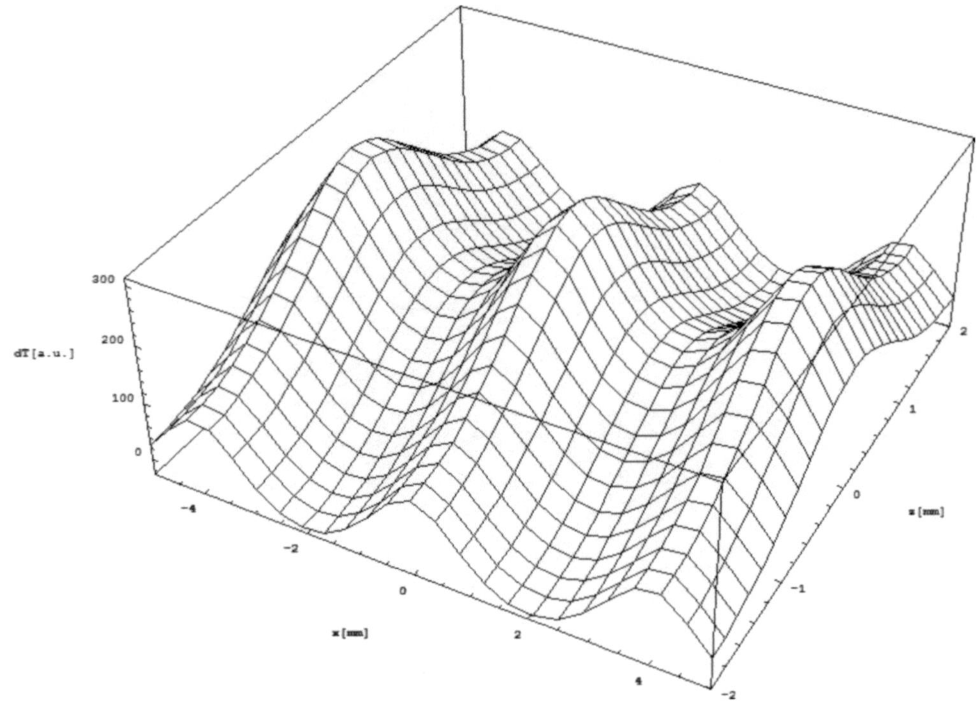

Figure 6.
The thermal field for a Gaussian electron beam when V = 0 mm/s, P = 1 KW and H = 2 mm. The substrate is supposed to be from Cu.

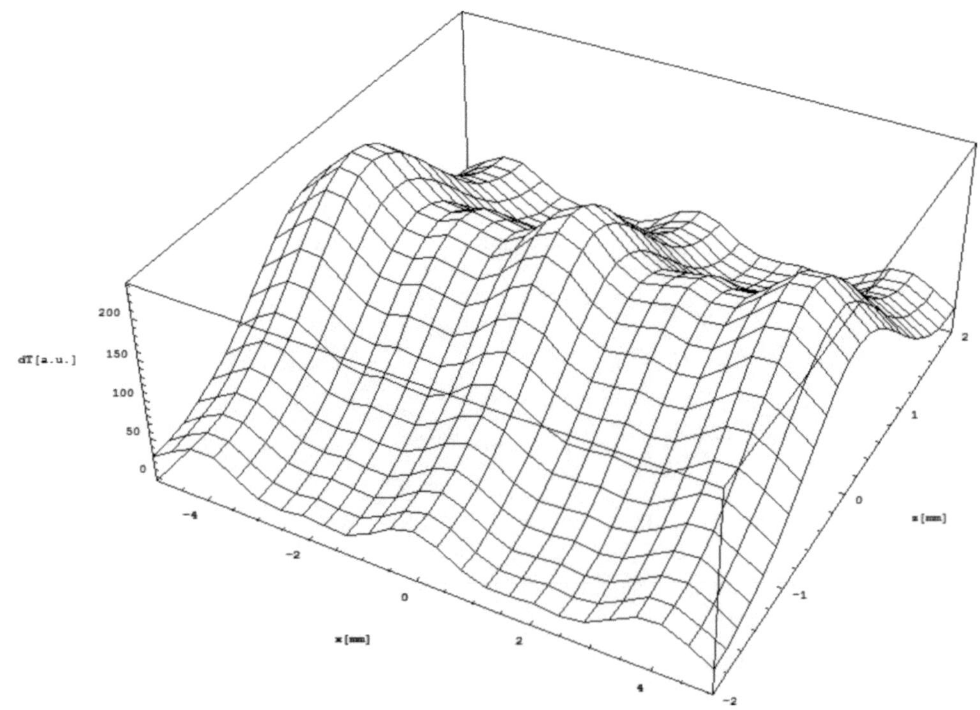

Figure 7.
The thermal field for a Gaussian electron beam when V = 10 mm/s, P = 1 KW and H = 2 mm. The substrate is supposed to be from Cu.

Figure no.	Beam type	Mode	Velocity [mm/s]	Incident power [kW]	Melted pool depth H [mm]	Thermal diffusivity γ [cm^2/s]	Thermal conductivity k [W/cmK]
Figure 1	Laser	TEM_{00}	0	1	2	1.14	3.95
Figure 2	Laser	TEM_{00}	10	1	2	1.14	3.95
Figure 3	Laser	TEM_{03}	0	2	3	1.14	3.95
Figure 4	Laser	TEM_{03}	10	4	4	1.14	3.95
Figure 5	Laser	TEM_{03}	100	10	3	1.14]	3.95
Figure 6	Electron	TEM_{00}	0	1	2	1.14	3.95
Figure 7	Electron	TEM_{00}	10	1	2	1.14	3.95

Table 1.
*The input parameters for the **Figures 1–7**.*

Element	Thermal diffusivity γ [cm^2/s]	Thermal conductivity k [W/cmK]
Cu	1.14	3.95
Au	1.22	3.15
Ag	1.72	4.28
Al	1.03	2.4

Table 2.
The thermal parameters for: Cu, Au, Ag and Al.

4. Conclusions

Our major conclusions are as follows: apart from Gaussian case, the increase in velocity of the other transversal modes does not affect too much the thermal profile; and second the large difference between the electron cladding and laser cladding is that in electron cladding an increase of beam velocity affects in an important amount the values of the thermal fields. The higher the velocity of electron beam, the lower the thermal fields at the surface sample. Our major conclusions are in good agreement with experimental data from literature; see, for example, references [11, 12]. On the other hand, it is known that there are some limitations in laser cladding, for example, high initial capital cost, high maintenance cost and presence of heat affected zone. For electron cladding, one can conclude that the cost is reduced as there are no involved mechanical cutting force, work holding and fixturing.

Acknowledgements

The authors acknowledge the support of MCI-OI under the contract POC G 135/2016.

Conflict of interest

The authors declare no conflict of interest.

Author details

Mihai Oane[1], Ion N. Mihăilescu[2] and Carmen-Georgeta Ristoscu[2*]

1 Electrons Accelerators Laboratory, National Institute for Lasers, Plasma and Radiation Physics, Măgurele-Ilfov, Romania

2 Lasers Department, National Institute for Lasers, Plasma and Radiation Physics, Măgurele-Ilfov, Romania

*Address all correspondence to: carmen.ristoscu@inflpr.ro

References

[1] Oane M, Ticos D, Ticoş CM. Charged Particle Beams Processing Versus Laser Processing. Germany: Scholars' Press; 2015. pp. 60-61, ISBN: 978-3-639-66753-0, monograph

[2] Oane M, Peled A, Medianu RV. Notes on Laser Processing. Germany: Lambert Academic Publishing; 2013. pp. 7-9, ISBN: 978-3-659-487-48739-2, monograph

[3] Oane M, Medianu RV, Bucă A. Chapter 16: A parallel between laser irradiation and electrons irradiation of solids. In: Radiation Effects in Materials. IntechOpen; 2016. pp. 413-430, ISBN: 978-953-51-2417-7, monograph

[4] Steen WM, Mazumder J. Laser Material Processing. London, Dordrecht, Heidelberg, New York: Springer; 2010, ISBN: 978-1-84996-061-8, monograph

[5] Cline HE, Anthony TR. Heat treating and melting material with a scanning laser or electron beam. Journal of Applied Physics. 1977;**48**(9):3895-3900. DOI: 10.1063/1.324261

[6] Picasso M, Marsden CF, Wagniere JD, Frenk A, Rappaz M. A simple but realistic model for laser cladding. Metallurgical and Materials Transactions B. 1994;**25**(B):281-291. DOI: 10.1007/BF02665211

[7] Mul D, Krivezhenko D, Zimoglyadova T, Popelyukh A, Lazurenko D, Shevtsova L. Surface hardening of steel by electron-beam cladding of Ti+C and Ti+B_4C powder compositions at air atmosphere. Applied Mechanics and Materials. 2015;**788**: 241-245. DOI: 10.4028/www.scientific.net/AMM.788.241

[8] Oane M, Toader D, Iacob N, Ticos CM. Thermal phenomena induced in a small graphite sample during irradiation with a few MeV electron beam: Experiment versus theoretical simulations. Nuclear Instruments and Methods in Physics Research B. 2014; **318**:232-236. DOI: 10.1016/j.nimb.2013.09.017

[9] Oane M, Toader D, Iacob N, Ticos CM. Thermal phenomena induced in a small tungsten sample during irradiation with a few MeV electron beam: Experiment versus simulations. Nuclear Instruments and Methods in Physics Research B. 2014;**337**:17-20. DOI: 10.1016/j.nimb.2014.07.012

[10] Bauerle D. Laser Processing and Chemistry. 2nd ed. Berlin: Springer; 1996; ISBN 10: 354060541X/ISBN 13: 9783540605416

[11] Wirth F, Eisenbarth D, Wegener K. Absorptivity measurements and heat source modeling to simulate laser cladding. Physics Procedia. 2016;**83**: 1424-1434. DOI: 10.1016/j.phpro.2016.08.148

[12] Meriaudeau F, Truchetet F, Grevey D, Vannes AB. Laser cladding process and image processing. Journal of Lasers in Engineering. 1997;**6**:161-187

Chapter 7

Optical Sensor for Nonlinear and Quantum Optical Effects

Antônio Carlos Amaro de Faria

Abstract

In this chapter, the main foundations for the conception, design, and the project of optical sensors that explore the effects of nonlinear and quantum optics are presented. These sensors have a variety of applications from the design of waveguides with self-selection of propagation modes to signal processing and quantum computing. The chapter seeks to present formal aspects of applied modern optics in a detailed, sequential, and concise manner.

Keywords: optical sensor, nonlinear optics, quantum optics, optical fiber, optical signal processing

1. Introduction

Classical electrodynamics is the basis for the analysis and formulation of electromagnetic waves. From the equations of Maxwell, it is possible to obtain the equation of the movement of the electric and magnetic fields whose solution describes the propagation of the electromagnetic wave. This formal treatment was originally developed by Maxwell [1], who verified that the electromagnetic wave propagated with the speed of light provided that the optics could be described from the electromagnetism. The medium through which the electromagnetic wave propagates responds in various ways to the electromagnetic field. This response depends on how the atoms and molecules are arranged spatially composing the constituent medium and how the interaction or scattering of the electromagnetic wave through the medium will occur. In other words, the way the medium responds to the electromagnetic excitation is contained in the middle polarization due to the propagation of the electromagnetic wave. It is in this context that some recent analyses have discovered some solutions from the nonlinear response of the medium to the propagation of the electromagnetic wave which may lead to an approach of some quantum effects from a nonlinear treatment of electromagnetism in the material medium.

The propagation of optical pulses through waveguides such as optical fibers can give rise to nonlinear optical effects and quantum effects. The appropriate modeling of these effects can be used for the development of sensors to the optical fiber whose resolution can be regulated properly. In addition, the method allows the selection of propagation modes by selecting the desired modes by knowing the band gap of the waveguide or the photonic crystal.

The development of sensors to the optical fiber is based on the propagation of optical pulses through waveguides like optical fibers and photonic crystals. The propagation of the pulses through waveguides can generate nonlinear and quantum effects as Raman and Brillouin effects. We have analytically modeled these effects from the Maxwell equations on dielectric media describing the propagation of these optical pulses by developing a model that can be implemented computationally for the processing and propagation of these optical signals. The optical sensor can select the modes of propagation of the optical beams through the natural conduction band of the photonic crystal since the function of our model, the so-called optical potential, describes the optical light scattering through the crystal [2].

2. Modeling the optical lattice

The modeling of an optical system is very important for the design and development of many applications of optics in electronics, photonics, integrated optics, and an array of devices based on the light. Nonlinear effects from the interaction of light with matter in waveguides and photonic crystals may be suitable for a variety of optical applications [3–5]. Quantum effects such as the Raman effect and the Brillouin effect can be conveniently dealt with by exploring the nonlinear aspects of optical wave propagation in waveguides. Effects, e.g., self-focusing and low dispersion, of a guided beam can be applied and exploited in several technologies that use waveguides and sensors to the optical fiber [6, 7]. Consider an optical field given by

$$E\left(\vec{r},t\right) = A\left(\vec{r}\right)\exp\left(i\beta_0 z\right). \tag{1}$$

In Eq. (1), β_0 is the propagation constant, and beam one propagates along the z direction and self-focuses along the transverse directions x and y. The function $A[r(x,y,z)]$ represents the evolution of the beam envelope. The nonlinearity refers to nonlinear medium polarization that is the change of the refraction index of the medium that is responsible by the self-focusing of the pulse.

Considering the Maxwell equations and the medium polarization, one can obtain the generalized nonlinear Schrödinger (NLS) equation that describes soliton solutions [8]:

$$i\frac{\partial u}{\partial z} + \frac{1}{2}\left(\frac{\partial^2 u}{\partial x^2} + \frac{\partial^2 u}{\partial y^2}\right) \pm F\left(|u|^2 u\right) = 0. \tag{2}$$

In Eq. (2), u is proportional to electric field. The term $F(|u|^2 u)$ represents a generalized term of nonlinear optical effects. The solutions of Eq. (2) can represent and describe the optical beam profile one propagating through an optical lattice and an optical fiber. It describes also quantum noise as Raman and Brillouin scattering in the optical system.

We will sequentially show how quantum effects from the interaction of an optical beam with the constituents of the waveguide or the crystal lattice through which it propagates can be described in the context of nonlinear electrodynamics. In other words, we will show an equivalence to quantum optics and nonlinear electrodynamics characterized by nonlinear polarization of the medium. We will demonstrate the equivalence between its properties and the properties of solutions from a dynamical action. This action can map optical systems, and the method is based on the variational principle whose solutions give the same from that of Eq. (2).

Considering the optical rays equation

$$\frac{d}{ds}\left(n\frac{d\vec{r}}{ds}\right) = \nabla n, \tag{3}$$

where n is the refraction index, s is a specific trajectory, and $\vec{r} = \vec{r}(s)$, the principle of least action may be written as

$$\delta \int nds = 0, \tag{4}$$

which is Fermat's principle for the paths of light rays. So the scalar wave equation in optical context can be written as

$$\nabla^2\phi - \frac{n^2}{c^2}\frac{d^2\phi}{dt^2} = 0, \tag{5}$$

that is satisfied by a plane wave solution

$$\phi = \phi_0 e^{i\left(\vec{k}.\vec{r} - \omega t\right)}. \tag{6}$$

The wave number k and the frequency ω are related by

$$k = \frac{2\pi}{\lambda} = \frac{n\omega}{c}. \tag{7}$$

So Eq. (6) can be written as

$$\phi = \phi_0 e^{i\vec{k}_0(nz - ct)}, \tag{8}$$

and it is assumed that $\vec{k}$ is in the z direction. The refractive index n can change very gradually in space, and a solution resembling the plane wave can be written as

$$\phi = e^{A\left(\vec{r}\right) + ik_0\left[L\left(\vec{r}\right) - ct\right]}, \tag{9}$$

and the quantities $A\left(\vec{r}\right)$ and $L\left(\vec{r}\right)$ are real functions.

Now we can establish a relation between this description of optical pulses, the NLS equation and the Schrödinger equation. This equivalence, in context of the classical level at least, between NLS equation solutions and the method that we propose in this work, can be constructed considering the relation

$$k_0(L - ct) = 2\pi\left(\frac{L}{\lambda_0} - \omega t\right). \tag{10}$$

In this sense, we are led to conclude that the energy E of the system and its frequency ω are proportional

$$E = h\omega. \tag{11}$$

The wave length and the frequency are related by

$$\lambda\omega = v, \tag{12}$$

and, now, considering that for a light ray one propagating in a certain medium with speed $v < c$, where c is the speed light in the vacuum

$$v = \frac{E}{p}, \tag{13}$$

and another simple relation

$$\lambda = \frac{v}{\omega}, \tag{14}$$

we come to an important relationship, using Eqs. (11) and (12):

$$\lambda = \frac{h}{p}. \tag{15}$$

So as an important consequence of this approach, we can write the wave Eq. (5) as

$$\nabla^2 \phi - \frac{1}{v^2} \frac{d^2 \phi}{dt^2} = 0, \tag{16}$$

where v is the wave velocity in the medium of index n. This equation can yet be rewritten, considering a temporal dependence of kind $e^{-i\omega t}$ in the form

$$\nabla^2 \phi + \frac{4\pi^2}{\lambda^2} \phi = 0, \tag{17}$$

that is the time-independent wave equation. So it is perfectly acceptable that a certain optical field Ψ satisfies an equation of the same form as

$$\nabla^2 \psi + k^2 \psi = 0, \tag{18}$$

and doing the identifications

$$p = \sqrt{2m(E - V)}, \tag{19}$$

with p being the momentum and the terms inside the square root stands by kinetic energy or the difference between total energy E and the potential energy V of particle, we get the Schrödinger equation [9]

$$\nabla^2 \psi + \frac{8\pi^2 m}{h^2} (E - V) \psi = 0, \tag{20}$$

with m being the particle mass and h the Planck constant.

3. Brief introduction on optical solitons

3.1 Nonlinear Schrödinger equation

The equation that describes optical fields in a nonlinear medium is known as the nonlinear Schrödinger equation. In this section, we succinctly present the origin of the NLS equation for a CW beam propagating inside a nonlinear optical medium. From Maxwell equations in a nonlinear medium, one gets the wave equation for the electric field [7, 10]:

$$\nabla^2 E - \frac{1}{c^2}\frac{\partial^2 E}{\partial t^2} = \frac{1}{\varepsilon_0 c^2}\frac{\partial^2 P}{\partial t^2} \tag{21}$$

with c being the speed of light in vacuum and ε_0 the vacuum permittivity. The total polarization $\vec{P}$ is

$$P(r,t) = P_L(r,t) + P_{NL}(r,t). \tag{22}$$

P_L is the linear part, and P_{NL} is the nonlinear part, given respectively by [8, 10]

$$P_L(r,t) = \varepsilon_0 \int \chi^1(t-t')E(r,t')dt' \tag{23}$$

and

$$P_{NL}(r,t) = \varepsilon_0 \int\int\int \chi^3(t-t_1, t-t_2, t-t_3) \times E(r,t_1)E(r,t_2)E(r,t_3)dt_1dt_2dt_3, \tag{24}$$

χ^1 and χ^3 are the first- and third-order susceptibility tensors. A general solution of Eq. (21) will be

$$E(r,t) = \frac{1}{2}\hat{x}\left[E(r,t)e^{-i\omega_0 t} + cc\right] \tag{25}$$

where $E(r,t) = A(r)e^{i\beta_0 Z}$ and $\beta_0 = k_0 n_0 \equiv \frac{2\pi n_0}{\lambda}$ are the propagation constants with the wavelength $\lambda = \frac{2\pi c}{\omega_0}$. The beam diffracts and self-focuses along the two transverse directions X and Y where X, Y, and Z are the spatial coordinates associated with r. The function A(X,Y,Z) is the evolution of the beam envelope; it would be a constant in the absence of nonlinear and diffractive effects. Nonlinear and diffractive effects and neglecting $\frac{d^2A}{dz^2}$ the beam envelope satisfy the following nonlinear parabolic equation:

$$2i\beta_0\frac{\partial A}{\partial Z} + \left(\frac{\partial^2 A}{\partial X^2} + \frac{\partial^2 A}{\partial Y^2}\right) + 2\beta_0 k_0 n_{nl}(I)A = 0. \tag{26}$$

Introducing the following variables

$$x = \frac{X}{\omega_0}, y = \frac{Y}{\omega_0}, z = \frac{Z}{L_d}, u = (k_0|n_2|L_d)^{1/2}A \tag{27}$$

where ω_0 is a transverse scaling parameter related to the input beam width and $L_d = \beta_0\omega_{02}$ is the diffraction length; Eq. (26) takes the form of a NLS equation:

$$i\frac{\partial u}{\partial z} + \frac{1}{2}\left(\frac{\partial^2 u}{\partial x^2} + \frac{\partial^2 u}{\partial y^2}\right) \pm |u|^2 u = 0. \tag{28}$$

Now one can consider the NLS equation in the form

$$i\frac{\partial u}{\partial z} + \frac{1}{2}\left(\frac{\partial^2 u}{\partial x^2} + \sigma_d\frac{\partial^2 u}{\partial y^2}\right) \pm |u|^2 u = 0, \tag{29}$$

and the y independent form of it

$$i\frac{\partial u}{\partial z}+\frac{1}{2}\frac{\partial^2 u}{\partial x^2}\pm|u|^2u=0. \tag{30}$$

Eq. (30) describes the spatial and temporal optical solitons as treated in Refs. [5,6, 11–15], respectively.

4. Nonlinear and quantum optical sensor principles

Nonlinear effects in optical fibers are common when, for example, increasing the power of the optical source. In this case, optical noise such as Raman effect and Kerr effect originate the interaction of optical fields with matter. These effects, depending on their technological application, may be undesirable. In our work, we discuss one set of solutions for optical fields whose nonlinear effects can be used to suppress certain propagation modes harmful to technological applications. On propagation of signals, for example, the Kerr effect can be a factor representing the loss of optical signal, and in this sense, using the solutions presented in this work, an optical network can be specially designed to suppress the distortion of the optical signal by the Kerr effect. The fiber or group of optical fibers can be designed so that the distortion of optical signals through nonlinear effects is eliminated. In another perspective, our method can be combined with space-division multiplexing (SDM) and nonlinear cancelation methods that offer the opportunity to reverse the effect of Kerr distortion [16]. The method developed in this work can be implemented to identify patterns of nonlinear modes which contribute to the distortion of the optical signal in the transmission system [17].

In this case, the propagation modes of the optical beam by the system will be conveniently selected and processed in the transmission link as shown in **Figure 1**. In this section, we will mathematically demonstrate how any optical signal can be transformed conveniently into appropriate optical pulse. In other words, any sign optical can be mapped to a field originally known.

It is appropriate to point out that the mathematical approach is necessary to detail the practical implementation of the recognition and processing of optical signals that can reach the optical beam level in the waveguide or any optical network.

4.1 Optical systems

The discovery of non-Hermitian observables with real spectrum in optical systems may provide an important and useful relationship between the crystal structure of an optical system and parity-time (*PT*) symmetry [18]. Optical

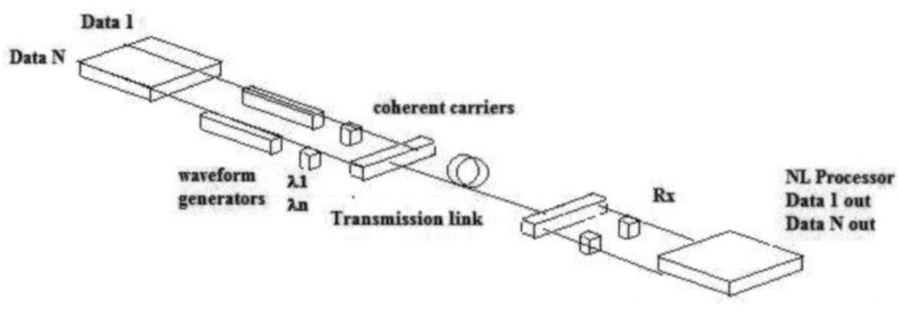

Figure 1.
*Architecture of a transmission system. The nonlinear (NL) processor identifies **λI** e **λn** modes and rectifies the optical distortions through the preprocessed patterns of non-Hermitian solitons diffracted by the optical network. Like any signal can be mapped, as will be presented in Section 4, the processor may recognize any optical pattern from the original signal, as a nonlinear distortion or optical soliton, and eliminate it, reverse it, or pass it on. In figure **Rx** is a signal identifier (basically an oscilloscope) with output to the Nonlinear Processor (NL Processor).*

potentials that have *PT* symmetry are potential that obey the following relation $V(x) = V^{*}(-x)$, where the superscript star denotes the conjugate complex. The symmetry PT may describe an optical system as a lattice structure or a waveguide. From the formal point of view, the light that propagates through a waveguide can be described by classical electrodynamics and the quantum mechanics, which seems to establish a deep relationship between them. However nonlinear optical phenomena lead to the same dynamic equations describing the propagation of quantum light in some materials media. We can explore the optical dynamics of a beam and waves propagating by an optical lattice which can be described by PT symmetric complex potentials. The optical beam propagation can be described by the Schröndinger-like equation:

$$i\frac{\partial\psi}{\partial z} + \frac{\partial^2\psi}{\partial x^2} + [V(x) + iW(x)]\psi + |\psi|^2\psi = 0. \tag{31}$$

Eq. (31) describes an optical beam one propagating in a self-focusing Kerr nonlinear PT symmetric potential as $\mathrm{V(x) = V(-x)}$ and $\mathrm{W(-x) = -W(x)}$. The solutions of Eq. (31) can be expressed as follows:

$$\psi(x,z) = \phi(x)\exp(i\lambda z), \tag{32}$$

where $\phi(x)$ is the nonlinear eigenmode and λ is the propagation constant. Then one obtains the following equation for the field $\phi(x)$:

$$\frac{\partial^2\phi}{\partial x^2} + [V(x) + iW(x)]\phi + |\phi|^2\phi = \lambda\phi. \tag{33}$$

In this point it is important to note that the optical system can be mapped by the band structure of the optical lattice or waveguide through the propagation constant λ. In fact the eigenvalue (λ) and the eigenmode $\phi(x)$ satisfy

$$\left[\frac{d^2}{dx^2} + F(x)\right]\phi = \lambda\phi, \tag{34}$$

whose solutions can be write in the form

$$\phi_n(k,x) = u_n(k,x)e^{ikx}, u_n(x+D,k) = u_n(x,k) \tag{35}$$

with $u_n(x,k)$ and satisfying

$$\frac{d^2u_n(x,k)}{dx^2} + 2ik\frac{du}{dx} + \left[F(x) - k^2\right]\phi = \lambda(k)\phi. \tag{36}$$

From Eq. (35) result important properties of optical systems described by symmetry *PT*, as the selection of propagation modes $\lambda(k)$.

4.2 Mapping optical systems

In order to show how the *PT* symmetric complex potentials can map optical lattice structures, consider Eq. (31) where

$$V(x) = V_0\,\mathrm{sec}\,h^2(x),\quad W(x) = W_0\,\mathrm{sec}\,h(x)\tanh(x), \tag{37}$$

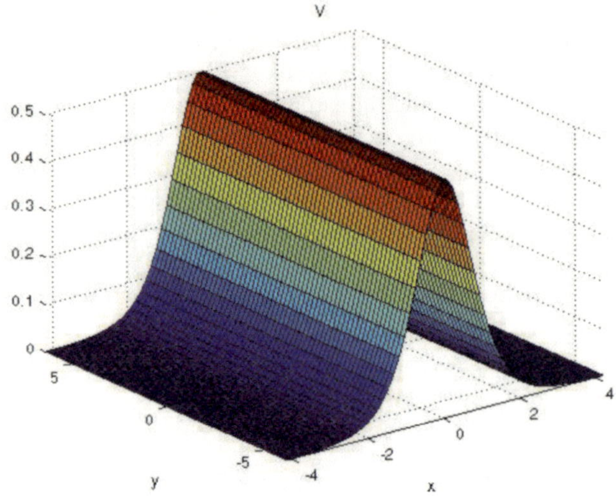

Figure 2.
Profile of potential V(*x*) *along the optical lattice.*

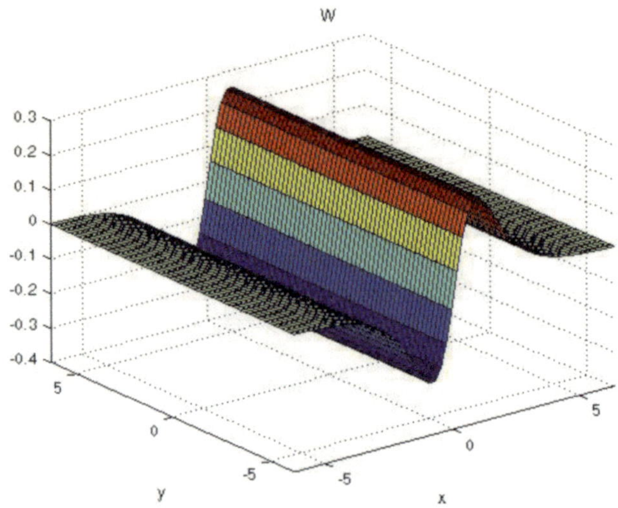

Figure 3.
Profile of potential W(*x*) *along the optical lattice.*

whose solution can be given by

$$\phi = \phi_0 \operatorname{sec}h(x) \exp\left\{i\mu \tan^{-1}[\sinh(x)]\right\}, \tag{38}$$

that represents a nonlinear mode with $\lambda = 1$, when $V_0 = 1$, $W_0 = 0.5$, and $\mu = W_0/3$. The profiles of the potentials *V(x)* and *W(x)* are shown in **Figures 2** and **3**.

5. New non-Hermitian optical systems

The application of non-Hermitian optical systems in the modeling of the optical lattices and waveguides [18] can be performed by the variational method proposed by [19]. In this approach, a generalization of the NLS equation (29)

$$i\frac{\partial u}{\partial z}+\frac{1}{2}\left(\frac{\partial^2 u}{\partial x^2}+\frac{\partial^2 u}{\partial y^2}\right)\pm F\left(|u|^2 u\right)=0, \tag{39}$$

can be obtained by a properly Lagrangian density L, introducing the action as

$$S=\int_0^z dz\int\int_{-\infty}^{\infty} L(x,y,z)dxdy. \tag{40}$$

In this approach, a mechanism to describe nonlinear and non-Hermitian optical systems can be obtained starting from an adequate action given by Eq. (40) with a Lagrangian

$$L=\frac{1}{2}(\nabla\phi)^2-V(\phi), \tag{41}$$

where ϕ is proportional to the electric field as in Eq. (32) and $V(\phi)$ is the optical potential that model the optical lattice of the system. In this sense, Eq. (33) can be written as

$$\frac{1}{2}\frac{d^2\phi}{dx^2}=\frac{dV(\phi)}{d\phi}, \tag{42}$$

and solved as follows

$$\int\frac{d\phi}{\sqrt{2V(\phi)}}=\pm(x-x_0). \tag{43}$$

On the other hand, these same solutions emerge from the energy calculation of optical system, which can be written as

$$E=\frac{1}{2}\int_{-\infty}^{\infty} dx\left[\left(\frac{d\phi}{dx}-W_\phi\right)^2+2W_\phi\frac{d\phi}{dx}\right], \tag{44}$$

where $\frac{dW(\phi)}{d\phi}=W_\phi=\sqrt{2V\phi}$.

The configuration that minimizes the energy of system, given by Eq. (19), is obtained by the following differential equation:

$$\frac{d\phi}{dx}=W_\phi. \tag{45}$$

5.1 New models and sensors

Nonlinear and non-Hermitian optical systems can be modeled now by the method based on Lagrangian. The motion equations will describe the propagation of the optical field through of a generic optical lattice.

Consider a model that is based on nonlinear and non-Hermitian optical potential

$$W_\phi=\phi^2+2ia\phi-b^2, \tag{46}$$

where a and b are optical parameters. The optical field solution, using, is

$$\phi(x) = -ia + \alpha \tanh(\alpha x), \tag{47}$$

where $\alpha = \sqrt{b^2 - a^2}$, and the energy has real spectrum that can be calculated by

$$E = \frac{4}{3}\alpha^3. \tag{48}$$

Another nonlinear and non-Hermitian optical potential is given by

$$W_\phi = \cos(\phi) + ia\sin(\phi) + b \tag{49}$$

whose solution for the optical field is

$$\phi(x) = 2\arctan\left[\frac{-ia - 2\beta\tanh(\beta x)}{b-1}\right], \tag{50}$$

where $2\beta = \sqrt{1 - (a^2 + b^2)}$, whose field configuration is plotted in **Figure 4**.

It is important to emphasize that a new class of non-Hermitian optical systems can be generated by Eq. (42) and from Eq. (43) follows that

$$\int \frac{d\phi}{\sqrt{2V(\phi)}} = x - x_0 = \int \frac{d\phi}{\sqrt{2V(\phi)}}. \tag{51}$$

In general one can rewrite the above equation simply as $\phi = f(\varphi)$ then we can try to generate a new non-Hermitian model with real energy from non-Hermitian model given by Eq. (46). In this case, we can use:

$$f(\phi) = \sinh(\phi), \tag{52}$$

getting the non-Hermitian potential

$$W(\phi) = \sinh(\phi) + 2ia\ln\left[\cosh(\phi) - 2(1+b^2)\tanh^{-1}[\tanh(\phi)]\right], \tag{53}$$

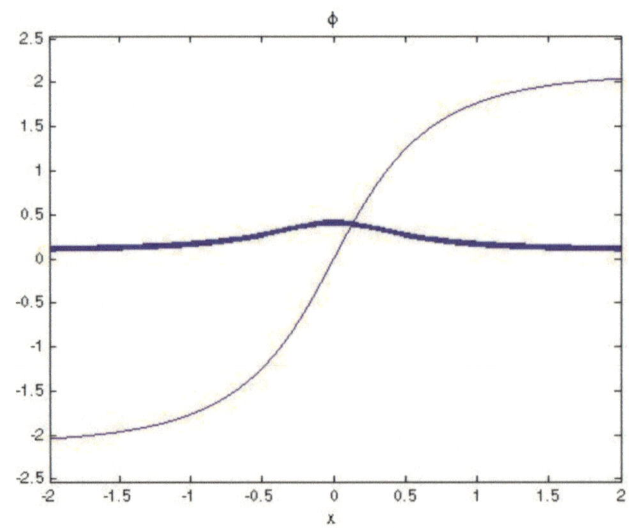

Figure 4.
The real (thin line) and imaginary (thick line) parts of optical field configurations of nonlinear and non-Hermitian optical potentials.

with the correspondent solution

$$\phi(x) = \sinh^{-1}[-ia + \alpha a \tanh(kax)]. \tag{54}$$

It is important to note that the optical field solution ϕ has a real component and an imaginary component. The real component represents the diffracted component of the optical field as shown in **Figure 4**. In our investigation, we have found that non-Hermitian optical systems can describe optical crystal structures and optical pulses propagating through such structures. We can observe that the modes of propagation, in Eq. (10), describe a particular conduction band of the optical lattice. In this sense, we can design a waveguide to act as a natural selector of certain wavelengths. In some photonic devices using sensors to optical fiber, for example, gyroscopes, optical noises from quantum scattering as the Raman effect is undesirable. So we can design a waveguide which suppresses the Raman propagation modes in the conduction band of the crystal lattice. Another important application of the modeling that we propose in this paper is that certain quantum noise from the interaction of light with matter can be treated very easily, as is the case for Raman scattering. In this case, the scattering propagation mode and its interaction with the crystal lattice can be modeled by the procedure that we have described. Thus we can simulate the propagation modes of an optical pulse by a crystalline network and mapping the crystalline structure whose information is present in the potential optical $V(x)$ and $W(x)$ of Eq. (12) by the symmetry parameters present in motion equations describing the propagation of an optical beam through an optical network. It is interesting to note that our procedure covers a wide range of nonlinear scatterings, which allow the numerical implementation of this method to recognize and model the patterns of quantum noise from the scattering of light with matter. Another interesting application of the method we have developed in soliton-based communications is that the crystal lattice can be designed such that the λ_k propagation modes can be used to adjust the bit size of the slot [14, 20]. This may be performed once for soliton train; it is possible to relate the soliton width T_0 with bit size B as

$$B = \frac{1}{T_B} = \frac{1}{2q_0 T_0}, \tag{55}$$

where T_B is the bit time slot and $2q_0 = T_B/T_0$ is the separation between neighboring soliton, as shown in **Figure 5**.

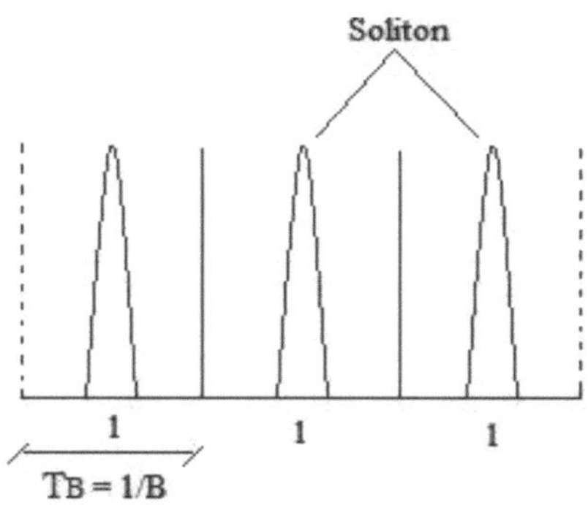

Figure 5.
Soliton bit stream. Each soliton occupying a bit slot can represent binary 1, and the absence of soliton can represent a binary 0.

6. Conclusions

In this chapter, we describe how nonlinear and quantum optical effects can be applied to the development of optical sensors. From Maxwell's equations, one can obtain an equation that describes the nonlinear behavior of an optical pulse that propagates through an optical system such as a waveguide. The nonlinear behavior of an optical pulse can be understood here as optical noise, the propagation of an optical pulse through a general optical system, or even quantum optical effects that can be properly described by a nonlinear second-order equation known as generalized nonlinear Schrödinger equation as described in Section 2. This equation has simple solutions called optical solitons. In Sections 3 and 4, we show how the term optical potential $F(|u|^2u)$ from NLS, where u describes the electric field, can model several optical systems through which a nonlinear optical pulse can propagate, generating physical information that can be transmitted and processed electronically. In Section 5, we show that nonlinear optical systems, which are optical systems capable of transmitting such nonlinear optical pulses, can be modeled simply by a first-order equation from non-Hermitian optical systems, which are systems that present a type of symmetry in spatial directions and in time, that is, $x \to -x$ and $t \to -t$. We apply these results by describing how different optical systems can be equivalent or mapped together. In principle, a particular optical system, such as an approximately one-dimensional waveguide, can be designed by reproducing the same optical effects as another waveguide. In the example, we apply these results although the beam optical profile is different in the measurement instrument data output, an oscilloscope, for example, the optical system which in this case may be a waveguide, can be mapped through its *design function* $W(\phi)$. The electric field amplitudes $\phi(x)$ can be appropriately combined to generate optical pulses in various optical systems and can be transmitted and processed from photonics and integrated optics and applied to the development of various emerging technologies such as quantum computing and quantum information.

Author details

Antônio Carlos Amaro de Faria[1,2]

1 Institute for Advanced Studies—IAS, São José dos Campos, SP, Brazil

2 Federal Technological University of Parana, Parana, Brazil

*Address all correspondence to: atoni.carlos@gmail.com

References

[1] Maxwell JC. Electrodynamics. London: Cambridge Press; 1889

[2] Taylor JT. Optical Solitons-Theory and Experiment. New York: Cambridge University Press; 1992

[3] Shen Y. Principles of Nonlinear Optics. New York: Wiley; 1984

[4] Boyd RW. Nonlinear Optics. San Diego, CA: Academic Press; 1992

[5] Agrawal GP. Applications of Nonlinear Fiber Optics. San Diego, CA: Academic Press; 2001

[6] Haus HA, Wong WS. Solitons in optical communications. Reviews of Modern Physics. 1996;**68**:423

[7] Kivshar YS, Agrawal GP. Optical Solitons from Fibers to Photonic Crystals. San Diego, CA: Academic Press; 2003

[8] Agrawal GP. Nonlinear Fiber Optics. San Diego, CA: Academic Press; 2001

[9] Born M, Wolf E. Principle of Optics. New York: Cambridge University Press; 1999

[10] Ablowitz MJ, Clarkson PA. Solitons Nonlinear Evolution Equations and Inverse Scattering. New York: Cambridge University Press; 1991

[11] Luther-Davis B, Stegman GI. Spatial Solitons. New York: Springer; 2001

[12] Hasegawa A, Tappert F. Transmission of stationary nonlinear optical pulses in dispersive dielectric fibers. Applied Physics Letters. 1973;**23**:142

[13] Mollenauer LF et al. Experimental observation of picosecond pulse narrowing and solitons in optical fibers. Physical Review Letters. 1980;**45**:1095

[14] Hasegawa H, Kodama Y. Solitons in Optical Communications. New York: Oxford University Press; 1995

[15] Mollenauer LF et al. Optical Fiber Telecommunications III. San Diego, CA: Academic Press; 1997

[16] Temprana E et al. Two-fold transmission reach enhancement enabled by transmitter-side digital backpropagation and optical frequency comb-derived information carriers. Optics Express. 2015;**23**:20774

[17] Temprana E et al. Beating the nonlinear capacity limit. Optics & Photonics News. 2016;**30**:32

[18] Musslimani ZH et al. Optical solitons in pt periodic potentials. Physical Review Letters. 2008;**100**: 030402

[19] Whitham GB. Linear and Nonlinear Waves. New York: Wiley; 1974

[20] Mollenauer LF, Smith K. Demonstration of soliton transmission over more than 4000 km in fiber with loss periodically compensated by Raman gain. Optics Letters. 1988;**13**:675

Chapter 8

Digital Sorting of Optical Vortices in Perturbed Singular Beams

Alexander Volyar, Mikhail Bretsko, Yana Akimova and Yuriy Egorov

Abstract

The chapter provides a brief overview of shaping and measuring techniques of the vortex spectra (squared amplitudes and initial phases of vortex modes) including radial indices. The main physical mechanisms causing the formation of laser beams with a complex vortex composition, in particular, in biological media, are indicated, and the need for a digital analysis of vortex spectra is substantiated. It is the analysis of vortex spectra that allows us to find the orbital angular momentum and informational entropy (Shannon's entropy) of perturbed laser beams in real time. In the main part of the chapter, we consider in detail a new approach for measuring vortex spectra without cuts and gluing of the wavefront, based on digital analyzing high-order intensity moments of complex beams and sorting the vortex beam in computer memory sells. It is shown that certain types of weak local inhomogeneities cause a vortex avalanche causing a sharp dips and bursts of the orbital angular momentum spectra and quick ups and downs of the informational entropy. An important object of analysis is also the vortex spectra of beams scattered by simple opaque obstacles such as a hole, a disk, and a sector aperture.

Keywords: optical vortex, moment's intensity, orbital angular momentum, medical optics

1. Introduction

As is well known, optical vortices [1] accompany light scattering processes due to both simple and complex medium inhomogeneities (see, e.g., [2] and references therein). Scattered light carries a huge array of information, both on the composition and structure of scattering (diffraction) centers, and on the structure of the initial light beam. At the same time, optical information can be read off both by analyzing the frequency spectrum [3–5] and the spectrum of optical vortices [6, 7]. An important aspect of this problem is the study of biomedical objects [8, 9], for example, the composition of the blood or the reflection of a vortex beam from the skin surface [10] for express diagnostics of skin diseases. The fact is that a vortex beam scattered by the skin is transformed into a speckle-like structure resembling one that occurs when light passes through a medium with weak turbulence [11, 12]. At the same time, the speckle structure is formed by the skeleton of optical vortex array [13–15]. In turn, the analysis of such a complex vortex structure is conveniently carried out on the basis of vortex fractal techniques [16] (see also [17] and references therein). There are a variety of approaches for the fractal vortex models of laser beams scattered by biological tissues [18, 19] based on the light scattering by nonspherical particles [20] that involves the representation of the wave field in terms of Legendre polynomials. However, a real optical experiment for measuring the vortex spectrum requires the use of particular approaches to the orthogonal basis for the scattered beam expansion in terms of special functions, which can differ significantly from the corresponding theoretical models.

The problem of using the properties of optical vortices in various areas of science and technology requires the development of reliable but simple techniques for measuring the spectrum of optical vortices in complex beams scattered by various objects. Therefore, the attention of many researchers is drawn to measuring the orbital angular momentum (OAM) of vortex beams [7, 21–30] that is directly related to the spectrum of optical vortices. Really, the complex amplitude $\Psi(x,y,z)$ of a composite paraxial beam can be written in the form of a superposition of the orthogonal vortex modes $\psi_m(x,y,z)$

$$\Psi(x,y,z) = \sum_{m=-N}^{N} C_m \Psi_m(x.y,z), \tag{1}$$

where $C_m = \langle \Psi | \psi_m \rangle$ stands for the mode amplitudes specified by the normalizing condition $\sum_{m=-N}^{N} |C_m|^2 = 1$, and $2N$ is a total number of vortex modes, Then the average on photon OAM of the composite beam is found by the formula [31].

$$\ell_z = \sum_{m=-N}^{N} m\,|C_m|^2. \tag{2}$$

The basis of this expression is an implicit connection between the topological charge m and the OAM ℓ_{zm} of a single mode in the beam. However, expression (2) includes the squared modulus $|C_m|^2$ of a complex number while we are talking about measuring the complex number C_m itself, i.e., its amplitude and the initial phase. The known for us approaches to measuring OAM sorting modes can be divided into four groups according to the methods used. All known for us approaches to measuring OAM spectrum and mode sorting can be conditionally divided into four groups according to the methods used and the effect on the internal beam structure. The most widely used method of holographic spatial filtering of the Laguerre-Gaussian, Hermite-Gaussian, Bessel-Gaussian, Airy-Gaussian and other mode types [22–24, 30]. At that, this method allows modes sorting, both by topological charge (azimuthal numbers m) and by radial numbers n. The holographic mode sorting resembles the effect of a "white light" decomposition into a color spectrum by means of a prism or diffraction grating. The second type of measurements requires the use of a so-called log-polar transparency and diaphragms [6, 7, 24] placed along the path of a complex beam. Such a structured transparency converts the beam into a set of horizontal or vertical fringes, the pattern analysis of which allows one to obtain the OAM spectrum. The direct application of digital processing of interference patterns of a composite vortex beam [32] and a collinear phase-shifting holography [33] can be considered as the third interference technique. However, measurements in the listed above approaches lead to complete or partial damage of the combined beam and losses of useful information. At the same time, an original approach presented in the papers [26, 27, 34] to measuring the OAM in the vortex beams with a fractional topological charge enables one to avoid the beam damage in the result of employing cylindrical lenses and analyzing the second order intensity moments [35]. Unfortunately, such a simple approach is not applicable for both sorting and measuring the spectrum of vortex modes. In [26, 27], the technique analyzes a fractional OAM [36] of the combined beam as a whole at the focal plane of the cylindrical lens without cutting and gluing the wavefront while the first two techniques distorts radically the initial beam structure. In the technique [26], the measurement of second-order intensity matrix elements was used that enables the authors to devoid breaking down the beam structure. However, it makes sense to use such an approach only if the modes of a complex beam have axial symmetry. The authors of [34] improved the technique by using two cylindrical lenses whose axes are rotated by the angle $\pi/2$ relative to each other. This made it possible to measure the fractional OAM for any type of the mode beam symmetry.

At the same time, the measurement of the fractional OAM does not give complete information about the combined beam content. As can be seen from expression (2), the same value of the OAM can correspond to a different composition of vortex modes with squared amplitudes C_m^2. For example, in a weakly turbulent medium [37] or optical fibers [38], there is an intense energy exchange between the vortex modes of the singular beam. It makes to change both the magnitude and the spectrum of the vortex modes including their initial phases. To measure the vortex spectrum, a special technique was developed based on the analysis of higher-order intensity moments [25] at the focal plane of a spherical lens. A characteristic feature of this technique is the ability to measure not only amplitudes but also the initial phases of the vortex modes. The point is that the intensity moments at the focal plane of the spherical lens are degenerate with respect to the sign of the vortex topological charge. Therefore, the authors of [25] note that the technique is applicable only for nondegenerate spectra of combined singular beams. In [39], the authors expanded the possibilities of measuring the vortex spectrum, also covering optical vortices with different signs of topological charges due to additional transformations of the Laguerre-Gauss beams (LG) into Hermite-Gauss ones (HG) via astigmatic conversions at a cylindrical lens [40]. However, the method of measuring optical vortices in LG beams of higher orders remained uncovered.

In this chapter, we consider in detail the technique of measuring the vortex spectrum based on the analysis of high-order intensity moments that excludes cuts and gluing of the beam without losses of information on initial mode phases. Unlike the method of holographic gratings that transforms a combined beam into vortex modes with different propagation directions, just as a prism converts "white" light into a spatial frequency spectrum, we will try to demonstrate how a perturbed singular beam can be sorted into vortex modes sited in computer memory cells and then reproduce the beam main characteristics: OAM, information entropy, and initial topological charge in real time. Moreover, knowing the digital spectrum, the beam can be recovered again, and by adjusting the spectral vortex amplitudes we can improve the structure of the transmitted field.

2. Preliminary remarks

As far as we know, the first theoretical and experimental studies of optical vortex arrays refer to 1991, when the authors of [41] succeeded in reproducing holographically individual letters and words due to ordering optical vortex array in typical phase skeleton on the base of the technique that had been developed back in the early 1980s [29]. However, only after the article [42] by Berry did the studies of the vortex array properties become widespread. As a result, it was shown that the diffracted beam turns into a combined beam containing a large number of optical vortices with integer topological charges. These are sometimes called the beams with fractional topological charges. Using Eq. (2), we can verify that beams with a fractional topological charge also have fractional OAM ℓ_z. However, in the process of shaping such beams, Eq. (2) requires careful use. Indeed, it is assumed that the functions Ψ_m of mode beams in Eq. (2) are normalized; otherwise, misunderstandings arise. For example, in [43] considers a Bessel beam diffracted by a holographic grating with a fractional topological charge (see the callout in **Figure 1**).

If the Bessel beam is represented in the form of a conical beam of plane waves with fractional phase bypath p, then after the inverse Fourier transform we get the non-normalized superposition of Bessel-Gauss modes, including vortices with both positive and negative topological charges:

$$\Psi = 2NGe^{ip\pi}\sin(p\pi)\sum_{m=-\infty}^{\infty} i^m J_m(KR)e^{im\varphi}/(p-m), \tag{3}$$

where $J_m(x)$ stands for a Bessel function of the first kind and m order, $R = r/w_0$, r and φ are polar coordinates, $N = \exp\left(iK^2/2kZ\right)$, $G = \exp\left(ikr^2/2Z\right)$, $Z = z - iz_0$, w_0 is a beam waist at $z = 0$, K is a scale parameter, k is a wavenumber. After the corresponding transformations, we find the OAM of the perturbed beam

$$\ell_z = \sum_{m=-\infty}^{\infty} mI_m\left(|K|^2 w_0^2/4\right)/(p-m)^2 / \sum_{m=-\infty}^{\infty} I_m\left(|K|^2 w_0^2/4\right)/(p-m)^2. \tag{4}$$

The result of the plotting is represented by the curve 1 in **Figure 1(a)**. The OAM oscillates at large values of the topological charge p. The integer values of the topological charge p correspond to sharp bursts of the OAM ℓ_z. However, a small deviation of the parameter p from the integer value causes sharp OAM dips. A completely different situation occurs if the elementary beams in Eq. (3) are normalized, as the authors of [44] do, the OAM oscillations disappear (see **Figure 1(b)** and curve 2 in **Figure 1(c)**). In this case, OAM $\ell_z(p)$ obeys a simple relation (see also [45])

$$\ell_z = p - \sin(2\pi p)/\pi. \tag{5}$$

The OAM oscillations disappear (see **Figure 1(c)**, curve 2). As the authors of [45] revealed, a gradual increase in OAM is observed only at small values of the fractional topological charge. In fact, we are dealing with different beams, although the basis for their shaping is the same physical process. As we will see later, the

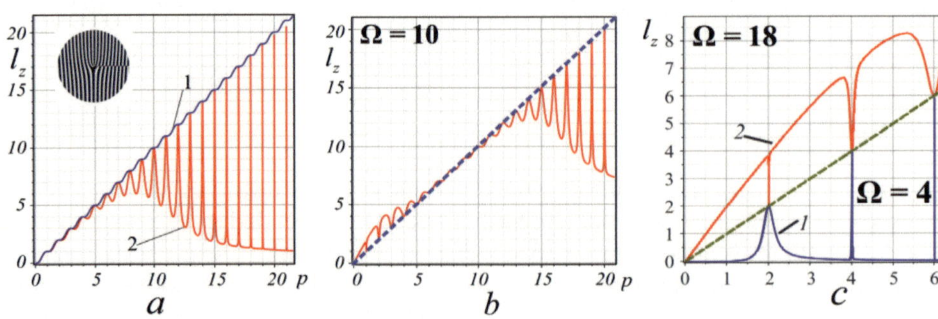

Figure 1.
(a) Dependence of the OAM ℓ_z on the fractional topological charge p: curve 1 is calculated according to Eq. (3) with $K = 2 \cdot 10^4 m^{-1}$, curve 2 presents Eq. (5), leader—diffraction grating for $p = 5/2$. (b and c) Computer modeling of the OAM curves $\ell_z(m)$ related with combined beams in Eq. (4): (b) total value of the OAM curves at $\Omega = 10$; (c) the characteristic spectral bursts of OAM; curve 1—the OAM bursts ($\Omega = 0.4$), curve 2—dips of the OAM ($\Omega = 14$).

choice of the normalization of vortex modes endows the combined beam with special properties.

3. Theoretical background of the digital vortex sorting and experimental results

3.1 Nondegenerate case

For a simplest model for our studies, we choose a scalar field of wave in the form of superposition LG_n^m of the lowest order with the N monochromatic beams, where $m > 0$ and $n = 0$ are the azimuthal and radial indices, respectively, so that all the vortex modes in superposition have the same waist radius w_0 at $z = 0$. Let us consider the wave field in the waist plane $z = 0$ in the form [25].

$$\Psi(r, \varphi, z = 0) = \sum_{m=0}^{N-1} C_m LG_{n=0}^{m} = \sum_{m=0}^{N-1} C_m r^m e^{i(m\varphi+\beta_m)} G(r, z)/N_m, \tag{6}$$

where $N_m = \sqrt{2^{-m-1} m!\pi}$ stands for the normalization factor, C_m is the mode amplitude, β_m is the initial mode phase. We consider the nondegenerate case of the field representation$(m \geq 0 \text{ or} m \leq 0)$ due to the axial symmetry; combined beams with different topological signs $\pm m$ are indistinguishable in intensity moments.

Our goal is to analyze the distribution of the wave field intensity $\Im(r, \varphi, z = 0)$ in such a way as to express the squared amplitudes and the initial phases of the vortex modes in terms of physically measured quantities in the region of the beam waist. For this, we make use of the intensity moments approach [35] in the form

$$J_{p,q} = \iint_S M_{p,q}(r, \varphi)\Im(r, \varphi) r\,dr\,d\varphi \Big/ \iint_S \Im(r, \varphi) r\,dr\,d\varphi, \tag{7}$$

where the intensity distribution is written as

$$\Im = \Psi^*\Psi = \sum_{m=0}^{N-1} \frac{C_m^2}{N_m^2} r^{2m} G(r)^2 + 2 \sum_{\substack{m, m'=0, \\ m > m'}}^{N-1} \frac{C_{m'} C_m}{N_m N_{m'}} r^{m+m'} \cos\left[(m - m')\varphi\right] \cos\beta_{m,m'} G(r)^2$$
$$-2 \sum_{\substack{m, m'=0, \\ m > m'}}^{N-1} \frac{C_{m'} C_m}{N_m N_{m'}} r^{m+m'} \sin\left[(m - m')\varphi\right] \sin\beta_{m,m'} G(r)^2, \tag{8}$$

$M_{p,q}(r, \varphi)$ stands for the intensity moments function, $p, q = 0, 1, 2, \ldots$. Now the problem is to choose a combination of intensity moments $J_{p,q}$ in such a way as to exclude all terms of the last two sums and leave only the first one in Eq. (7). The equations for the squared amplitudes and the initial phases are separated if the moment function is written as $M_p(r, \varphi) = r^p$. Then the equations for the squared amplitudes take the form

$$\sum_{m=0}^{N-1} \frac{(N-1+m)!}{m!} C_m^2 = \sum_{j=0}^{p} \binom{p}{j} J_{2j,2(p-j)}, p \leq N-2,$$
$$\sum_{m=0}^{N-1} \frac{m+1/2}{4\sqrt{2}m!} \Gamma(m+1/2)\, C_m^2 = \sqrt{J_{2,0}+J_{0,2}}. \tag{9}$$

The right-hand sides of the linear Eq. (9) are easy to calculate, and the left-hand sides are directly measured experimentally using the Eq. (7) that gives the desired values of the squared amplitudes C_m^2.

As intensity of array of the beams depends only on the difference of phase between two pairs of modes, in computation of the initial phases we will assume that one of beams phase is given (let us say $\beta_0 = 0$). Now it is important to select the combination of moments of intensity $J_{p,q}$ that so either the second or the third sum in Eq. (2) is considered in the calculation. It is natural to calculate not all differences of phase $\beta_{m,m'}$, but only $\beta_{0,m}$. However, we did not find such combinations $J_{p,q}$ that could filter out only terms with $\beta_{0,m}$. Generally speaking, to calculate the phase difference $\beta_{m,m'}$ it is required to find the M equations for variables $X_{m,m'} = C_m C_{m'} \cos\beta_{m m'}$ and $Y_{m,m'} = C_m C_{m'} \sin\beta_{m,m'}$ in Eq. (2) the second and third sum, number of which is equal to the number of 2-combinations of N elements. It turns out that the number of equations can be significantly decreased if using the moments $J_{2(p+1),1}$ for $X_{m,m'}$ variables and $J_{1,2(p+1)}$ for $Y_{m.m'}$ variables. The system of linear equations for the phase difference are written in the form

$$J_{2(p+1),1} = \frac{1}{2^{2(p+1)}} \sum_{m=0,}^{N-1} \sum_{k=0}^{p} \binom{2p}{k} \left[\frac{M_{m,n_1}}{N_n N_{n_1}} X_{m,n_1} + \frac{M_{m,n_2}}{N_m N_{n_2}} X_{m,n_2} \right], \tag{10}$$

$$J_{1,2(p+1)} = \frac{1}{2^{2(p+1)}} \sum_{m=0,}^{N-1} \sum_{k=0}^{p} (-1)^{p-k} \binom{2p}{k} \left[\frac{M_{m,n_1}}{N_m N_{n_1}} Y_{m,n_1} + \frac{M_{m,n_2}}{N_m N_{n_2}} Y_{m,n_2} \right]. \tag{11}$$

where $n_1 = |m-2(p-k) \pm 3|, n_2 = |m-2(p-k) \pm 1|, M_{m,n_{1,2}} = \int_0^\infty r^{m+n_{1,2}+1} G^2 dr$. The number of linear equations in each system Eqs. (10) and (11) is $K = 3(N-3),\ N \geq 6$. It is noteworthy that Eqs. (10) and (11) contain only the terms with an odd difference of indices, including $\beta_{0,m}$, so that a finite solution enables us to obtain all phases of the partial beams in the form $\tan\beta_{m,m'} = Y_{m,m'}/X_{m,m'}$.

A key element of the experimental setup in **Figure 2(a)** was a spatial light modulator SLM (Thorlabs EXULUS-4K1), which converted the fundamental TEM_{00} mode of the He-Ne laser (wavelength $\lambda = 0.6328\ mcm$ and power 1 mW) into a composite beam with mode amplitudes C_m and initial phases β_m. To make this possible, the laser beam was additionally filtered by the system FF. The beam splitter BS formed two working arms of the experimental setup. The beam in the first arm projected by a spherical lens onto the input pupil of the CCD1 camera (Thorlabs DCC1645) is subjected to the image computer processing. The result of the computer processing is a digitization of the beam intensity distribution and calculations of high-order intensity moments J_{pq}. Additional computer software made it possible to compose a system of linear Eqs. (9)–(11) and calculate both the mode amplitudes C_m and their initial phases β_m in real time. The second arm was in use for the control measurement of the OAM of the composite beam. For this purpose, the beam was focused by a cylindrical lens CL onto the input pupil of the

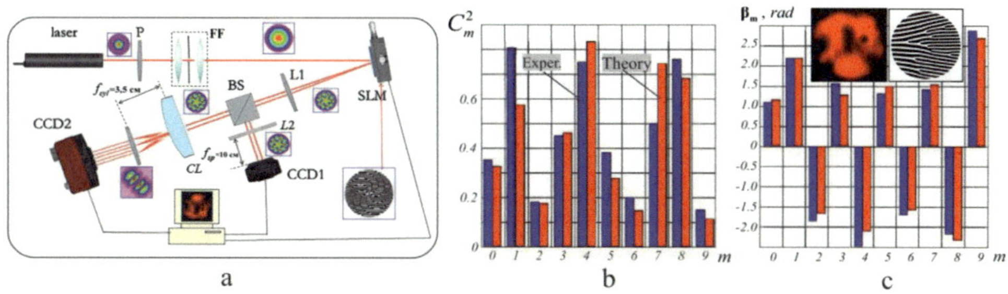

Figure 2.
(a) Model of experimental setup for real-time vortex measurement and value of the OAM, P—polarizer, FF—light filter of space, SLM—space light modulator, L1, L2—spherical lenses with focal length f_{sh}, BS—beam splitter, CL—cylindrical lens with a focal length f_{cyl}, CCD1,2—CCD-color cameras; (b and c) the vortex spectrum of the beam array with $N = 10$: (b) $C_m^2(m)$, (c) $\beta_m(m)$ the OAM: theory $\ell_z = 0.39$, experiment $\ell_z = 0.42$; callout: the intensity distribution and corresponding diffraction grating.

second CCD2 camera (Thorlabs DCC1645) then the second-order intensity moment J_{xy} was measured and the average OAM per photon was calculated in accordance with the technique described in detail in the paper [26].

Before proceeding to the current measurements, it was necessary to adjust the experimental setup. For this, a number of calibration measurements were carried out. The digitization of the intensity distribution $\Im(x, y)$ required normalization of the transverse coordinates x, y in units of the Gaussian beam waist at the plane of the CCD camera. To measure the initial beam waist radius w_0, the second-order intensity moments J_{12} were used in accordance with the beam quality measurement method [46]. In order to ensure calibration measurements, a special computer program was developed that assigned random values of the amplitudes and initial phases of the vortex modes in Eq. (1) due to the random-number generator. Then, an appropriate diffraction grating was formed on the liquid crystal element of the SLM modulator and the calibration beam was restored. A typical intensity distribution of the calibration beam is shown in the callout of **Figure 2(c)**. The corresponding spectra of the squared amplitudes $|C_m|^2$ and initial phases β_m are presented in **Figure 2(b)** and **(c)**. Red and blue colors in **Figure 2(a)** and **(b)** present experimental and theoretical data, respectively. Note that the OAM measured in the second arm was $\ell_z = 4.2$ and the first arm calculated according to Eq. (2) was $\ell_z = 4.5$, while the theory gives $\ell_z = 4.9$. The measurement error for $N = 10$ beams did not exceed 3–4% for amplitudes and 5–6% for phases. The most interesting effect is observed in the region of OAM oscillations in **Figure 1(a)** with a characteristic perturbation of the holographic grating in **Figure 2(c)**. This effect is accompanied by a sharp restructuring of the vortex spectrum. For small perturbations of the singular beam, additional vortex modes arise only near the initial value of the topological charge $m = M$ (see [47]). However, in general case of the hologram perturbation, vortex modes with opposite signs of topological charges appear. Such a vortex spectrum reconstruction cannot be detected using the above approach. It is required to expand the measurement technique.

3.2 Degenerate case and vortex avalanche

We consider the case when the initial LG beam with a zero radial index $n = 0, m \neq 0$ is subjected to local perturbation at the central region of the holographic grating. As a result of the perturbation, vortex modes with various types of topological charges appear in the beam. The method of the beam expansion in a series over the vortex modes depends on the type of the grating perturbation. We suppose (and then prove) that the chosen perturbation does not excite LG modes with radial indices $n = 0$, but only modes with different topological charges m appear [39]. Then we write the complex amplitude of the perturbed beam in the form

$$\Psi(R,\varphi,z) = \sum_{m=-N}^{N} \frac{C_m R^{|m|} e^{im\,\varphi} G(R)}{M_m \sigma^{|m|+1}(z)}, \tag{12}$$

where M_m is a normalized factor and $\sigma(z) = 1 - iz/z_0$.

It is important to notice the intensity moments $J_{p.q}$ are degenerate with respect to the sign of the vortex topological charge m in the case of axial symmetry of the modes in Eq. (12). To remove degeneracy of the intensity moments, it is important to change symmetry of singular beams, but without destroying of structure beams so that the vortex modes with opposite topological charges in sign have distinct geometric contours. Such a requirement provides by the astigmatic transformation with a cylindrical lens of the Laguerre-Gauss beam, which was considered detail in [34] (the general theory of paraxial astigmatic transformations can be found in [41]). We suppose that the cylindrical lens with a focal length f is located at the plane $z = 0$. A paraxial beam with complex amplitude (12) is projected at the lens input. The plane of the beam waist is matched with the plane of the lens. Besides, we restrict our study to zero initial phases of mode beams, i.e., amplitudes C_m are real values. Then, following [34], we write the complex amplitude of a combined beam at the wave diffraction zone as

$$\Psi_{CL}(x,y,z) = \sum_{m=-N}^{N} A_m H_m(F) \exp(\Phi), \tag{13}$$

where $A_m = -i\frac{z_0}{z}\frac{1}{qq_0}\left(-\frac{w_0}{q_0}\right)^m\left(1-\frac{q_0^2}{q^2}\right)$, $\Phi = i\frac{k}{2z}(x^2+y^2) - \left(\frac{z_0}{zw_0q}\right)^2 x^2 - \left(\frac{z_0}{zw_0q_0}\right)^2 y^2$, $F = \frac{z_0}{w_0z}\frac{i\frac{q_0}{q}x - \frac{q}{q_0}y}{\sqrt{q^2-q_0^2}}$, $q_0^2 = 1 - i\frac{z_0}{z}, q^2 = 1 + i\frac{z_0}{z_1}, z_1 = \frac{zf}{z-f}$..

Let us consider new normalized coordinates so that the beam axes are directed by an angle $\pi/4$ relative to the axes of the cylindrical lens: $u = (x+y)/w_0, v = (x-y)/w_0$, we will consider field of paraxial beam in double focus plane $z = 2f$, and also require that $z_0/2f = 1$. Then distribution of the beam intensity is written as

$$\Im = |\Psi_{CL}|^2 = \left\{ \sum_{m,n=0}^{N} C_{-n}C_{-m}\frac{A_{m,n}}{N_{m,n}} H_n\left(\frac{u}{\sqrt{2}}\right) H_m\left(\frac{u}{\sqrt{2}}\right) + \sum_{m,n=1}^{N} C_n C_m \frac{A_{m,n}}{N_{m,n}} H_n\left(\frac{v}{\sqrt{2}}\right) H_m\left(\frac{v}{\sqrt{2}}\right) \right.$$
$$\left. + \sum_{\substack{n=0,\\ m=1}}^{N} C_{-n}C_m \frac{A_{m,n}}{N_{m,n}} H_n\left(\frac{u}{\sqrt{2}}\right) H_m\left(\frac{v}{\sqrt{2}}\right) + \sum_{\substack{n=1,\\ m=0}}^{N} C_n C_{-m}\frac{A_{m,n}}{N_{m,n}} H_n\left(\frac{v}{\sqrt{2}}\right) H_m\left(\frac{u}{\sqrt{2}}\right) \right\} \exp\left(-\frac{u^2+v^2}{2}\right) \tag{14}$$

where $H_n(x)$ is Hermite polynomials, $A_{m,n} = (-1)^{|n|+|m|} e^{i\frac{|m|-|n|}{4}\pi}$, $N_{n,m}^2 = \frac{\pi}{4}\left(\frac{w_o}{2}\right)^{n+m} 2^{n+m+1} n!m!$. As can be seen from Eq. (14), the terms corresponding to optical vortices with positive and negative topological charges are only partially divided. The last two terms characterize the cross terms $C_{-n}C_m$ and C_nC_{-m}. Besides, the first two terms also contain cross amplitudes $C_{-n}C_{-m}$ and $C_nC_m(|n| \neq |m|)$. On the other hand, our task is to measure only the squared amplitudes C_m^2. It is important to note, that the factors $A_{m,n} = -A_{n,m} = i$ with the difference of the indices $|m-n|=2$ in Eq. (14) are imaginary, therefore the cross terms of the beam intensity distribution with such indices difference disappear.

It seems that for solving this problem it is reasonable to choose intensity moment functions in the form of Hermite polynomials Eq. (14) as $M_{p,0}(u)=H_p\left(u/\sqrt{2}\right)$ and $M_{0,q}(v)=H_q\left(v/\sqrt{2}\right)$. Next, one can use the orthogonality condition of the HG beams and write a system of linear equations, the number of which is equal to the number of variables, as was done above for a non-degenerate vortex array. However, as shown by the assessed computations, as the indices p and q increase along with the $2N$ variables C_m^2, the $2N-6$ cross terms $C_{\pm n}C_{\pm m}$ and $C_{\pm n}C_{\mp m}$ are also added. The system of Eq. (14) cannot be closed. However, one can act otherwise. First, we make two Fourier transforms of the intensity distribution $\Im(x,y)$ in the form

$$\Im(\xi)=\int\limits_{-\infty}^{\infty}\int\limits_{-\infty}^{\infty}\cos\left(\xi\frac{u}{\sqrt{2}}\right)\Im(x.y)\,dx\,dy,\quad \Im(\eta)=\int\limits_{-\infty}^{\infty}\int\limits_{-\infty}^{\infty}\cos\left(\eta\frac{v}{\sqrt{2}}\right)\Im(x.y)\,dx\,dy. \tag{15}$$

In particular, the terms $\Im_{-m.-n}$ in the intensity distribution $\Im$ are

$$\begin{aligned}\Im_{-m.-n}(\xi)&=\int\limits_{-\infty}^{\infty}\int\limits_{-\infty}^{\infty}\cos\left(\xi\frac{u}{\sqrt{2}}\right)\Im(x.y)\,dx\,dy=\sum_{n=0}^{N}C_{-n}^2\frac{2^{n+1/2}}{N_{n,n}}\xi^{2n}L_n\left(\frac{\xi^2}{2}\right)e^{-\frac{\xi^2}{4}}\\&+\sum_{\substack{m,n=0,\\|m-n|\neq 0,2}}^{N}C_{-n}C_{-(n+2m)}\frac{(-1)^{n+2m}2^{n+1/2}}{N_{m,n}}\xi^{2m}L_n^{2m}\left(\frac{\xi^2}{2}\right)e^{-\frac{\xi^2}{4}},\end{aligned} \tag{16}$$

Thus, after the Fourier transform the distribution of intensity $\Im(\xi,\eta)$ contains all the squared amplitudes, and, also, the cross terms with the difference $|m|-|n|=\pm 4$. The cross terms $\Im_{-m,m}$ and $\Im_{m,-n}$ refer to vortices with positive and negative charges. Besides, the terms $\Im_{-m,-(n+2m)}(\xi)$ includes amplitudes C_{-n}^2 with the similar coefficients, and the member $\Im_{m,(n+2m)}(\xi)$ includes amplitudes C_n^2 with only different coefficients. The expressions are received for the terms, that $\Im_{-m,-(n+2m)}(\eta)$ and $\Im_{m,(n+2m)}(\eta)$, if we swap the signs of the indices.

The equations for squared amplitudes $C_{\pm n}^2$ will be found, if we will make use the full basis of the Laguerre-Gaussian modes for expansion of the intensity moments:

$$J_{p,2q}^{(+)}=\int\limits_{-\infty}^{\infty}\xi L_p^{2q}\left(\frac{\xi^2}{2}\right)\Im(\xi)\,d\xi/J_{00},\quad J_{p,2q}^{(-)}=\int\limits_{-\infty}^{\infty}\eta L_p^{2q}\left(\frac{\eta^2}{2}\right)\Im(\eta)\,d\eta/J_{00}, \tag{17}$$

where J_{00} is a total beam intensity and $L_p^q(x)$ is Laguerre polynomials. Then, using [39], we find a system of linear equations

$$J_{p,q}^{\mp}=\left(\sum_{n=1}^{N}C_{\mp n}^2+\frac{(2q+p)!(1/2)^p}{p!4^{-(2q+p+1)}}\sum_{m,n=1}^{N}\frac{(2m+n)!(1/2)^n}{n!4^{-(2m+n+1)}N_{nm}}C_{\pm n}C_{\mp m}\right)/J_{00}, \tag{18}$$

The system of Eq. (18) enables us to find the values of the squared amplitudes via the measured values of the intensity moments $J_{p,q}^{\pm}$. Each equation contains $2N$ terms for amplitudes C_m^2 and $N-3$ cross terms C_mC_n. Increasing the value of the indices p and q does not change the number of variables, so the system of Eq. (18) can be closed. Note that each of the Eq. (18) can be solved independently of each other and find the amplitudes C_m^2 and C_{-m}^2. The total number of linear equations is relative to signs $(\pm)$, so that the systems can be easy solved for finding the squared amplitudes in real time using modern computer software.

For measurements, the experimental setup shown in **Figure 2(a)** was used. However, in contrast to the non-degenerate case, the main experimental data were measured in the second arm, provided that the detection plane was located at the double focus region of the cylindrical lens, while the first arm was used for general adjustment [39].

We will implement the above technique for a new type of combined beams with conspicuous dips and bursts in the OAM spectrum similar to that shown in **Figure 1(a)** and **(b)**. Let us write the complex amplitude of the combined vortex beam in the form

$$\Psi(r,\varphi,z) = \sin(p\pi)e^{iP\pi}\sum_{m=-N}^{N}\begin{pmatrix}\cos(m\pi/2)\\ \sin(m\pi/2)\end{pmatrix}\frac{\Omega^{|m|}}{|m|!(M-m+\delta p)}\frac{R^{|m|}e^{im\varphi}}{\sigma(z)^{|m|+1}}\exp\left(-R^2\right), \tag{19}$$

where Ω is a scale parameter, δp stands for the perturbation of the holographic grating responsible for shaping the beam, M is a topological charge of the non-perturbed forked grating. Factors $\cos(m\pi/2)$ or $\sin(m\pi/2)$ thin out the vortex spectrum remaining the modes with either even or odd $m-$ indices. This makes the dips in the OAM spectrum more distinct. The complex amplitude (12) corresponds to the OAM in the form

$$\ell_z = \sum_{m=-N}^{N} m\begin{pmatrix}\cos^2(m\pi/2)\\ \sin^2(m\pi/2)\end{pmatrix}\frac{\Omega^{2|m|}}{|m|!^2(M-m+\delta p)^2}\Big/\sum_{m=-N}^{N}\begin{pmatrix}\cos^2(m\pi/2)\\ \sin^2(m\pi/2)\end{pmatrix}\frac{\Omega^{2|m|}}{|m|!^2(M-m+\delta p)^2}. \tag{20}$$

Figure 1(b) and **(c)** displays the case of even vortices beams in the curves of $\ell_z(p)$. The spectrum of OAM in **Figure 1(b)** characterizes the general view of resonant bursts and dips. The Ω parameter of scale admits displace along the axis of the resonant regions $p = M + \delta p$. **Figure 1(c)** (curves 1 and 2) describes their characteristic features. In the even vortex modes of integer topological charge are located resonant dips. Swaps of the cosine with the sine in Eq. (19) displaces along the axis p of the resonant regions by one. Increase of the topological charge M is accompanied by resonances compression, until their depth gets become approximately equal to M at $\Omega = 18$ and $p < 6$. For large parameters $\Omega >> 20$, the OAM is doubled $\ell_z = 2M$ in range of the initial topological charges M. As was shown in [47, 48], a perturbation δp causes a local distortion of the forked grating defect for the large values parameter p shown in **Figure 3(a)**.

As the perturbation grows the area of the hologram defect increases (**Figure 3(b)**). Small variations in the holographic grating structure lead to a cardinal reconstruction of the vortex spectrum (**Figure 4(a)**). Weak perturbations $\delta p \sim 10^{-2}$ of the grating cause an avalanche of optical vortices. The satellites appear in the OAM spectrum, the maxima of which has topological charges less than the initial M. The energy of the basic mode M flows partially into the spectral satellites, that are generated both in the region of positive $m > 0$ and negative $m > 0$ topological charges.

For large perturbations $\delta p \sim 0.1$, a vortex avalanche picks almost all the energy out of the basic mode. The vortex avalanche at small deviations δp of the topological

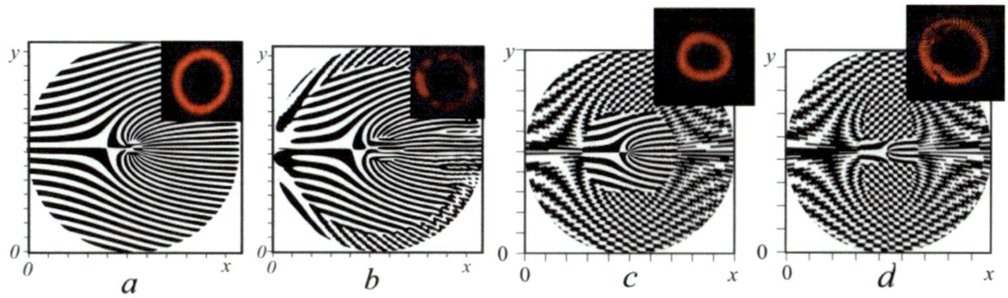

Figure 3.
Forked defect perturbations of the holographic gratings with (a and b) $M = 20$, $\Omega = 2.1$, (c and d) $M = 4$, $\Omega = 14$: (a) $\delta p = 10^{-3}$, (b) $\delta p = 0.1$, (c) $\delta p = -10^{-3}$ (d) $\delta p = -0.5$, callouts: intensity distributions of the beams restored by the perturbed holograms.

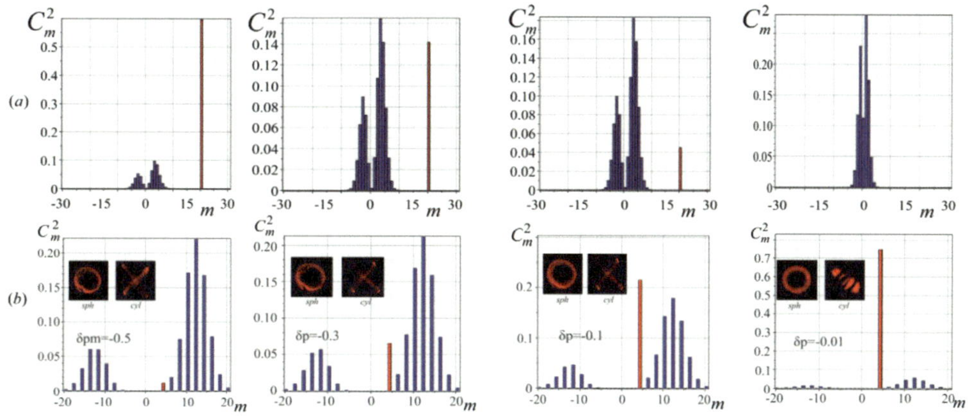

Figure 4.
Experimental vortex spectra $C_m^2(m)$ for perturbed singular beams with (a) $M = 20$, $\Omega = 3.5$ and (b) $\Omega = 18, M = 4$; callouts: intensity distributions at the beam waist after spherical (sph) with focal length $f = 0.1$ m and cylindrical (cyl) lens with $f_{cyl} = 0.5$ m; red color—an initial mode.

charge from integer values M immediately causes a resonance dip in the OAM spectrum (**Figure 5(a)**). With increasing M the depth of the dip decreases while its shape is distorted (see **Figure 5(b)**).

Absolutely other situation arises when the holographic grating with $M > 10$ and a forked defect is subjected to a small perturbation. As shown in **Figure 3(c)**, even a weak perturbation $\delta p = -0.001$ changes drastically the entire relief of the holographic grating although the intensity distribution. In the result, even small differences in the grating relief affects inevitably the shape of the vortex spectrum

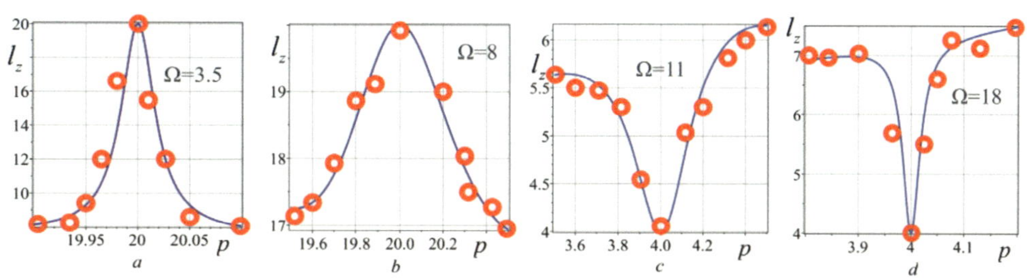

Figure 5.
Dips and bursts in the OAM spectrum $\ell_z(p)$ at various beam parameters Ω and M, $p = M + \delta p$: (a and b) $M = 20$, (c and d) $M = 4$; solid lines—theory, circles—experiment.

$|C_m(m)|^2$ and $\Omega > 10$ induce the vortex avalanche (see **Figure 4(b)**). However, the avalanche type differs significantly from that in the previous case. Indeed, as well as in the above case, the perturbation provokes appearing resonance satellites with both positive and negative topological charges (TC). However, their values are always greater than those of the initial TC. In this case, energy is also pumped from the basic mode to the satellites but with higher topological charges under the condition of the resonance $m = p$. In the OAM spectrum, it looks like the OAM burst at the resonance as shown in **Figure 5(c)** and **(d)** but the maxima ℓ_z are at the region of fractional values p. As the scale parameter increases, the height of the flash decreases, and the resonance circuit broadens (**Figure 5(d)**).

Further increasing the parameter p ($\Omega = const$) is accompanied by smoothing the curve $\ell_z(p)$ so that $\ell_z|_{\max} \to 2p$ except for the OAM dips at the integer values $m = p$.

4. OAM, informational entropy and topological charge of truncated vortex beams

A special case is represented by natural changes of the vortex spectrum due to external influences on the laser beam (diffraction by opaque obstacles, external interference in the beam, etc.). The simplest opaque obstacles are the sector, circular and annular apertures where we stop our attention using a simple technique of the vortex digital sorting.

4.1 Sector aperture

Let us consider propagation a scalar beam of Laguerre-Gaussian LG_0^m with a zero radial index $p = 0$ and an azimuthal index (a topological charge) m through the rough regular sector aperture obstacle, that an angle shown in **Figure 1**. The edge of sector touches axis of the beam. The field of beam at the initial plane $z = 0$ can be represented as [49].

$$\Psi_m(r, \varphi, \alpha) = (\rho/w)^{|m|} e^{i m \varphi} e^{-\rho^2/w^2} = r^{|m|} e^{i m \varphi} e^{-r^2}, \quad \alpha < \varphi < 2\pi - \alpha, \tag{21}$$

where $r = \rho/w$. w is a beam waist radius at the plane $z = 0$, ρ and φ represent polar coordinates. We rewrite the beam field (1) formed by the rigid-edges aperture with the half angle α as a sum of non-normalized Laguerre-Gauss beams LG_p^m in the form

$$\Psi_m(r, \varphi, \alpha) = \sum_{n=-\infty}^{\infty} C_{m,n}(\alpha) LG_0^n(r, \varphi) = \sum_{n=-\infty}^{\infty} C_{m,n}(\alpha) r^{|n|} e^{i n \varphi} e^{-r^2}, \tag{22}$$

where the beam amplitudes are

$$C_{m,n}(\alpha) = (-1)^{m-n} \Gamma\left(\frac{|m| + |n|}{2} + 1\right) \frac{\sin[(m-n)(\pi - \alpha)]}{m - n} \Big/ \left(\pi 2^{\frac{|m|-|n|}{2}} |n|!\right), \tag{23}$$

where $\Gamma(x)$ is a Gamma function. The terms in the series (23) with radial indices $p \neq 0$ disappear due to orthogonality of the LG modes.

The complex amplitude (22) can describe the spatial evolution of the perturbed beam if we replace $r \to r/\sigma$ and multiply the sum by σ where $\sigma = 1 - iz/z_0$.

Figure 6(b) and **(c)** shows a computer simulation (**Figure 6(b)**) and experiment (**Figure 6(c)**) of the intensity distribution $\Im(r, \varphi, z) = \Psi\Psi^*$ along the vortex beam length with $m = 15$ modulated by the sector aperture with $\alpha = \pi/4$. It is important to note, that the cut out dark sector from the beam intensity distribution does not lead to its self-healing. When the beam propagates, the dark sector rotates synchronously. We showed, both theoretically and experimentally, that beam self-healing does not occur at any beam length z even at a very small sector angle α.

4.1.1 The vortex spectrum

We consider vortex beams spectra for topological charge that most clearly reflect properties of the sector perturbation. Computer simulation and experimental data of typical vortex spectra is shown in **Figure 7**. We revealed a clear maximum of vortex mode intensity $C^2_{m,n}$ for the initial topological charges $n = m = 5$ with $\alpha = 45°$. The short part of the beam intensity redistributed symmetric at the neighboring vortices orders. But, if we increase the angle α, we see a violation of the symmetric distribution of intensity among the vortex modes. There formed a second maximum in the spectrum range of negative topological charges $n < 0$. It is important to note, that the authors of Ref. [50], taking into consideration the optical uncertainty principle, also plotted the vortex spectrum for a topologically neutral beam $m = 0$ and a beam with a small topological charge $m = 2$ at the angle of aperture $\alpha = 45°$. The authors did not detect a maximum of spectra in the negative region of topological charges. As we presented above, the emergence of the second

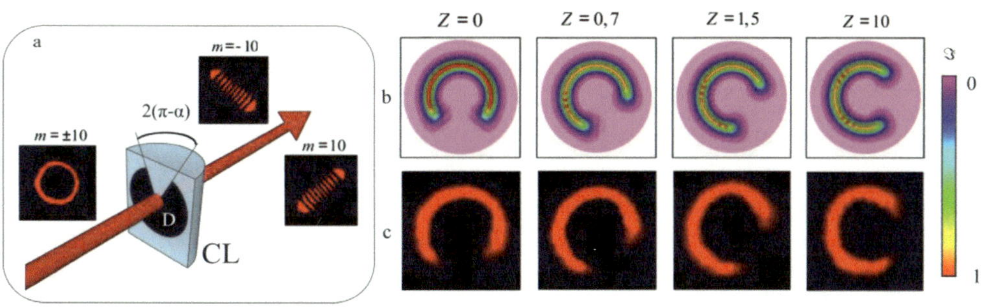

Figure 6.
(a) Sketch of the hard-edged aperture (D) installed at the cylindrical lens plane (CL). Images LG and HG beams illustrate the astigmatic transformation of a single vortex beam. (b and c) Intensity distribution $\Im(r, \varphi, z)$ of the vortex beam with $m = 15$ along length $Z = z/z_0$ perturbed by a hard-edged sector diaphragm with $\alpha = \pi/4$: (b) theory, (c) experiment.

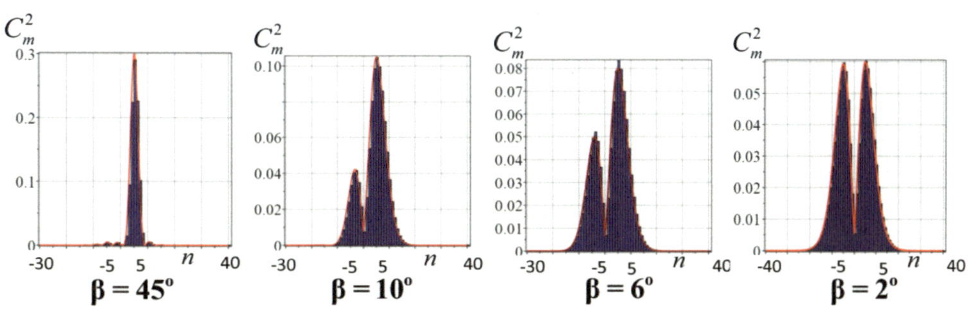

Figure 7.
Distribution of vortex beams spectrum $C^2_{m,n}$ with charges $m = 5$ perturbed by the hard-edged aperture with the sector adjacent angle $\beta = \pi - \alpha$, the solid line is spectrum outline.

intensity maximum is possible only at sufficiently large angles of overlap of beam α and also large topological charges m of the initial singular beam.

4.1.2 The orbital angular momentum

The OAM per photon of a complex perturbed beam can be calculated in accordance with Eq. (23). The mode amplitudes are given by the normalized field Ψ_m of the Laguerre-Gauss beams so that the squared amplitudes $C^2_{m,n}$ and $\overline{C}^2_{m,n}$ are obeyed a simple relation $\overline{C}^2_{m,n} = 2^{-|n|-2}|n|!C^2_{m,n}$. The changes of OAM $\ell_z(\alpha, m)$ after increasing the aperture angle α (decreasing the adjacent angle $\beta = \pi - \alpha$) is presented by **Figure 7(a)** for topological charges $m = 5, m = 10$ and $m = 15$ of the initial vortex beam.

The OAM is practically unchanged a wide range of angles and remains almost equal to the initial OAM, despite the rapid increase in the number of vortex states (see **Figure 8(a)**). After the second spectral maximum is formed in the negative region of topological charges **Figure 7**, there is a sharp decrease of the OAM. The OAM is equal to zero already at the angle $\beta \approx 2°$.

4.1.3 Informational entropy (Shannon's entropy)

The normalized squared amplitude $C^2_n \in (0, 1)$ in the Eq. (23) can be treated as a conditional probability $P(n/m)$ of finding out a vortex beam in the state $|n\rangle$ among of $2N$ states (see, e.g., [51] and references therein). This approach to counting a number of vortex states can be used for the Shannon Entropy [51] and is written as

$$H_I = -\sum_{n=0}^{N} P(n/m)\log_2 P(n/m) = -\sum_{n=0}^{N} C^2_n(\alpha, m)\log_2 C^2_n(\alpha, m) > 0. \qquad (24)$$

The Shannon entropy (24) characterizes the amount of uncertainty (randomness) that arises when a perturbation acts on a vortex beam. For example, in the case of the same amplitudes of the mode beams, the Shannon entropy of $H_I = 2$ means that to remove the uncertainty it is necessary to expend 1 bit of information.

In **Figure 8(b)** presents the dependence of the Shannon Entropy H_I on the angle α for various of number topological charges m. In a broad range of aperture angles $0 < \alpha < 7\pi/8$, the informational entropy H_I increases equally for any topological charges m while the OAM ℓ_z does not change (see **Figure 8(a)**) which corresponds to the same changes in the spectrum of vortex states in **Figure 7**.

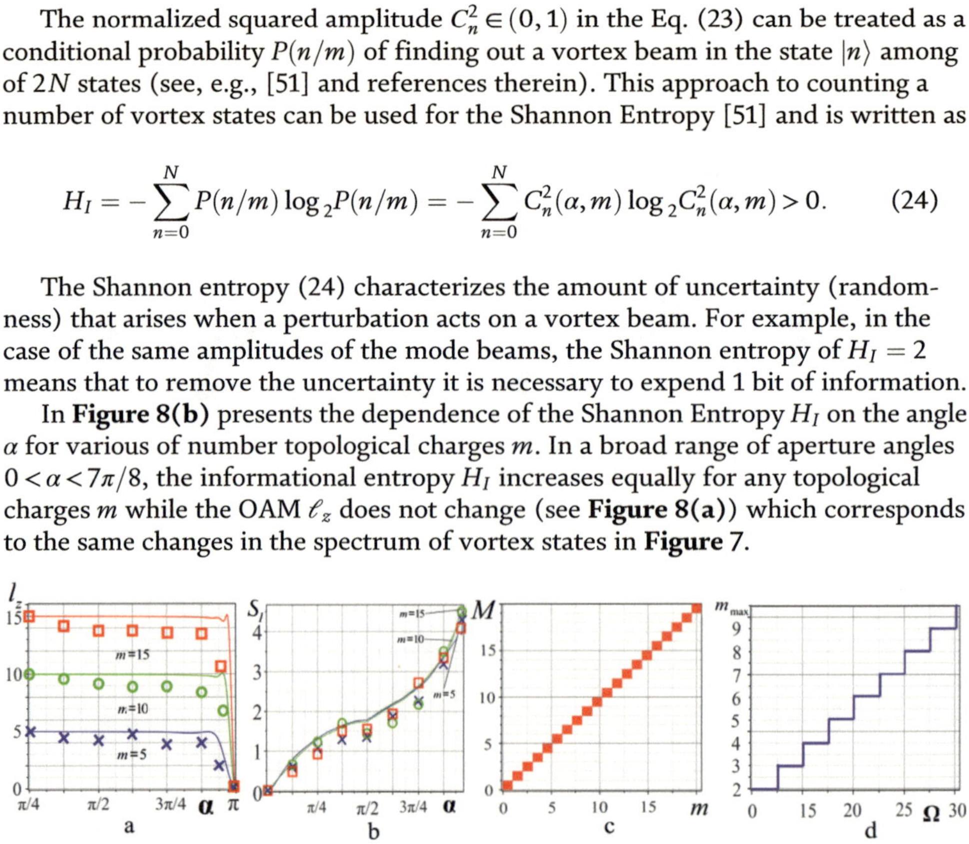

Figure 8.
Computer simulation of (a) the OAM $\ell_z(\alpha, m)$ and (b) information entropy $H_I(\alpha, m)$ for the initial topological charges, $m = 5, 10, 15$; (c) transformation of the vortex TC Mof the perturbed beam via variation TC of the initial beam; (d) maximal TC m_{max} of the spectral vortex satellites via perturbation Ω of the perturbed holographic gratings; (solid lines); crosslets ($\times$), circlets ($\odot$) and squares ($\square$) are correspondent experimental data.

4.1.4 The topological charge

According to Berry [42], the topological charge of the vortex array is defined as the difference of the fluxes of vortex trajectories through the beam cross section taking into account vortex directions and "weights," and is calculated as

$$M = \frac{1}{2\pi} \mathrm{Im} \left\{ \lim_{r \to \infty} \int_0^{2\pi} \partial_\varphi \Psi_m / \Psi_m d\varphi \right\}. \tag{25}$$

Berry showed [42] that the vortex beam TC has always integer values when the spiral phase plate is perturbed equal to the integer value of the unperturbed plate step. We investigated the changes of the TC under the beam sector perturbations [52] and perturbation of the holographic grating [47]. Note that all further calculations are based on the requirement that the perturbation does not introduce changes in the mode phases. We performed a series of computer TC estimations of the perturbed beam (20) for various initial TC. The following restrictions were used. As the spectra of vortices in **Figure 7** show, the squared mode amplitudes quickly tend to zero as their TC increases. Therefore, we can restrict ourselves to a finite mode numbers $C_{mn}^2 \sim 10^{-3}$. The half-width beam radius can be estimated at the maximum intensity of the initial vortex beam $r_M = \sqrt{|m|/2}$. Therefore, in Eq. (25) we replaced the limit $r \to \infty$ by $r \to 3r_N$. The results of computer simulation are shown in **Figure 8(c)**. It can be seen from the figure that the total topological charge M at $\alpha = 3\pi.4$ remains equal to the initial TC of the unperturbed beam $M = m$. Computer calculations for sector aperture angles $\alpha = \pi/4, \pi/2, 30\pi/31$ show similar results. Therefore, we assume that, in the general case of arbitrary angles, the initial TC will be preserved under sector perturbations. We also examined the constancy condition TC under the perturbation of the holographic grating considered above and made sure that this condition is strictly satisfied. Moreover, we traced the maxima displacement of the spectral satellites in the vortex avalanche in **Figure 4** [47] and convinced (see **Figure 8(d)**) that their TC cannot take fractional values.

4.2 Circular and annular apertures

The problem of the birth and annihilation of phase singularities has been considered as far back as at the beginning of the last century in connection with the peculiarities of light diffraction at the edges of the half-plane or lenses and micro-objectives of telescopes and microscopes (see [53] and references therein). As a rule, the discussion came down to the technique of suppressing the corresponding aberrations. In this section, we focus on the digital vortex sorting after beam diffraction by the circular and annular apertures addressing the Shannon entropy problem of the diffracted combined vortex beams.

Note also that recently, special attention has been paid to studies on increasing the information capacity of optical channels due to LG_n^m modes with various radial indices n, but a constant topological charge m [54]. Such modes are sorted using the holographic grating techniques and *a single phase screen* [55]. In this section, we will focus on the digital sorting of LG modes with various radial indices.

We consider the perturbation of the vortex LG_0^m beam at the $z = 0$ plane with a ring aperture so that its complex amplitude is written in the form (see also **Figure 10a**)

$$\Psi_m = r^{|m|} e^{i m \varphi} \exp\left(-r^2\right), \quad R < r < R + h. \tag{26}$$

If the axis of the annular aperture coincides with the beam axis then the perturbation excites only LG beams with the same TC ($m = const$) but different radial indices n (this follows from the condition of orthogonally of LG beams). The perturbed beam field can be represented as an expansion over LG beams

$$\Psi_m = \sum_{n=0}^{\infty} C_{m,n} r^m L_n^m \left(2r^2\right) e^{im\varphi} e^{-r^2}. \tag{27}$$

The LG vortex modes amplitudes restricted by the ring of thickness h, we find for the field difference the beams passing simultaneously through the circular aperture of radius $R + h$ and the opaque disk of radius R. As a result we obtain

$$\begin{aligned} C_{m,n}(R) &= \int_0^{\infty} \Psi_m \left(LG_n^m\right)^* r\,dr \\ &= \left(2(R+h)^2\right)^{m+1} e^{-2(R+h)^2} {}_1F_1\left(1-n, m+2; -2(R+h)^2\right)/(m+1)m! \\ &\quad - \left(2R^2\right)^{m+1} e^{-2R^2} {}_1F_1\left(1-n, m+2; -2R^2\right)/(m+1)m!, \end{aligned} \tag{28}$$

where ${}_1F_1$ is a confluent hypergeometric function and we used the integrals from [56]. The expression (28) together with Eq. (27) allows covering three cases: (1) a circular aperture, $R = 0, h = R_0$; (2) an opaque disk $R = R_0, h = 0$; and (3) an annular aperture $R, h \neq 0$. An important property of mode amplitudes (28) is that the perturbation of a singular beam with a defined TC via these types of axial apertures does not excite vortex modes with other TC. This means that the OAM ℓ_z does not change due to such a perturbation process. Does this mean that the perturbed vortex beam completely restores its initial properties during propagation, i.e., possesses the self-healing effect? We will peer into this process carefully.

Recall that under the action of axial perturbation, vortex modes with new topological charges do not appear in the perturbed beam. Therefore, the digital sorting of vortex modes in a perturbed beam can be carried out in accordance with Eq. (15) for the nondegenerate case (see Section 3.1). In this case, as a function of moments $M_{p,q}(r, \varphi)$ should choose the forms $M_{p,q} = \sin rL_p^k(2r^2)$ or $M_{p,q} = \cos rL_p^k(2r^2)$ while in the intensity distribution $\Im(r, \varphi)$ use the complex amplitude (26). Variation of the indices p and k enables us to obtain a closed system of linear equations for the squared amplitudes $C_{m,n}^2$.

The experimental results of measuring the vortex spectra are shown in **Figure 9**, where the average values of the squared amplitudes are plotted along the ordinate axis. A characteristic feature of the dependences $C_{m,n}^2(n)$ is long spectral tails, which are omitted in the figures, but which make a significant contribution to the calculation of information entropy. Truncation of a topologically neutral beam ($m = 0$) with a circular aperture in **Figure 9(Ia)** leads to overlapping many side rings, resulting in a wave-like form of the spectral tail. In the perturbed vortex beams shown in **Figure 9(Ibc)**, a broadening of the vortex spectrum and a decrease in the tail amplitudes are observed. Such characteristic features of the vortex spectra insert significant uncertainty into information carried by the vortex beam, which is represented as the dependence of information entropy H_I on the aperture radius R_0 in **Figure 9(Id)**. It is noteworthy that even small variations in the aperture radius lead to changes in the entropy H_I that grow with increasing TC. Similar changes in the entropy $H_I(h)$ shown in **Figure 9(IId)** occur when a vortex beam is perturbed by a annular aperture, which are the result of transformations in the vortex spectra in **Figure 9(a–c)**. The presented results show that any external interference in the beam immediately affects the uncertainty of the vortex beam state, the magnitude of which can be estimated by measuring informational entropy.

Another interesting feature of the axial aperture action is manifested under the combined vortex beam perturbation consisting of two vortex beams with the same values but different TC signs (m and $-m$). If the amplitudes of the beams are the same, then the complex amplitude of such a combined beam is described by Eq. (28) when replacing the phase factor $\exp(im\varphi) \to 2\cos m\varphi$. **Figure 10(b–d)** illustrates the intensity distribution of such perturbed beams at the measurement plane. Each of these beams receives the same perturbations, regardless of the sign of their TC. Even if the beam amplitudes are different, they receive the same amount of the vortex state uncertainty. However, this apparent indistinguishability of the modes can be easily detected experimentally due to opposite phase circulation of the fields, and the modes can be sorted out in different memory cells.

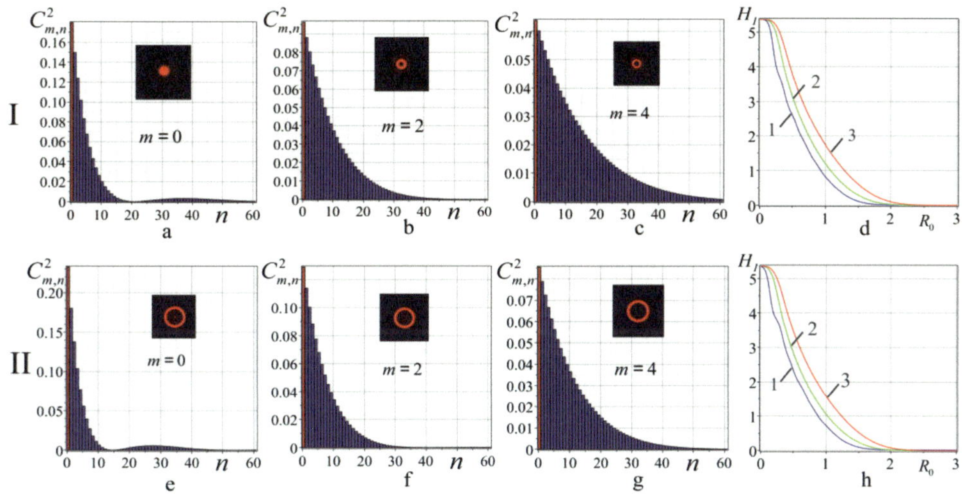

Figure 9.
The vortex spectra $C^2_{m,n}$ and the Shannon entropy $H_I(h)$ of the perturbed singular beams with different TC m: (I) the circular aperture; (II) the annular aperture; (Id) the Shannon entropy H_I via the aperture radius R_0; (IIh) the Shannon entropy H_I via the annular width, curves 1 and 3 correspond to the TC $m = 0, m = 2, m = 4$, respectively; callouts: corresponding intensity distributions.

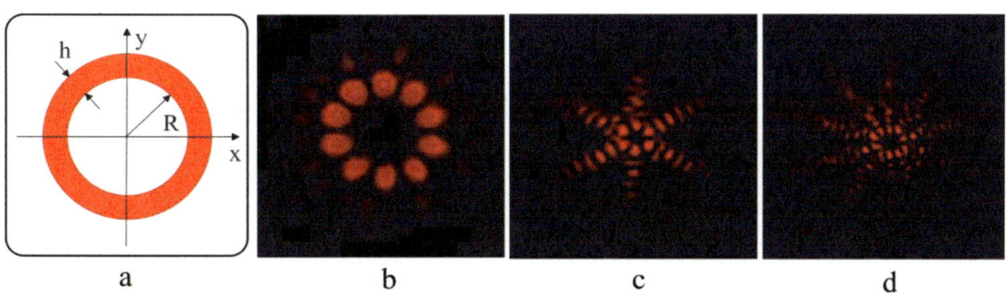

Figure 10.
(a) Sketch of the annular aperture. (b–d) Intensity distributions of the singular beams with TC $m = 5$ and $m = -5$ perturbed by the circular aperture with the radius (b) $R_o = 0.5$, (c) $R_o = 0.1$ and (d) the annular aperture with $R_o = 0.5, h = 0.2$.

5. Conclusions

We examined the technique of digital sorting of vortex modes that makes it possible to measure in real time the vortex spectrum (squared amplitudes and initial phases) including radial indices, OAM, and informational entropy of perturbed singular beams. The considered approach is based on the measurement of intensity moments of higher orders and a digital solution of a linear equations system that eliminates the cuts and gluing of the beam wavefront without losing information on the modes initial phases. Moreover, the digital vortex spectrum also enables us to restore the initial combined beam and, correcting parameters of the spectral modes, to improve its characteristics.

The digital approach has been tested on vortex beams free of wave defects perturbed both by local defects of holographic gratings responsible for the beam generation and by the sectorial, circular and annular aperture. We revealed that a local perturbation of the holographic grating near the central forked defect causes bursts and dips in the OAM spectrum. The depth and height of the spectral dips and bursts are controlled by the parameters of the holographic grating and can vary over a wide range. The perturbation inserted by the sector aperture is regulated by the sector angle. Over a wide range of sector angles, the beam OAM remains almost unchanged. However, when the sector angle is relatively large, so that most of the light flux is cut off by the aperture, the optical uncertainty principle begins to act, and the OAM sharply decreases to almost zero that is accompanied by a rapid growth of the Shannon entropy. At the same time, the beam topological charge remains unchanged for any sectorial perturbations. The axial perturbation via a circular and annular aperture does not change either the OAM or topological charge. However, a wide range of Laguerre-Gauss modes with the same topological charges but different radial indices leads to a rapid increase in information entropy as the pupil of the circular aperture or the ring thickness of the annular aperture decreases. This allows not only to estimate the noise level in the optical information transmission line, but also to record external interference in the information flow. We also note that the digital sorting of optical vortices in a perturbed light flux opens up broad prospects for its employment for medical express-diagnostics of skin diseases, since, for example, this allows us to detect slight changes in the vortex spectrum of a laser beam scattered by inflamed or dehydrated skin areas.

Acknowledgements

The authors are grateful to E. Abramochkin (Samara Branch of the Lebedev Physical Institute, Russian Academy of Sciences, Samara, Russia) for a useful discussion of the mathematical approach. The reported study was funded by RFBR according to the research project № 19-29-01233.

Author details

Alexander Volyar*, Mikhail Bretsko, Yana Akimova and Yuriy Egorov
Physics and Technology Institute, V.I. Vernadsky Crimean Federal University, Simferopol, Republic of Crimea, Russia

*Address all correspondence to: volyar@singular-optics.org

References

[1] Allen L, Beijersbergen MW, Spreeuw RJC, Woerdman JP. Orbital angular momentum of light and the transformation of Laguerre-Gaussian laser modes. Physical Review A. 1992; **45**(11):8185-8189. DOI: 10.1103/PhysRevA.45.8185

[2] Berry M. Nature's optics and our understanding of light. Contemporary Physics. 2015;**56**(1):2-16. DOI: 10.1080/00107514.2015.97162

[3] Mendenhall MJ, Nunez AS, Martin RK. Human skin detection in the visible and near infrared. Applied Optics. 2015;**51**(2):10559-10570. DOI: 10.1364/AO.54.010559

[4] Michel AP, Liakat S, Bors K, Gmachl CF. In vivo measurement of mid-infrared light scattering from human skin. Biomedical Optics Express. 2013;**4**(4):520-530. DOI: 10.1364/BOE.4.000520

[5] Chorvat D, Chorvatova A. Multi-wavelength fluorescence lifetime spectroscopy: A new approach to the study of endogenous fluorescence in living cells and tissues. Laser Physics Letters. 2009;**6**:175-193. DOI: 10.1002/lapl.200810132

[6] Gbur GJ. Singular Optics. New York: CRC Press; 2017. DOI: 10.1201/9781315374260

[7] Lavery MPJ, Berkhout GCG, Courtial J, Padgett MJ. Measurement of the light orbital angular momentum spectrum using an optical geometric transformation. Journal of Optics. 2013; **13**:064006. DOI: 10.1088/2040-8978/13/6/064006

[8] Shen Y, Yang X, Qi R, Wan Z, Fu X, Gon M. Recent advances on tunable vortex beam devices for biomedical applications. BJSTR. 2018;**9**(3):2/5-4/5. DOI: 10.26717/BJSTR.2018.09.001801

[9] Otaka H, Shimakura H, Motoyoshi I. Perception of human skin conditions and image statistics. Journal of the Optical Society of America A. 2019; **36**(9):1609-1616. DOI: 10.1364/JOSAA.36.001609

[10] Ney M, Abdulhalim I. Does human skin truly behave as an array of helical antennae in the millimeter and terahertz wave ranges? Optics Letters. 2010; **35**(19):3180-3182. DOI: 10.1364/OL.35.003180

[11] Gbur G, Korotkova O. Angular spectrum representation for the propagation of arbitrary coherent and partially coherent beams through atmospheric turbulence. Journal of the Optical Society of America. A. 2007;**24**: 745-752. DOI: 10.1364/JOSAA.24.000745

[12] Soifer VA, Korotkova O, Khonina SN, Shchepakina EA. Vortex beams in turbulent media. Review Computer Optics. 2016;**40**(5):605-621. DOI: 10.18287/2412-6179-2016-40-5-605-624

[13] Majumdar A, Kirkpatrick SJ. Statistical studies on optical vortices in dynamic speckle fields. Journal of Biomedical Photonics & Engineering. 2018;**4**(2):020301-1-020301-9. DOI: 10.18287/JBPE18.04.020301

[14] Reddy SG, Prabhakar S, Kumar A, Banerji J, Singh RP. Higher order optical vortices and formation of speckles. Optics Letters. 2014;**39**(15):4364-4367. DOI: 10.1364/OL.39.004364

[15] Kirkpatrick SJ, Khaksari K, Thomas D, Duncan DD. Optical vortex behavior in dynamic speckle fields. Journal of Biomedical Optics. 2012; **17**(5):050504-1-050504-3. DOI: 10.1117/1.JBO.17.5.050504

[16] Machado FJ, Monsoriu JA, Furlan WD. Fractal light vortices. In:

Perez-de-Tejada H, editor. Vortex Dynamics and Optical Vortices. Rijeka: IntechOpen; 2017. DOI: 10.5772/66343

[17] Segev M, Soljačić M, Dudley JM. Fractal optics and beyond. Nature Photonics. 2012;**6**(4):209-210. DOI: 10.1038/nphoton.2012.71

[18] Sheppard CJR. Fractal model of light scattering in biological tissue and cells. Optics Letters. 2007;**32**(2):142-144. DOI: 10.1364/OL.32.000142

[19] Xu M, Alfano RR. Fractal mechanisms of light scattering in biological tissue and cells. Optics Letters. 2005;**30**(22):3051-3053. DOI: 10.1364/OL.30.003051

[20] Si K, Gong W, Sheppard CJR. Model for light scattering in biological tissue and cells based on random rough nonspherical particles. Applied Optics. 2009;**48**(6):1153-1157. DOI: 10.1364/AO.48.001153

[21] Khonina SN, Kotlyar VV, Soifer VA, Paakkonen P, Turunen J. Measuring the light field orbital angular momentum using DOE. Optical Memory and Neural Networks. 2001;**10**(4):241-255

[22] Khonina SN, Kotlyar VV, Soifer VA, Jefimovs K, Turunen J. Generation and selection of laser beams represented by a superposition of two angular harmonics. Journal of Modern Optics. 2004;**51**: 761-773. DOI: 10.1080/09500340408235551

[23] Khonina SN, Kazanskiy NL, Soifer VA. In: Yasin M, Harun SW, Arof H, editors. Optical Vortices in a Fiber: Mode Division Multiplexing and Multimode Self-Imaging, Recent Progress in Optical Fiber Research. Rijeka: IntechOpen; 2012. DOI: 10.5772/28067

[24] Berkhout GCG, Lavery MPJ, Courtial J, Beijersbergen MW, Padgett MJ. Efficient sorting of orbital angular momentum states of light. Physical Review Letters. 2010;**105**: 153601. DOI: PhysRevLett.105.153601

[25] Volyar A, Bretsko M, Akimova Y, Egorov Y. Measurement of the vortex spectrum in a vortex-beam array without cuts and gluing of the wavefront. Optics Letters. 2018;**43**(22): 5635-5638. DOI: 10.1364/OL.43.005635

[26] Alperin SN, Niederiter RD, Gopinath JT, Siements KE. Quantitative measurement of the orbital angular momentum of light with a single, stationary lens. Optics Letters. 2016;**41**: 5019-5022. DOI: 10.1364/OL.41.005019

[27] Alperin N, Siemens ME. Angular momentum of topologically structured darkness. Physical Review Letters. 2017; **119**:203902. DOI: PhysRevLett.119.203902

[28] Kotlyar VV, Kovalev AA, Porfirev AP. Methods for determining the orbital angular momentum of a laser beam. Computer Optics. 2019;**43**(1): 42-53. DOI: 10.18287/2412-6179-2019-43-1-42-53

[29] Soifer VA, Golub MA. Laser Beam Mode Selection by Computer-Generated Holograms. Boca Raton: CRC Press; 1994. p. 224. ISBN: 978-0-8493-2476-5

[30] Gibson G, Courtial J, Padgett MJ, Vasnetsov M, Pas'ko V, Barnett SM, et al. Free-space information transfer using light beams carrying orbital angular momentum. Optics Express. 2004;**12**:5448-5456

[31] Berry MV. Paraxial beams of spinning light. In: Proceedings of the SPIE International Conference Singular Optics. Vol. 3487. 1998. pp. 6-11. DOI: 10.1117/12.317704

[32] D'errico A, D'amelio R, Piccirillo B, Cardano F, Marrucc L. Measuring the complex orbital angular momentum spectrum and spatial mode

decomposition of structured light beams. Optica. 2017;**4**:1350-1357

[33] Andersen JM, Alperin SN, Voitev AA, Holtzmann WG, Gopinath JT, Simens ME. Characterizing vortex beams from a spatial light modulator with collinear phase-shifting holography. Applied Optics. 2019;**58**: 404-409. DOI: 10.1364/AO.58.000404

[34] Kotlyar V, Kovalev A, Porfirev A. Calculation of fractional orbital angular momentum of superpositions of optical vortices by intensity moments. Optics Express. 2019;**27**(8):11236-11251. DOI: 10.1364/OE.27.011236

[35] Flusser J, Suk T, Zitová B. Moments and Moment Invariants in Pattern Recognition. John Wiley & Sons, Ltd: Chichester; 2009. p. 312. DOI: 10.1002/9780470684757

[36] Alexeyev CN, Egorov YA, Volyar AV. Mutual transformations of fractional-order and integer-order optical vortices. Physical Review A. 2017;**96**:063807. DOI: 10.1103/PhysRevA.96.063807

[37] Fu S, Gao C. Influences of atmospheric turbulence effects on the orbital angular momentum spectra of vortex beams. Photonics Research. 2016; **4**:B1-B4. DOI: 10.1364/PRJ.4.0000B1

[38] Kaiser T, Flamm D, Schroter S, Duparre M. Complete modal decomposition for optical fibers using CGH-based correlation filters. Optics Express. 2009;**17**:9347-9356. DOI: 10.1364/OE.17.009347

[39] Volyar A, Bretsko M, Akimova Y, Egorov Y. Measurement of the vortex and orbital angular momentum spectra with a single cylindrical lens. Applied Optics. 2019;**58**(21):5748-5755. DOI: 10.1364/AO.58.005748

[40] Abramochkin E, Razueva E, Volostnikov V. General astigmatic transform of Hermite–Laguerre–Gaussian beams. Journal of the Optical Society of America A. 2010;**27**: 2506-2513. DOI: 10.1364/JOSAA.27.002506

[41] Abramochkin E, Volostnikov V. Beam transformations and non-transformed beams. Optics Communications. 1991;**83**(12):123-135. DOI: 10.1016/0030-4018(91)90534-K

[42] Berry MV. Optical vortices evolving from helicoidal integer and fractional phase steps. Journal of Optics A. 2004;**6**: 259-268. DOI: 10.1088/1464-4258/6/2/018

[43] Fadeyeva TA, Rubass AF, Aleksandrov RV, Volyar AV. Does the optical angular momentum change smoothly in fractional-charged vortex beams? Journal of the Optical Society of America B: Optical Physics. 2014;**31**(4): 798-805. DOI: 10.1364/JOSAB.31.000798

[44] Götte B, O'Holleran K, Preece D, Flossmann F, Franke-Arnold S, Barnett SM, et al. Light beams with fractional orbital angular momentum and their vortex structure. Optics Express. 2008;**16**:993-1006. DOI: 10.1364/OE.16.000993

[45] Gutiérrez-Vega JC, López-Mariscal C. Nondiffracting vortex beams with continuous orbital angular momentum order dependence. Journal of Optics A. 2008;**10**:015009. DOI: 10.1088/1464-4258/10/01/015009

[46] Nemes G, Siegman A. Measurement of all ten second-order moments of an astigmatic beam by the use of rotating simple astigmatic (anamorphic) optics. Journal of the Optical Society of America. A. 1994;**11**:2257. DOI: 10.1364/JOSAA.11.002257

[47] Volyar A, Bretsko M, Akimova Y, Egorov Y. Vortex avalanche in the perturbed singular beams. Journal of the

Optical Society of America A. 2019; **36**(6):1064-1071. DOI: 10.1364/ JOSAA.36.001064

[48] Volyar A, Egorov Y. Super pulses of orbital angular momentum in fractional-order spiroid vortex beams. Optics Letters. 2018;**43**(1):74-77. DOI: 10.1364/OL.43.000074

[49] Volyar A, Bretsko M, Akimova Y, Egorov Y. Orbital angular momentum and informational entropy in perturbed vortex beams. Optics Letters. 2019; **44**(23):5687-5690. DOI: 10.1364/ OL.44.005687

[50] Yao E, Franke-Arnold S, Courtial J, Barnett S, Padgett M. Fourier relationship between angular position and optical orbital angular momentum. Optics Express. 2006;**14**:9071-9076. DOI: 10.1364/OE.14.009071

[51] Yu FTS. Entropy and Information Optics. Boca Raton: CRC Press; 2000. 360 p. DOI: 10.1201/9781315214498

[52] Volyar A, Bretsko M, Akimova Y, Egorov Y, Milyukov V. Sectorial perturbation of vortex beams: Shannon entropy, orbital angular momentum and topological charge. Computer Optics. 2019;**43**(5):722-733. DOI: 10.18287/ 2412-6179-2019-43-5-723-734

[53] Born M, Wolf E. Principles of Optics. 7th ed. London: Cambridge University Press; 1999. 952 p. DOI: 10.1017/CBO9781139644181

[54] Karimi E, Giovannini D, Bolduc E, Bent N, Miatto FM, Padgett MJ, et al. Exploring the quantum nature of the radial degree of freedom of a photon via Hong-Ou-Mandel interference. Physical Review A. 2014;**89**:013829-1-013829-5. DOI: 10.1103/PhysRevA.89.013829

[55] Bouchard F, Herrera-Valencia N, Brandt F, Fickler R, Huber M, Malik M. Measuring azimuthal and radial modes of photons. Optics Express. 2018; **26**(24):31925-31941. DOI: 10.1364/ OE.26.031925

[56] Prudnikov AP, Brychkov YA, Marichev OI. Integrals and Series. Special Functions. New York: Gordon and Breach; 1986. p. 798. ISBN: 2881240909 9782881240904

Chapter 9

Nonlinear Optical Responsive Molecular Switches

Mahadevappa Y. Kariduraganavar, Radha V. Doddamani, Balachandar Waddar and Saidi Reddy Parne

Abstract

Nonlinear optical (NLO) materials have gained much attention during the last two decades owing to their potentiality in the field of optical data storage, optical information processing, optical switching, and telecommunication. NLO responsive macroscopic devices possess extensive applications in our day to day life. Such devices are considered as assemblies of several macroscopic components designed to achieve specific functions. The extension of this concept to the molecular level forms the basis of molecular devices. In this context, the design of NLO switches, that is, molecules characterized by their ability to alternate between two or more chemical forms displaying contrasts in one of their NLO properties, has motivated many experimental and theoretical works. Thus, this chapter focuses on the rational design of molecular NLO switches based on stimuli and materials with extensive examples reported in the literature. The factors affecting the efficiency of optical switches are discussed. The device fabrication of optical switches and their efficiency based on the optical switch, internal architecture, and substrate materials are described. In the end, applications of switches and future prospectus of designing new molecules with references are suitably discussed.

Keywords: nonlinear optical materials, optical switches, chromophores, simulation, DFT

1. Introduction

There is a strong demand for the advancement of high-performance NLO materials, owing to the emergence of photonic technologies in the area of telecommunication where the information is coded, transported, and routed through optically. Thus, NLO materials have created enormous interest among the scientific community due to their role in optical switching, sensor protection, light modulators, memory storage devices, information processing devices [1–3].

A considerable number of NLO materials have been developed over the last two decades. As a new method of communication, optical signal achieves faster network speed and broader bandwidth, and thus, the optical switch is an essential device for optoelectronic and integrated optics. Optical switches play the role of protecting, testing, monitoring, and management of the network. In this context, the design of NLO switches based molecules, which are characterized by their ability to alternate between two or more chemical forms displaying contrasts in one of their NLO properties, has motivated much experimental and theoretical work [4, 5].

Understanding these, researchers have been tried to synthesize and characterize the several molecular NLO switches that differ by the stimulus used to provoke the commutation. These stimuli can include light irradiation, redox reaction, pH variation, ion recognition, and others. The design of multistate and multifunctional systems has also provoked a large amount of work, aiming for the enhancement of the storage capacity of the optical memories or the diversification of the relevant devices [6, 7].

Generally, most of the molecular switches can be divided into two categories, namely stereoisomerism and structural isomerism. Other mechanisms are covered in the literature, such as dipole, charge, or spin switching, but these processes generally include some kind of structural change as well [8, 9]. As possible candidates for integration into responsive materials, researchers have been intensely studying the organic, organometallic molecules and other highly advanced materials like supramolecules, carbon nanotubes, etc. with switchable hyperpolarizability over the past two decades [10–13].

Realizing the importance of the factors as mentioned above, in this chapter, we illustrate the design and categorize the highly efficient NLO molecular switches based on stimuli and materials. After a critical overview, we provide numerous case studies, including organic, organometallic molecules, supramolecules, carbon/graphene materials, nanoscale structures, polymers, and crystalline solids [14, 15]. Finally, we illustrate the influence of the environment on the NLO switching properties with examples. The rational design of molecular NLO switches with recent device fabrication is also discussed. The chapter also explains the experimental and the theoretical approaches, which have reached maturity in obtaining NLO responsive molecular switches. The prime objective of the chapter was to examine the different systems with particular reference to their photophysical properties as well as the switching mechanism at the molecular level.

2. Factors affecting the efficiency of switches

Many scientists have been working for the last three decades, to develop the optical switches, in which one light beam can switch another beam on and off, the same way that one electrical signal can switch another on and off in an electrical transistor [16]. This type of light switch could be the basic building block for optical circuits and even for optical computers. In principle, such a device could work at terahertz frequencies or higher, and the dissipation of power could be far less. For an ideal device of an optical switch, there could be five crucial factors. They are: (1) amplification (a weak beam is used to switch a strong beam); (2) compact size, preferably small enough to make millions on a chip; (3) fast response time; (4) possibility of cascading devices (the output of one device can act as the input for a subsequent device) and (5) room-temperature operation [17–19].

3. Molecular switches based on stimuli

Functional materials, which are exhibiting distinct functionalities in response to a specific stimulus, are highly desirable for the fabrication of advanced devices with superior dynamic performances [20]. The stimuli can include light irradiation, temperature variation, redox reaction, induced ion recognition, and so on.

3.1 Light-induced optical switches/photoswitches

The photochemical isomerization of an olefinic bond is one of the fundamental processes in the vision. The structural perturbations induced by the *cis/trans* isomerization of the retinal moiety results in conformational changes in the protein molecule, and through a cascade of reactions, the neural pulse responsible for vision is generated. This principle was mimicked to design molecular systems that can undergo light-induced structural changes for signaling applications.

Light mainly offers several advantages over the chemical stimuli, viz., its spatial and temporal control, ease of implementation, remote addressability, and low toxicity. These optimistic attributes have led to the development of various photoswitching moieties for the applications in photo-pharmacology, sensing, energy production, and actuation [21].

The azobenzene moiety enjoys widespread popularity, and thus a number of artificial systems based on the azobenzene functionality have been investigated. Owing to its ease of synthesis, reasonable quantum yields, large extinction coefficients, and so azobenzenes exist in two forms, *E* (*trans*) and *Z* (*cis*) forms and their interconversions can be induced by irradiation with appropriate wavelengths [22]. For instance, upon photochemical excitation, structural changes occur in molecular systems having azobenzene functionality, and such changes were utilized for modulating the host-guest properties, phase changes in liquid crystals, photo-triggered reversible gel formation and photoinduced switching of conductivity. However, thermal reactions prohibit the use of azobenzene based molecular systems from their use in information storage systems. Chen et al. [23] reported dual-mode optical switch by conjugating helicene and bis-azobenzene moieties in proximity with highly diastereoselective photochromic and chirochromic switching both in solution as well as in nematic liquid crystals with tunable pitches and handedness by modification of four different metastable states. In 2013, Roldan et al. [24], for the first time, investigated the conductance properties of a photoswitchable dimethyl-dihydro pyrene derivative in single-molecule junctions using the mechanically controllable break junction technique. They studied the reversible photo-thermal switching of a bipyridine appended dimethyl-dihydro pyrene/cyclophane-dienephotochromic system. The photo-thermally triggered conductance switching between these two isomers is fully reversible, stable over more than five sequential cycles, with on/off ratio up to 104. This study represented highly promising candidates for constructing nanoscale molecular building blocks. Deng et al. [25] designed and synthesized two nonlinear optical chromophores based on dithienylethene as the ultraviolet NLO switches. They investigated open/close behavior using the UV–Vis spectra. The dipole moments, first and second hyperpolarizabilities, etc. have been calculated and analyzed for both opening and closure states of the chromophores using quantum chemical calculations. The switching ability of the NLO chromophores was studied through monitoring the SHG at the opening/closure states. The results suggested that indicate that the maximal SHG changes were observed at energy density equal to about 90 J/m^2. The resulting induced second-order susceptibilities remained up to 150 h without decreasing, even after switching off of the external UV light, which indicated that the chromophores have the potential application as the NLO switches widely applied in photonics. Another example where the purple merocyanine exhibits strong absorption at 563 nm and can completely block an incoming signal, the colorless spiropyran allows the beam to pass, depicted through the structures in **Figure 1**. Through these types of systems, the building of logic operations possible, which also lay an attractive foundation for digital processing at the molecular level [26].

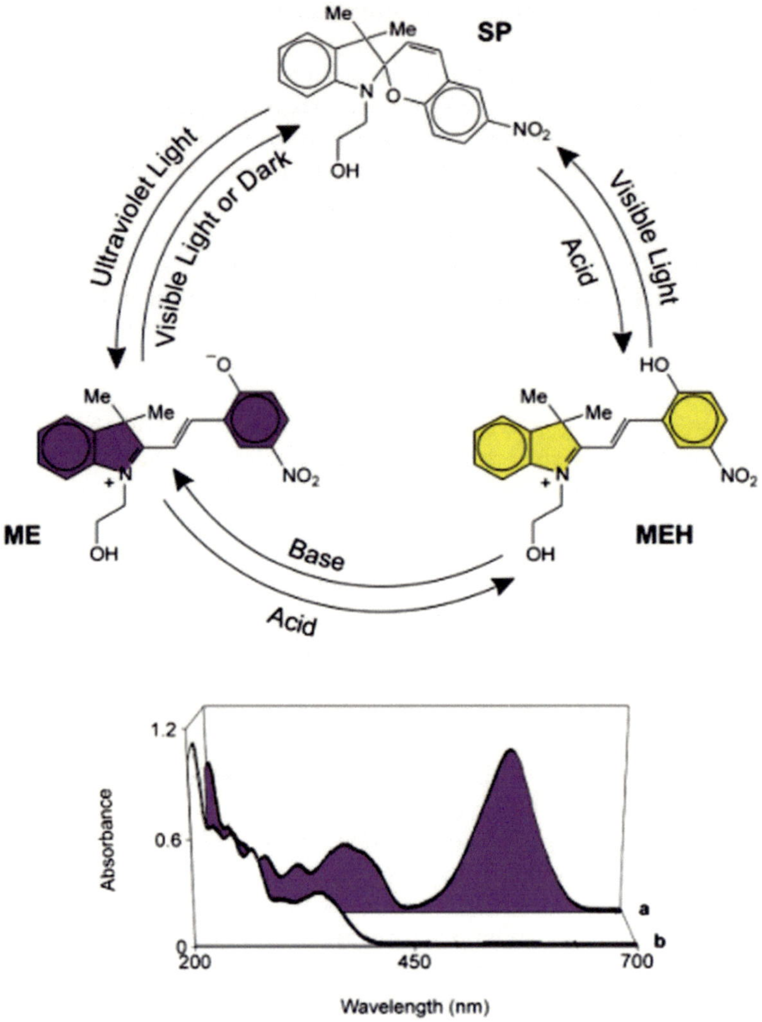

Figure 1.
The UV light, visible light, and H^+ induce the interconversion between the three states of SP, ME, and MEH. Absorption spectra of SP (in MeCN, 10^{-4} M, 25°C) recorded after (a in lower) and before (b in lower) irradiation at 254 nm for 5 min (Adopted from [26] with consent; National Academy of Sciences, 2002).

Similarly, it revealed [8] that five-membered and cyclic biradicals developed based on 15 elements are robust and efficient photochemical switches that can be activated by a red light with the quantum yield of the photoisomerization as high as 24.6%, and the thermal equilibration of the photo-activation product proceeds swiftly at room temperature, which is completely reversible. As demonstrated, the biradical character could be completely turned on and off, so that the system could be employed to control the chemical equilibria that involve activation products of the cyclic biradicals. Similarly, Li et al. [20] designed and synthesized two novel light-driven chiral fluorescent molecular switches that are able to exhibit unprecedented reversible *Z*/*E* photoisomerization behavior along with tunable fluorescence intensity in both isotropic and anisotropic media are described in **Figure 2**. Szukalski et al. [27] introduced all-optical switches that are based on deoxyribonucleic acid (DNA) in the form of electrospun fibers, where DNA is semi-intercalated with a push-pull, luminescent nonlinear pyrazoline derivative. Hansel et al. [28] have studied the switching ability and the nonlinear optical contrast

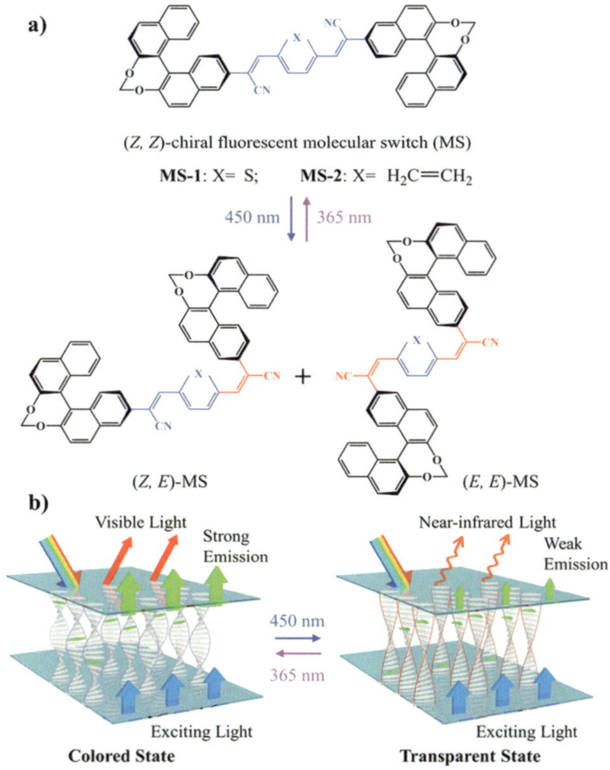

Figure 2.
(a) The chemical structures and reversible photoisomerization process of light-driven chiral fluorescent molecular switches MS-1 and MS-2. (b) The schematic illustration of the self-organized helical superstructures in the cholesteric LCs tuned by visible/UV light, reflecting different wavelengths, and fluorescing with different intensity when excited by blue light (Adopted from [20]; Springer Nature, 2012).

between different switching states of three unequal fulgimide-functionalized self-assembled monolayers (SAMs) on Si(111), facilitated a very efficient reversible photoisomerization, using second harmonic generation, density functional theory, and wave function method calculations.

In order to obtain high NLO contrasts, controlled variations of the linker system and the fulgimides were performed. In all the fulgimide-functionalized interfaces, a photoinduced reversible ring-opening/closure reaction demonstrated via the NLO response. The closed-form of the fulgimide demonstrated a higher SHG signal amplitude than that of open-form, and this is mainly because of the extended π-electron system of the closed state. Also, by chemical modifications, large cross-sections for the light-induced reversible switching process in the order of 10^{-18} cm^{-2} have been observed, which are similar to the values known for fulgimides in solution (shown in **Figure 3**). Based on the results, it is concluded that chemical modifications of both the molecular switch and the linker system may even enhance the NLO contrast between the open and closed fulgimides. From the theoretical study, it is predicted that the switching unit and the extended peripheral π-system are the potential targets for enhancing NLO contrasts, in particular. In general, the generation of two-dimensional NLO switching materials using functionalized photoswitchable silicon interfaces may pave the way for applications in photonic and optoelectronic devices [29].

Ping et al. [30] employed density functional theory (DFT) and time-dependent density-functional theory (TDDFT) methods in solution to study the effect of

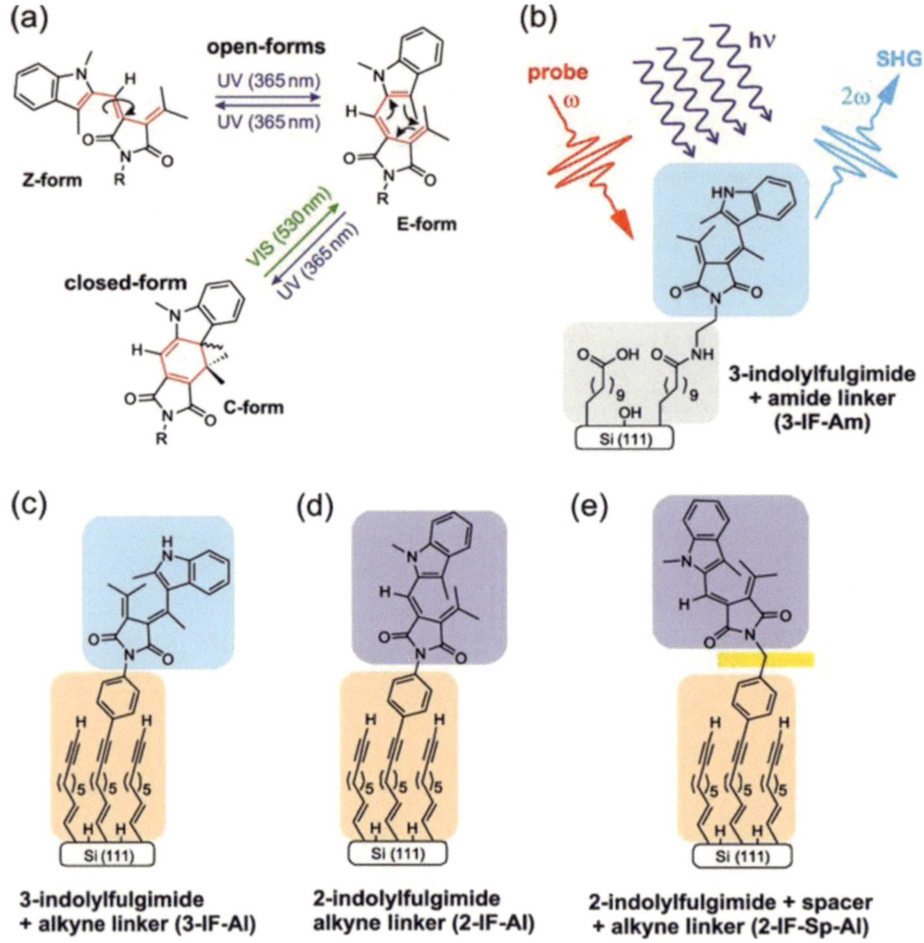

Figure 3.

(a) Photochromism in fulgimide upon irradiation with UV (λ = 365 nm) and visible (λ = 530 nm) light. Fulgimides yield a reversible reaction between the closed-form and a mixture E/Z-isomers of the open-form. (b) the scheme of the SAM containing fulgimide moiety, immobilized via an amide (Am) linker, called as 3-indolylfulgimide (3-IF-Am). SHG is used to study the switching state via the NLO response. (c)–(e) Fulgimide-containing SAMs on Si(111) (c) 3-indolylfulgimide (3-IF-Al), (d) 2-indolylfulgimide (2-IF-Al), and (e) 2-indolylfulgimide with a $-CH_2$*-spacer group (2-IF-Sp-Al) connected via an alkyne (Al) linker to the Si substrate (Adopted from [28]; ACS Publications, 2018).*

photoisomerization for dithiazolylarylene. The weak interaction of S···N and hydrogen bond of CH···N restrain the rotation of side-chain thiazolyl ring in open isomer 1a, the higher stability of which favors showing a high quantum yield of photoisomerization.

The UV–Vis spectrum at around 320 nm for open-isomer 1a is bathochromically shifted to 647 nm for closed-isomer 1b, in good agreement with the experimental photochromic phenomenon. The electron transition in electron CD (ECD) spectra for closed isomer 1b with two chiral carbon atoms is dominated by intramolecular charge transition (ICT) and local excitation (LE). Which attributes to one negative cotton effect (650 nm), one positive (440 nm) and one negative cotton effect (650 nm) where the two chiral carbon atoms play a slight role in these transitions. The potential energy surface (PES) in the S1 and S0 states, respectively, suggest that the cyclization reaction is allowed from open-isomer 1a to closed-isomer 1b in the photoexcited state with high-conversion quantum efficiency. At the same time, it is prohibited in the thermodynamic process. Besides, the second-order NLO response for closed-isomer 1b is nearly six times higher than that of open-isomer 1a. Further, it is confirmed that the photoirradiation excite the photoisomerization property to show a significant difference in the second-order NLO response, which could be employed in designing the photochromic materials and reversible NLO switches.

3.2 Thermal optical switches

The thermal-optical switch (TO), as the essential part of the optical switch, acts as a pivotal part of low-speed fiber-optic communication systems. Currently, the wavelengths of the reported thermal-optical switches are usually at 1550 and 650 nm, and the thermal-optical switch working at 650 nm is mainly used in a system of short-range communication [31]. The switchable NLO materials are essential for photonic as well as optoelectronic technologies. One crucial concern for NLO photoswitching, the most studied physical switching strategy, is how to enhance the switching contrast of SHG in crystals because the known values are generally below three times [32].

These devices are polymer-based to obtain flexibility. In order to make the molecular switch to be useful in the future technological applications, the system should retain its properties when incorporated in a processable matrix, to offer a photoswitchable film for data storage. Flexible and wearable optoelectronic devices are the current trend for an active lifestyle [33].

In this context, Sun and coworkers [34] developed waveguide optical switches, as shown in **Figure 4** for a flexible photonic integrated circuit using poly(methyl methacrylate) (PMMA) polymer. The optical field of the device was accelerated by beam propagation to optimize the position of the electrode. The properties of switching of the flexible optical switch at different working conditions, such as contact to the polymer, silicon, and skin, were studied using simulation. The device was constructed based on the photo curved polymer and lithography method. Using an excimer laser, the end faces of the flexible film device were processed with optimized parameters of 15 Hz and 28 mJ/cm^2. The measured response rise time and fall time on the substrate were respectively 1.98 and 2.71 ms. The consumption of power and the ratio of extinction were 16 mW and 11 dB, respectively. The properties indicated that the resulting flexible optical waveguide structure could be used in the light control area of a wearable device.

Further, novel polymer (fluorinated photopolymer) based thermo-optic switch arrays were successfully designed and fabricated based on dielectric-loaded surface plasmon polariton waveguide (DLSPPW) structure by Zheng and research group [35]. Briefly, the DLSPPW model included highly fluorinated

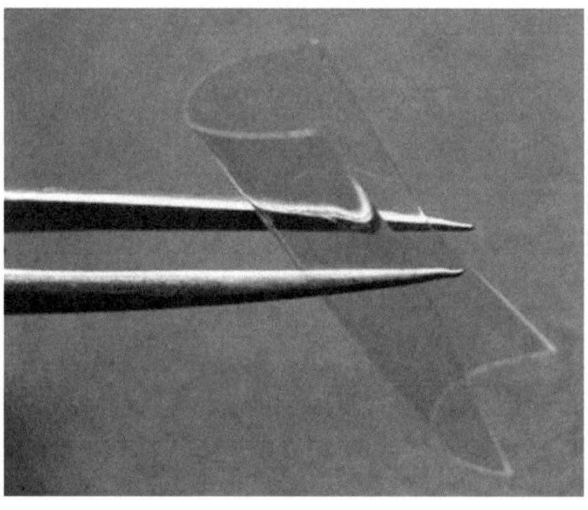

Figure 4.
Flexible polymeric film (Adopted from [34]; OSA Publishing, 2018).

low-loss photopolymers and organic–inorganic grafting materials that were used as the waveguide core and cladding and with PMMA substrate. The operation of the device at signal wavelengths was controlled via the thermo-optic effect by electrically heating the gold stripes of dielectric-loaded surface plasmon polariton waveguides. At 1550 nm wavelength, the propagation loss of 4 μm wide straight DLSPPW was performed as 0.55 dB/cm. The insertion loss of the device was around 4.5 dB. The fall time and switching rise of the device measured by 200 Hz square-wave voltage was found to be 370.2 and 287.5 μs, respectively. The switching power was about 5.6 mW, and the extinction ratio was about 13.5 dB. The flexible low-loss multifunctional waveguide switch arrays were realized to be suitable for large-scale optoelectronic integrated circuits. Zhen et al. [36] summarized the fundamental realization principles, novel configurations, superior materials, improved performance indexes, and ameliorated trigger methods of ultrafast optical switching.

Thermoswitching has shown impressive high SHG-switching contrasts, but the fast decay of thermally induced states demands constant heat sources to maintain the specific SHG intensities. Xing and coworkers [32] synthesized a photochromic and thermochromic bistable acentric compound, β-[(MQ)ZnCl3] (MQ+ = N-methyl-4,4′-bipyridinium), which represents the first crystalline compound with both heat- and photoinduced SHG switching behavior and the first example of a thermo-switchable NLO crystal that can maintain its expected second-order NLO intensity without any heat source. The SHG-switching contrast could reach about eight times after laser irradiation or two times after thermal annealing. The former value happens to be the highest recorded for photoswitchable NLO crystals. This work evidences that the higher SHG switching contrasts could be obtained by increasing the electron-transfer efficiency, self-absorption, and variation of permanent dipole moment.

3.3 Chemically activated optical switches

In the field of Biology, chemically responsive switches and machines are used to regulate the myriad of processes, which include the creation of protein folding, proton gradients, and viral entry and translocation [37]. Sakata et al. [38] reported the design, synthesis, and characterization of a family of thiol-reactive optical switches for labeling the proteins and other biomolecules. The site-selective introduction of photochromic probes within biomolecules is being employed as part of a new approach for optical control of bio-molecular activities and interactions within the cells. The thiol-reactive photochromic probes presented in this report cover spironaphthoxazine and five spirobenzopyrans. The location of the thiol-reactive group on the spirobenzopyran is different for each probe, and this feature could be used to control the geometry of the optical switch within a bio-conjugate. As indicated in **Figure 5**, the photochromes undergo rapid and reversible, and optically driven transitions between the states of a colorless spiro (SP) and a brightly colored

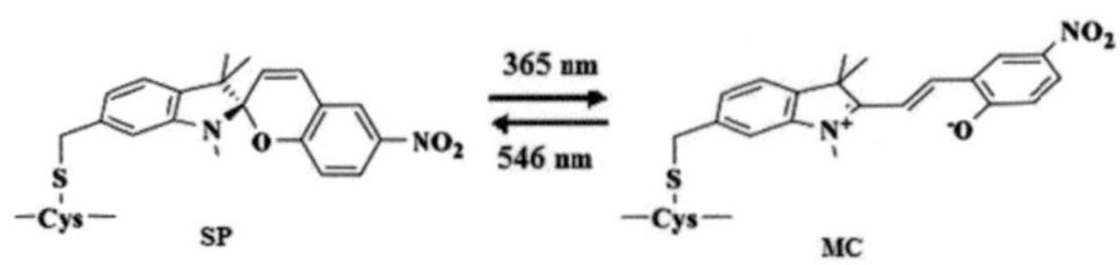

Figure 5.
Schematic representation of the reversible optical switching between the SP and MC states of a spirobenzopyran linked to a cysteine residue in a protein (Reused from [38]; American Chemical Society Publication, 2005).

merocyanine (MC). The absorption of MC of a spironaphthoxazine conjugate was red-shifted by more than 100 nm compared to the equivalent spirobenzopyran, which could be exploited for the independent control of the transition of MC to SP for up to two different spironaphthoxazine and spirobenzopyran conjugates within the sample.

4. Molecular switches based on materials

4.1 Optical switches based on organic chromophores

4.1.1 Design of chromophores

One of the major advantages of employing organic materials is that their physical and chemical properties can be customized through specific and well-defined synthetic modifications of the molecular structures. The properties such as positions and strengths of the low-energy one-photon and two-photon electronic resonances, the solubility, the aggregation propensity, the nonlinear polarizabilities, the strength of ion-pairing interactions in case of molecular salts dissolved in polymer hosts or in solution, the compatibility with various polymer host materials. Moreover, the photochemical and thermal stability of the NLO molecules can be logically controlled [39].

In NLO switches, most of the systems synthesized to date are π-conjugated donor-acceptor molecules, which consist of strong electron-donor (D) and electron-acceptor (A) groups anchored onto a π-conjugated linker. Coe [40] proposed to categorize the several systems based on the part of a molecule that could be altered while performing the reaction, i.e., the donor part (type I), the acceptor part (type II), or the conjugated bridge (type III), as shown in **Figure 6**. The NLO switches of type I and type II rely on redox switching, where the electron-donating capability of the donor group is reduced by oxidation, or on acido-switching, wherein the capability of the electron-donating group is reduced by protonation. The NLO switching based on redox interconversion has been established in a wide range of organometallic complexes. Most of the NLO switches of type III are photochromic compounds characterized by different commutation processes, including intramolecular proton transfer as in aniline derivatives, cis-trans photoisomerization in azobenzenes, and opening/closure of a conjugated ring in diarylethenes or spiropyrans [6].

Hales et al. [39] designed a class of chromophores using a relatively simple model, namely cyanine-like polymethines. The challenges associated with changing these microscopic properties of the chromophores into macroscopic properties for the suitable device applications are discussed in detail, along with the molecular engineering approaches to overcome these drawbacks.

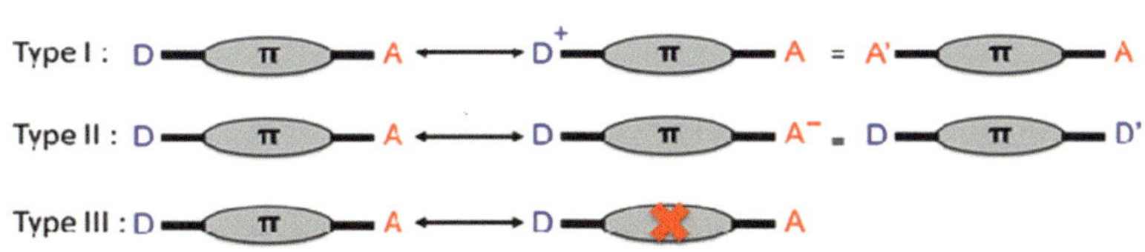

Figure 6.
The different types of NLO switching in dipolar D-π-A molecules, wherein the push-pull character is tuned upon switching, Type I = redox/proton transfer on D; TYPE II = redox/proton transfer on A; Type III = alteration of the bridge π-conjugation (Adopted from [6]; American Chemical Society, 2013).

Prabhu et al. [41] synthesized and crystallized a new chalcone compound (2E)-3-(3-methyl phenyl)-1-(4-nitrophenyl)prop-2-en-1-one based on the results obtained by Hirshfeld analysis, and nonlinear optical studies showed that the material is of a potential candidate for optical device applications such as optical limiters and optical switches. Tomas et al. [42] synthesized two molecules namely, 2,2′;-(1,4-phenylene-bis(ethyne-2,1-diyl))bis(N,N-diethyl-4-nitroaniline) (PDNA) and 2,2′-(1,4-phenylene bis(ethyne-2,1-diyl))bis(4-nitroaniline) (PNA), built up from two p-nitroaniline "push-pull" NLO chromophores slightly conjugated through a 1,4-phenylene bis(ethyne-2,1-diyl) bridge. The crystal structure of PNA reveals that the ground (off) state conformation corresponds to the centrosymmetric anti-conformer in which the molecular hyperpolarizability is equal to zero. Also, a computational investigation performed by the group reveals that once submitted to an intense electric field, the two p-nitroaniline subunits are gradually aligned in the direction of the field with the appearance of a metastable (on) state, stabilized by conjugation. During the on/off cycle, a loop of hysteresis was observed and thereby formed molecular bistability. This behavior was evidenced by investigating the NLO response of PNA at different field intensities by the electric field induced second-harmonic (EFISH) technique.

Kumar et al. [43] in their work proposed an optical switch based on nonlinear one-dimensional photonic bandgap material with polystyrene and SiO_2 as chosen constituent materials. The propagation of optical waves through these multilayered structures was studied for different intensities. The simulation of the device and its performance was carried out by transfer matrix method. The results suggest that the polystyrene/SiO_2 based nonlinear photonic bandgap material performed here can work as an optical switch. Also, the proposed device could be useful in optical time-division multiplexing (OTDM), demultiplexing, all-optical information processing, etc. This device could be a strong candidate for the variety of applications in the field of integrated optoelectronics, owing to its simple geometric structure and operating principle.

The physical properties of phase transition materials are directly linked to the transformation of structural moieties. In this context, Jing et al. [44] reported a new H-bonded binary crystal of cyclohexylaminium chlorodifluoroacetate, consisting of two distinct types of the anion-cationic structural moieties, which shows an order–disorder phase transition around 153 K (phase transition point, Tc). Thermal analyses, dielectric and second-order nonlinear optical measurements confirm the symmetry transformation from the point group mmm to mm2. It is noteworthy that cyclohexylaminium chlorodifluoroacetate exhibits a coexistence of switchable dielectric behaviors and bistable NLO effects. It exhibits NLO-activity below Tc, while its quadratic NLO effects disappear completely above Tc. Such an effect reveals the bistable feature of its quadratic NLO properties. The temperature variable-structure analyses reveal that the emergence of NLO effects below Tc is as to the partial ordering of anions as well as the small-angle reorientation of cations. This mechanism is different from that of the precedent NLO-switching materials, as it offers a potential way to design new electric-ordered molecular compounds as multifunctional materials.

4.2 Optical switches based on metallochromophores

The last two decades have been in the great boom in the development of metal–organic frameworks and coordination compounds. The resulting materials have proved to be excellent sorbents for catalysts, gases, objects for storing hydrogen as a fuel, elements of analytical devices for separating substances, as well as demonstrating unique physical properties in the field of nonlinear optics and ferroelectricity magnetism. Recently, unique features such as strong optical anisotropy in the visible region have also been studied in addition to the existing scope of applications, and this is mainly due to a vast surface area [45].

In this context, Green et al. [46] reviewed the progress in employing metal alkynyl complexes as switchable NLO systems. The origin of NLO effects in molecules is summarized, molecular switching stimuli and read-outs were briefly reviewed, and switching NLO properties using other organic, inorganic, and organometallic molecules were also briefly discussed. NLO properties using other organic, inorganic, and organometallic molecules were also briefly summarized. Sarbjeet et al. [47], in their review, the aspects of new ferrocene-based chromophores have presented, and structure–property correlations have been studied, as the ferrocene unit behaves as an extremely strong donor or acceptor in its chromophore. This property has been used extensively in fine-tuning the electronic properties of the organometallic chromophores.

Juan et al. [48] reported three organic–inorganic hybrid analogs, (1) $[Et_3(n\text{-}Pr)P][Mn(dca)_3]$, (2) $([Et_3(CH_2CHCH_2)P][Mn(dca)_3])$ and (3) $([Et_3(MeOCH_2)P][Mn(dca)_3])$ (dca = dicyanamide, $N(CN)_2$-), and showed similar 3-dimensional perovskite frameworks, wherein the cavities are occupied by the guest phosphonium cations. At room temperature, the analogs 1 and 3 correspond to the orthorhombic noncentrosymmetric space group P212121, while analogue 2 crystallizes in the monoclinic centrosymmetric space group P21/c. Both DSC and dielectric measurements confirmed the phase transitions in 1–3, where subtle structural distinctions of guest cations affect the crystal lattices, phase transition temperature, and physical properties. The analogs 1 and 3 are SHG active at room temperature and which can be used as NLO switches tuned by guest cations.

A series of dipolar bis(dibutyl-amino styryl)bipyridine M(II) dichloride complexes (M = Co, Ni, Cu, Zn) and octupolar tris(bipyridine) M(II) coordination compounds of D_3 symmetry (M = Fe, Ru, Os, Ni, Cu, Zn) was studied by Bozec et al. [49]. The quadratic NLO susceptibilities of these complexes were evaluated with the harmonic light scattering technique and found to be at 1.91 mm. From the study, it was clear that the nature of the metallic centers strongly influenced the optical and NLO properties, and the first hyperpolarizability could be tuned by changing the nature of the metal ion. Thus, this could be a competent approach to the design of switchable NLO materials. This is the reason; the photochromic metal complexes yielded an efficient on/off reversible switching $\mu\beta$ values as determined by EFISH at 1.91 mm.

Qiang et al. [50] successfully synthesized and grown block crystals of $[trimethyl(chloromethyl)ammonium]_2[ZnCl_4]$ $[(TMCM)_2\text{-}ZnCl_4]$, which undergoes two reversible solid-state phase transitions at 337 K (Tc_1) and 258 K (Tc_2), respectively, with coexisting switchable nonlinear optical and dielectric properties. Exceptionally, variable-temperature x-ray single-crystal structural analyses also revealed that the unusual space group changes between Tc_2 and Tc_1 and above Tc_1. As expected, the dielectric value of $(TMCM)_2\text{-}ZnCl_4$, indicated that it exhibits a low dielectric state, an intermediate dielectric state, and a high dielectric state. The transitions of $(TMCM)_2\text{-}ZnCl_4$ could be intrinsically ascribed to the sequential order–disorder transformations of both the cations and anions. In particular, the second harmonic generation efficiency shows a sharp step-like increase from almost zero to 0.47. All these results demonstrate its potential application as a switchable and tunable molecular dielectric and NLO material. By Ming and coworkers [51] proposed and numerically investigated all-optical switching based metal–insulator–metal (MIM) waveguide resonator with an outer portion smooth bend structure containing Kerr nonlinear materials. In the structure, pump light locates at mixed-modes (the dips overlap), and then resonance in the resonator, which supports for the improvement of nonlinear response. Light signal locates at split modes, which provides the full width at half-maximum of dips decrease significantly, and increase the difference of signal light transmittance; as a result, when pump light is on or off. This was also verified by the finite-difference time-domain simulations, with tiny optical bistability with signal light appear by varying the control-light intensity, and the switching time is at approximately the picosecond level.

Further, Benoit et al. [52] developed a robust and multi-usage detection tool that could recognize cations by molecular switches with a significant contrast of the second-order NLO optical properties. The concept of proof is given by evidencing by *ab initio* calculation, the ability of spiropyran/merocyanine systems to detect alkali, alkaline earth, and transition metal cations selectively. The complexes can have a broadband spectral and temporal response and strong reverse saturable absorption and two-photon absorption in the visible and near-IR region. Accordingly, Sun et al. [53] Pt (II) provided tridentate or bidentate complexes containing heterocyclic ligands with excellent nonlinear optical properties. As such, the complexes were useful for organic light-emitting diodes and optical-switching or sensing devices. To switch quadratic nonlinear optical effect has become an exciting branch of material science. However, solid-state molecular crystals exhibiting tunable and switchable NLO behaviors remain scanty. Herein, Adnan et al. [54] reported the picrate-based binary organic molecular crystal, triethylammonium picrate (TEAP), which undergoes a phase transition at Tc = 319 K above-room-temperature, being solidly confirmed by the dielectric and thermal measurements. A sizeable thermal hysteresis of ~7 K reveals the first-order feature for its phase transition. The quadratic NLO effects of TEAP can be switched in the vicinity of Tc, i.e., TEAP exhibits NLO-active response of ~1.5 times as large as that of KDP below Tc (NLO-on state); however, its NLO effects completely disappear above Tc (NLO-off state). The structure analyses reveal that the order–disorder transformations of triethylammonium cations and picrate anions together contribute to its both switchable NLO behavior and phase-transition. This work provides a new direction to the designing and assembling of stimuli-responsive materials. Efficient NLO properties of topological switches can be obtained from the push-pull chromophores with π-conjugated spacer and strong electron-withdrawing and releasing groups located on opposite sides of the skeleton ring. Sucarrat et al. [55] achieved this, as they generated unique Huckel-to-Mobius topology switches, with the macrocyclic ring of expanded porphyrins, which presents a conformational versatility that leads to original structural motifs. These systems could act as optoelectronic materials, and their range of applicability depends on the values of the nonlinear optical properties as well as the on substantial differences between the Huckel and Mobius structures.

Ting et al. [56] studied spirooxazine molecular switches, a photochromic material, which transform into open-form of metallic merocyanine by molecular switching and giving rise to significant contrasts in its second-order NLO properties as selective cation sensors. The resulting merocyanine derivatives exhibit 10–21 fold higher static second-order NLO responses. The large first hyperpolarizability relies on the alkaline earth metals present, causing values to increase nearly 21 fold, as evidenced by the larger charge distribution, lower transition energy, and separation distribution of the first hyperpolarizability. Therefore, spiropyran-to-merocyanine molecular switching could be employed to distinguish alkaline earth metals and determine the efficiency of cation detection.

4.3 Optical switches based on supramolecular systems

Chemically, photochemically and electrochemically controllable conformational motion of interlocked molecules such as catenanes, rotaxanes (as shown in

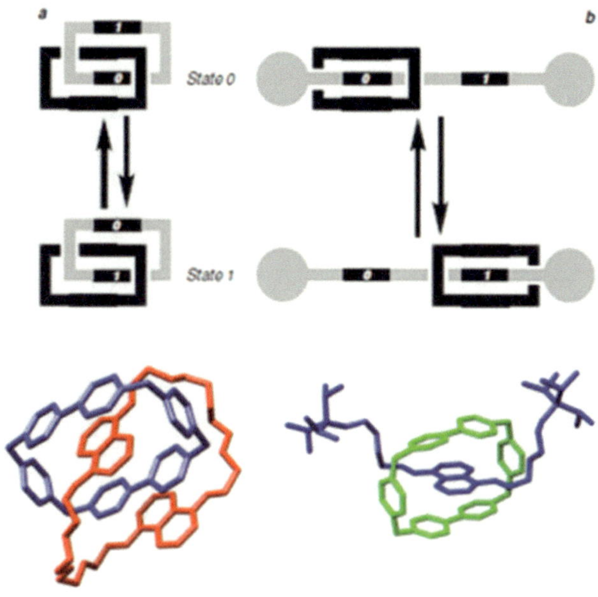

Figure 7.
Interlocked molecules: (a) Catenanes and (b) Rotaxanes (Adopted from [57]; American Chemical Society, 2001).

Figure 7), and pseudorotaxanes have been extensively investigated and reviewed [58]. The stimuli induced molecular motions in these systems may undoubtedly play a key role in the development of optoelectronic devices.

The nonlinear optical properties of phthalocyanines (Pcs) depend not only on the chemical structure of component monomers but also on their packing arrangement, which provides the possibility to incorporate switchability into the third NLO behavior of such materials at supramolecular level. Based on this, Hui et al. [59] rationally designed three series supramolecular third-order NLO photochromic moiety (switching materials) containing Pcs. By utilizing the photoinduced changes in the geometry and electronic delocalization distributions of photochromic moieties under the alternative illumination of suitable UV and visible light, the self-assemblies of Pcs were successfully photo-controlled in an efficient manner. Besides, they also showed large molecular second hyperpolarizabilities at the order of 10^{-30} esu both in “on” and “off” states. The results suggest that these materials would be of potential in emerging optoelectronics and photonic technologies.

Important features of the rotaxane system derived from noncovalent interactions between a component that contains complementary recognition sites. Such interactions, which are also responsible for the efficient template-directed synthesis of rotaxanes, include electron donor-acceptor ability, hydrogen bonding, hydrophobic–hydrophilic character, π-π stacking, electrostatic forces, metal-ligand bonding. Rotaxanes have, therefore, both molecular and supramolecular character: molecular because the components are held together mechanically and can be unlinked only by breaking strong covalent bonds; supramolecular because of the presence of weak noncovalent interactions [60].

Similarly, solid-solid-state reversible phase transitions are widely used in the switchable dielectric, ferroelectric, piezoelectrical, and pyroelectrical and NLO materials. For instance, Yi [61] reported a new crown ether clathrate, $[Habf\text{-}(18\text{-}crown\text{-}6)1.5]^{+}$ $[PF6]^{-}$ (Habf = p-ammonium benzene formamide) synthesis scheme as shown in **Figure 8**, which showed an interesting reversible phase transition and

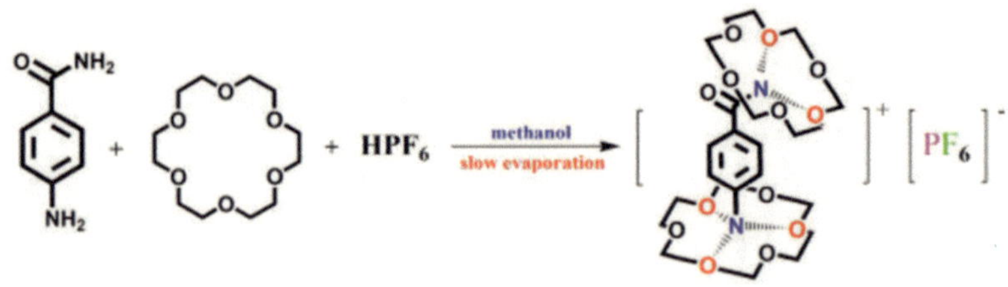

Figure 8.
The preparation of [Habf-(18-crown-6)1.5]⁺ [PF6]⁻ (Habf = p-ammonium benzene formamide) (Adopted from [61]; Royal Society of Chemistry, 2013).

switchable dielectric behaviors near room temperature (263 K) which was triggered by rotation and order–disorder motions of crowns. The synergetic rotation motion between the 18-crown-6 crown ethers and the order–disorder motions of both hexafluorophosphate guest anions and 18-crown-6 crown ether host molecules was analyzed by X-ray single-crystal diffraction. The apparent step-like dielectric change around the phase transition indicates that it has potential applications in dielectric switches.

Khan et al. [62] synthesized a novel molecular phase transition material, 1-methylpiperidinium perchlorate [18-crown-6] monohydrate, which exhibited reversible switchable dielectric anomalies near room temperature (260 K). The phase transition is from noncentrosymmetric to noncentrosymmetric, exhibiting a second harmonic generation response of ~0.8 times high than that of potassium dihydrogen phosphate. The temperature-dependent single-crystal X-ray diffraction reveals that the origin of its phase transition was attributed to the order-disorder transformation of perchlorate anion, the Me group of cation and the torsional angular change in the crown molecule; i.e., all three components contribute to the appearance of this phase change. The results of the work suggest that the synthesized compound could be conceived as a potential switchable dielectric and NLO material. Work also provided a practical approach to design novel NLO switch materials.

4.4 Optical switches based on graphene/carbon materials

The optoelectronic properties of graphene attracted much interest in recent years. Shiramin and Van [63] reported a comprehensive study, investigating the different design trade-offs involved in realizing, in particular, graphene-based modulators and switches. This involved the comparative study of representative hybrid graphene-waveguide configurations, focusing on optimizing their dimensions, the gate-oxide thickness, the polarization, the operating wavelength, and contact definition. Minmin et al. [64] fabricated MoS_2/graphene nanocomposite films with high optical homogeneity and large scale. Using a Z-scan technique with a femto-second laser at 800 nm, the saturable absorption of MoS_2/graphene nanocomposite films was enhanced as compared to pristine MoS_2 film and graphene film. The nonlinear absorption coefficient of MoS_2/graphene nanocomposite film measured at the same condition was close to −1217.8 cm/GW, and this is higher than those of MoS_2 film (−136.1 cm/GW) and graphene film (−961.6 cm/GW) [45].

In the same line, Amos et al. [65] reviewed on the optical properties of carbon nanotube (CNT) and their most relevant applications, including saturable absorption for the passive mode-locking of fiber lasers and optical switching for wavelength conversion. Several demonstrations of integrated graphene-based modulators, switches, detectors, and nonlinear devices have been reported [63].

4.5 Optical switches based on nanoparticles

Recent studies have shown an NLO response in inorganic nanocomposite materials with enhanced SHG coefficient. Some studies have reported the preparation and NLO properties of inorganic nanocomposite materials with noble metals [66]. Herein, Sun et al. [67] reviewed on recent theoretical progress in understanding the nonlinear optical response of noble metal nanocomposite materials with an enhanced third-order nonlinear susceptibility application on optical switchers and optical limiters. The enhanced third-order nonlinear susceptibility results from the surface plasmon resonance of noble metal nanoparticles, which strongly depends on irradiation light wavelength, concentration, and surrounding medium of noble metal particles. Furthermore, their applications as optical switchers with ultrashort time response and optical limiters of intense laser radiation were also reviewed and discussed.

Jagannath et al. [68] fabricated gold nanoparticles (NPs) embedded in sodium antimony borate glass system, and their optical, physical, structural, and ultrafast nonlinear optical properties were studied and analyzed. UV–visible spectroscopy measurements revealed the characteristic surface plasmon resonance of Au NPs at 565 nm and attributed to interband transition. The ultrafast NLO properties of gold doped glasses were studied by the Z-scan technique at a non-resonant wavelength (λ = 800 nm, 80 MHz repetition rate) with femtosecond Ti: sapphire laser pulses. The Z-scan data demonstrated that the investigated glasses are potential materials for the applications in NLO devices, particularly in optical switching devices.

5. Device fabrication of optical switches

The molecular switch is a multifunctional device, which identifies in telecommunication networks to even human body muscles as a modern optical device. In this context, light is a critical element of optical switches, which influences the switching mechanism to bring on/off state [69]. Fabrication of such optical switches mainly depends on the functions of the optical switch, internal architecture, and substrate materials. Most of the optical switches are based on the silicon platform; in some cases, SiGe is also used. The deposition of the material on the substrate is broadly classified into two classes, viz., physical vapor deposition and chemical vapor deposition. To reduce the errors in device fabrication, generally, the chemical vapor deposition technique preferred.

Light guided through couplers by varying the coupling ratio comes under the section of electro-optic switches. In telecommunication networks, they operate between the wavelengths 850 and 1550 nm, and the widely used materials are based on LiNbO, III-V group conductors, GaAs or InP, polymers, silicon. However, these switches always suffer from fabrication errors [70–78]. Qiao et al. [70] developed optical switches with a complementary metal-oxide-semiconductor (CMOS) and microfabrication techniques having low power consumption, high-speed cross talk, high insertion loss, nanosecond response speed and with excellent long-term stability which implies to be suitable for lightwave circuit applications. Abel et al. [79] designed and developed the high speed ultrafast low power consumption optical switch with CMOS combined monolithic integrated fabrication technique.

Wavelength multiplexing divisions-based operation in optical switches is preferably carried out with a variation of the refractive index of the material by thermo-optic coefficient. The phase changing materials are generally preferred in the thermo-optic switches as they have thermo optic effect. These switches are classified into interferometric and digital optical switches [78]. Thermo optical switches are relatively slow, and the biggest problem is in tuning the material properties with the limiting switch speed and thermo isolation problem [80–84]. Zequin et al. [81] designed with low energy consumption Mach Zehnder Interferometer (MZI) based switch with improved compatibility for configurable and high-bandwidth wavelength division multiplexing application. The fabrication techniques used were CMOS and Silicon-on-Insulator (SOI), which are proved to be the best. Thermo optic switches are highly applicable for MZI, as can have suitable configuration, operating between wavelengths 850 and 1500 nm.

A combination of optical switches based on reflecting surface-micromirror operation called micromechanical electro-mechanical system (MEMS) optical switches which have replaced various optical switches to overcome the various issues. These are the most dominant and promising switches in switching technology, available in 2D, 3D configurations as they possess advantages of good potential in high volume and with low cost with its compactness and optical transparency. It is well-known that silicon is the dominant material for the fabrication process. Surface micromachining, which sounds similar to surface milling, whereas one can only subtract the material from a device, with exterior micromachining, one can both add and subtract material, or one can match layers of material more precisely. **Figure 9** [82] gives a detailed schematic of the process from a plain silicon substrate to an entire surface assembled micromotor. The patterning is performed using a lithography mask, a thin film of metal wrapped within the glass with the fancy pattern on it (in this case, the shape of a micro-cantilever). Ultraviolet light is then illuminated through the holes in the mask, which suppresses some parts of the photoresist, allowing it to be quickly washed away. For instance, Sattari et al. [85] designed and experimentally demonstrated 64 × 64 silicon photonic switches with 3.7 dB on-chip losses, 0.91 μs switching times, digital operation and broadband response from 1460 to 1580 nm. These are scalable when compared to previously reported silicon photonic switches, through its MEMS-actuated adiabatic couplers with high on/off ratios, with no loss in the off state. In its present embodiment, this switch integrated 4096 MEMS-actuated adiabatic couplers on a 7 mm × 7 mm area, as reported is the largest silicon photonic integrated circuit so far demonstrated.

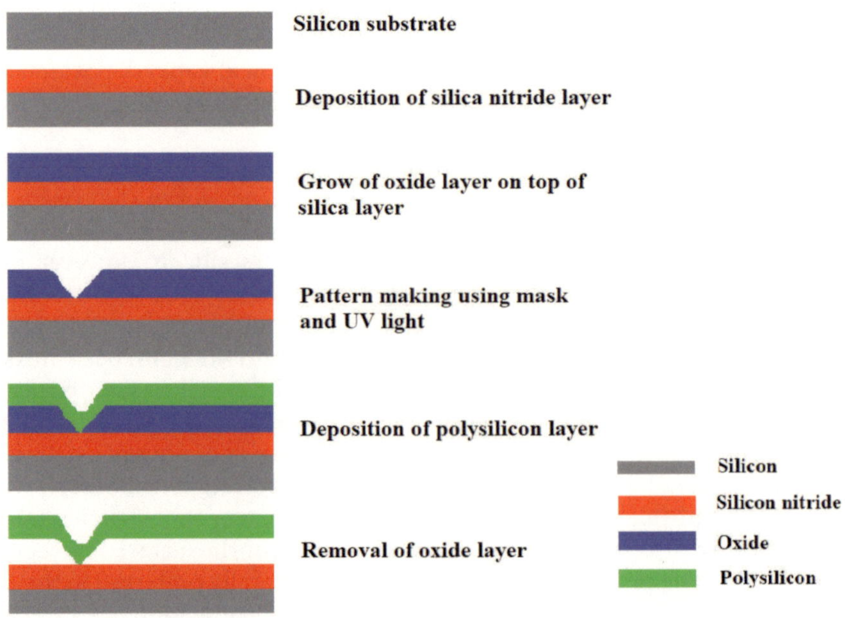

Figure 9.
Simple process for fabricating a micro-cantilever using surface micromachining technique (Reused from [82]).

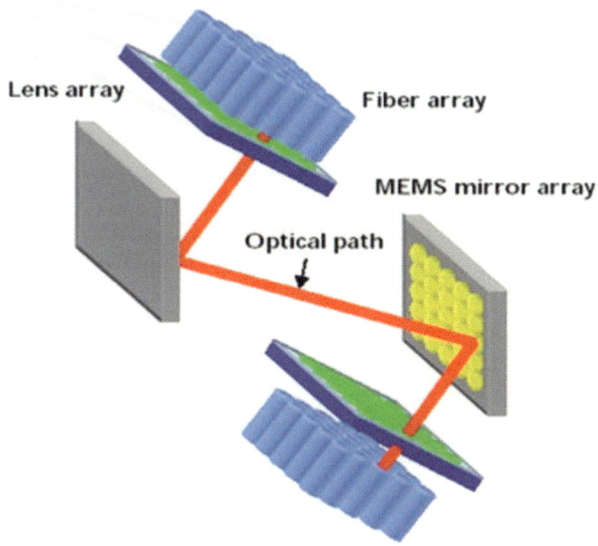

Figure 10.
Ideal MEMS optical switch (Reused from [86]; OSA Publishing).

MEMS optical switches are promising with low cost, lightweight, high speed, high resolution, and high sensitivity [6, 81]. **Figure 10** shows the operation process of the MEMS switching [86]. Based on the various advantages, the MEMS-based optical switches are the dominant and promising switches in switching technology. Primarily switches are fabricated with silicon platform and CMOS fabrication technique, which is compatible with other techniques. The internal architecture also matters in the efficiency of the performance of the device.

6. Applications of switches and future prospectus of designing new molecules

This chapter addresses the molecular design strategies to engineer new SHG materials, which are quite advanced and well employed. It is hoped that the recently introduced algorithm by which the SHG-relevant D-π-A motif can be identified and assessed its strength, which will help to move this field towards a more rational approach to materials discovery of dipolar SHG chromophores. Future challenges depend on extending the investigated objects to supramolecular (e.g., rotaxanes) architectures, which involves many NLO-responsive units, in order to exploit their cooperative effects for enhancing the NLO responses and contrasts [87].

The discovery of more materials is essential before sufficient data are available for establishing the generic relationships between the structure and SHG properties. Nevertheless, the SHG materials have exciting prospects for the NLO applications that employ SHG in a more 3D fashion, in particular to optical switches. Thus, this chapter focused on numerous examples of discrete molecules, as reflects the natural progress of the field [88]. Through these representative examples, one could demonstrate that the rational design of molecular NLO switches, which combines experimental and theoretical approaches.

In order to compare the performance of memory devices, the specifications are yet to be determined. The literature does not fulfill these requirements. Then one should ask whether any structure, with the right materials, is eventually competent in achieving the expected performance. To reply this, it is essential to know the mechanisms accountable for switching and conductance as the active mechanisms are still the subjects of debate, complicated by the fact that the experimental results are not always reproducible, even within the same laboratory. Therefore, it is not easy to study the trade-offs between, for example, in the case of switching and retention times or on-state current and retention time. Endurance may vary in a system that involves mass transport like electrochemistry, electromigration, and molecular reorientation as compared to the one in which only charge transports [89].

Author details

Mahadevappa Y. Kariduraganavar[1*], Radha V. Doddamani[1], Balachandar Waddar[2] and Saidi Reddy Parne[2]

1 Department of Chemistry, Karnatak University, Dharwad, India

2 Department of Physics, National Institute of Technology, Goa, India

*Address all correspondence to: mahadevappayk@gmail.com; kariduraganavarmy@kud.ac.in

References

[1] Marder SR, Sohn JE, Stucky GD. Materials for Nonlinear Optics: Chemical Perspectives. Vol. 455. Washington, DC: American Chemical Society; 1991

[2] Peyghambarian N, Koch SW, Mysyrowicz A. Introduction to Semiconductor Optics. Englewood Cliffs: Prentice-Hall; 1993

[3] Doddamani RV, Tasaganva RG, Inamdar SR, Kariduraganavar MY. Synthesis of chromophores and polyimides with a green chemistry approach for second-order nonlinear optical applications. Polymers for Advanced Technologies. 2018;**29**(7):2091-2102

[4] Asselberghs I, Clays K, Persoons A, Ward MD, McCleverty J. Switching of molecular second-order polarisability in solution. Journal of Materials Chemistry. 2004;**14**:2831-2839

[5] Delaire JA, Nakatani K. Linear and nonlinear optical properties of photochromic molecules and materials. Chemical Reviews. 2000;**100**:1817-1845

[6] Castet F, Rodriguez V, Pozzo JL, Ducasse L, Plaquet A, Champagne B. Design and characterization of molecular nonlinear optical switches. Accounts of Chemical Research. 2013;**46**(11):2656-2665

[7] Stegeman GI, Stegeman RA. Nonlinear optics: Phenomena, materials, and devices. In: Boreman G, editor. Wiley Series in Pure and Applied Optics. Hoboken, NJ: John Wiley & Sons; 2012

[8] Bresien J, Kröger-Badge T, Lochbrunner S, Michalik D, Müller H, Schulz A, et al. A chemical reaction controlled by light-activated molecular switches based on hetero-cyclopentanediyls. Chemical Science. 2019;**10**(12):3486-3493

[9] Zhang JL, Zhong JQ, Lin JD, Hu WP, Wu K, Xu GQ, et al. Towards single molecule switches. Chemical Society Reviews. 2015;**44**:2998-3022

[10] Hollemann AF, Wiberg E, Wiberg N. Lehrbuch der Anorganischen Chemie. Berlin, Germany: Walter de Gruyter; 2007

[11] Sabnis RW, Ross E, Köthe J, Naumann R, Fischer W, Mayer WD, et al. Ullmann's Encyclopedia of Industrial Chemistry. Weinheim, Germany: Wiley-VCH; 2009

[12] Feringa BL, van Delden RA, Koumura N, Geertsema EM. Chiroptical molecular switches. Chemical Reviews. 2000;**100**:1789-1816

[13] Sun Z, Snoke DW. Optical switching with organics. Nature Photonics. 2019;**13**(6):370-371

[14] Zasedatelev AV, Baranikov AV, Urbonas D, Fabio S, Ullrich S, Thilo S, et al. A room-temperature organic polariton transistor. Nature Photonics. 2019;**13**:378-383

[15] Berini P, De Leon I. Surface plasmon–polariton amplifiers and lasers. Nature Photonics. 2012;**6**:16-24

[16] Liebman PA, Parker KR, Dratz EA. The molecular mechanism of visual excitation and its relation to the structure and composition of the rod outer segment. Annual Review of Physiology. 1987;**49**(1):765-791

[17] Li J, Bisoyi HK, Tian J, Guo J, Li Q. Optically rewritable transparent liquid crystal displays enabled by light-driven chiral fluorescent molecular switches. Advanced Materials. 2019;**31**(10):1807751

[18] Harris JD, Moran MJ, Aprahamian I. New molecular switch architectures.

Proceedings of the National Academy of Sciences. 2018;**115**(38):9414-9422

[19] Snoke D. Microcavity polaritons: A new type of light switch. Nature Nanotechnology. 2013;**8**:393-395

[20] Ray D, Foy JT, Hughes RP, Aprahamian I. A switching cascade of hydrazone-based rotary switches through coordination-coupled proton relays. Nature Chemistry. 2012;**4**:757-762

[21] Tatum LA, Foy JT, Aprahamian I. Waste management of chemically activated switches: Using a photoacid to eliminate accumulation of side products. Journal of the American Chemical Society. 2014;**136**:17438-17441

[22] Jones IM, Hamilton AD. Designed molecular switches: Controlling the conformation of benzamido-diphenylacetylenes. Organic Letters. 2010;**12**:3651-3653

[23] Chen CH, Chen YK, Sha CK. Enantioselective total synthesis of otteliones A and B. Organic Letters. 2010;**12**(7):1377-1379

[24] Roldan D, Kaliginedi V, Cobo S, Kolivoska V, Bucher C, Hong W, et al. Charge transport in photoswitchable dimethyldihydropyrene-type single-molecule junctions. Journal of the American Chemical Society. 2013;**135**(16):5974-5977

[25] Deng G, Chun X, Li Z, Zhang X, Yang Y, Huo F, et al. Synthesis and features of nonlinear optical switches based on dithienylethene unit. Tetrahedron Letters. 2018;**59**(37):3448-3452

[26] Raymo FM, Giordani S. All-optical processing with molecular switches. Proceedings of the National Academy of Sciences. 2002;**99**(8):4941-4944

[27] Szukalski A, Moffa M, Camposeo A, Pisignano D, Mysliwiec J. All-optical switching in dye-doped DNA nanofibers. Journal of Materials Chemistry C. 2019;**7**(1):170-176

[28] Hänsel M, Barta C, Rietze C, Utecht M, Rück-Braun K, Saalfrank P, et al. Two-dimensional nonlinear optical switching materials: Molecular engineering toward high nonlinear optical contrasts. The Journal of Physical Chemistry C. 2018;**122**(44):25555-25564

[29] Wirsing S, Hänsel M, Belova V, Schreiber F, Broch K, Engels B, et al. Excited-state dynamics in perylene-based organic semiconductor thin films: Theory meets experiment. The Journal of Physical Chemistry C. 2019;**123**(45):27561-27572

[30] Song P, Gao AH, Zhou PW, Chu TS. Theoretical study on photoisomerization effect with a reversible nonlinear optical switch for dithiazolylarylene. The Journal of Physical Chemistry A. 2012;**116**(22):5392-5397

[31] Lv J, Cao Y, Lin B, Yang Y, Sun Y, Li S, et al. Polymer MZ thermal optical switch at 532-nm based on wet etching and UV-writing waveguide. Polymers. 2019;**11**(6):995

[32] Xing XS, Sa RJ, Li PX, Zhang NN, Zhou ZY, Liu BW, et al. Second-order nonlinear optical switching with a record-high contrast for a photochromic and thermochromic bistable crystal. Chemical Science. 2017;**8**(11):7751-7757

[33] Schwarz JA, Contescu CI, Putyera K, editors. Dekker Encyclopedia of Nanoscience and Nanotechnology (Vol. 3). Boca Raton: CRC Press; 2014

[34] Sun Y, Cao Y, Wang Q, Yi Y, Sun X, Wu Y, et al. Polymer thermal optical switch for a flexible photonic circuit. Applied Optics. 2018;**57**(1):14-17

[35] Zheng Y, Chen CM, Gu YL, Zhang DM, Cai ZZ, Shi ZS, et al.

Fluorinated photopolymer thermo-optic switch arrays with dielectric-loaded surface plasmon polariton waveguide structure. Optical Materials Express. 2015;**5**(9):1934-1948

[36] Chai Z, Hu X, Wang F, Niu X, Xie J, Gong Q. Ultrafast all-optical switching. Advanced Optical Materials. 2017;**5**(7):1600665

[37] Knipe PC, Thompson S, Hamilton AD. Ion-mediated conformational switches. Chemical Science. 2015;**6**(3):1630-1639

[38] Sakata T, Yan Y, Marriott G. Family of site-selective molecular optical switches. The Journal of Organic Chemistry. 2005;**70**(6):2009-2013

[39] Hales JM, Barlow S, Kim H, Mukhopadhyay S, Brédas JL, Perry JW, et al. Design of organic chromophores for all-optical signal processing applications. Chemistry of Materials. 2013;**26**(1):549-560

[40] Coe BJ. Molecular materials possessing switchable quadratic nonlinear optical properties. Chemistry - A European Journal. 1999;**5**(9):2464-2471

[41] Prabhu SR, Jayarama A, Chandrasekharan K, Upadhyaya V, Ng SW. Synthesis, growth, structural characterization, Hirshfeld analysis and nonlinear optical studies of a methyl substituted chalcone. Journal of Molecular Structure. 2017;**1136**:244-252

[42] Guerrero T, Santillan R, García-Ortega H, Morales-Saavedra OG, Farfán N, Lacroix PG. Bis (4-nitroanilines) in interactions through a π-conjugated bridge: Conformational effects and potential molecular switches. New Journal of Chemistry. 2017;**41**(20):11881-11890

[43] Kumar A, Kumar V, Nautiyal A, Singh KS, Ojha SP. Optical switch based on nonlinear one dimensional photonic band gap material. Optik. 2017;**145**:473-478

[44] Zhang J, Sun Z, Zhang W, Ji C, Liu S, Han S, et al. Hydrogen-bonded switchable dielectric material showing the bistability of second-order nonlinear optical properties. Crystal Growth & Design. 2017;**17**(6):3250-3256

[45] Mingabudinova LR, Vinogradov VV, Milichko VA, Hey-Hawkins E, Vinogradov AV. Metal–organic frameworks as competitive materials for non-linear optics. Chemical Society Reviews. 2016;**45**(19):5408-5431

[46] Green KA, Cifuentes MP, Samoc M, Humphrey MG. Metal alkynyl complexes as switchable NLO systems. Coordination Chemistry Reviews. 2011;**255**(21-22):2530-2541

[47] Kaur S, Kaur M, Kaur P, Clays K, Singh K. Ferrocene chromophores continue to inspire. Fine-tuning and switching of the second-order nonlinear optical response. Coordination Chemistry Reviews. 2017;**343**:185-219

[48] Geng FJ, Zhou L, Shi PP, Wang XL, Zheng X, Zhang Y, et al. Perovskite-type organic–inorganic hybrid NLO switches tuned by guest cations. Journal of Materials Chemistry C. 2017;**5**(6):1529-1536

[49] Le Bozec H, Guerchais V, Akdas-Killig H, Maury O, Ordronneau L, Boucekkine A, et al. Nonlinear optics. Quantum Optics. 2012;**43**(1-4):187-196. ISSN: 1543-053

[50] Liao WQ, Gao JX, Hua XN, Chen XG, Lu Y. Unusual two-step sequential reversible phase transitions with coexisting switchable nonlinear optical and dielectric behaviors in $[(CH_3)_3NCH_2Cl]_2$ $[ZnCl_4]$. Journal of Materials Chemistry C. 2017;**5**(45):11873-11878

[51] Tian M, Lu P, Chen L, Liu D, Lv C. All-optical switching in MIM waveguide resonator with an outer portion smooth bend structure containing nonlinear optical materials. Optics Communication. 2012;**285**(21-22):4562-4566

[52] Champagne B, Plaquet A, Pozzo JL, Rodriguez V, Castet F. Nonlinear optical molecular switches as selective cation sensors. Journal of the American Chemical Society. 2012;**134**(19):8101-8103

[53] Wenfang S, Zhang B, Liu R, Shao P, Ji Z, Li Z. Emissive and broadband nonlinear absorbing metal complexes and ligands as OLED, optical switching or optical sensing materials. USA Patent Information: US 20120100628, Classifications: US 436169000 Accession Number: 2012:609374; 2012

[54] Asghar MA, Zhang J, Han S, Sun Z, Ji C, Zeb A, et al. Triethylammonium picrate: An above-room-temperature phase transition material to switch quadratic nonlinear optical properties. Chinese Chemical Letters. 2018;**29**(2):285-288

[55] Torrent-Sucarrat M, Navarro S, Marcos E, Anglada JM, Luis JM. Design of Hückel–Möbius topological switches with high nonlinear optical properties. The Journal of Physical Chemistry C. 2017;**121**(35):19348-19357

[56] Ye JT, Wang L, Wang HQ, Chen ZZ, Qiu YQ, Xie HM. Spirooxazine molecular switches with nonlinear optical responses as selective cation sensors. RSC Advances. 2017;**7**(2):642-650

[57] Feringa BL. In control of motion: From molecular switches to molecular motors. Accounts of Chemical Research. 2001;**34**(6):504-513

[58] Credi A, Semeraro M, Silvi S, Venturi M. Redox control of molecular motion in switchable artificial nanoscale devices. Antioxidants & Redox Signaling. 2011;**14**(6):1119-1165

[59] Chen ZH, Zhang FS. Supramolecular third-order nonlinear optical switches based on the photo-controllable assemblies of phthalocyanines. Image Technology. 2011;**23**:21-25

[60] Mena-Hernando S, Pérez EM. Mechanically interlocked materials. Rotaxanes and catenanes beyond the small molecule. Chemical Society Reviews. 2019;**48**(19):5016-5032

[61] Liu Y, Zhou HT, Chen SP, Tan YH, Wang CF, Yang CS, et al. Reversible phase transition and switchable dielectric behaviors triggered by rotation and order-disorder motions of crowns. Dalton Transactions. 2018;**47**(11):3851-3856

[62] Khan T, Asghar MA, Sun Z, Zeb A, Ji C, Luo J. A supra-molecular switchable dielectric material with non-linear optical properties. Journal of Materials Chemistry C. 2017;**5**(11):2865-2870

[63] Shiramin LA, Van Thourhout D. Graphene modulators and switches integrated on silicon and silicon nitride waveguide. IEEE Journal of Selected Topics in Quantum Electronics. 2016;**23**(1):94-100

[64] He M, Quan C, He C, Huang Y, Zhu L, Yao Z, et al. Enhanced nonlinear saturable absorption of MoS_2/graphene nanocomposite films. The Journal of Physical Chemistry C. 2017;**121**(48):27147-27153

[65] Martinez A, Yamashita S. Carbon nanotube-based photonic devices: Applications in nonlinear optics. Carbon Nanotubes Applications on Electron Device. 2011;**1**:367-386

[66] Gao L, Zhen YL. Third order nonlinear optical response of metal

dielectric composites. Journal of Applied Physics. 2000;**87**:1620

[67] Sun YY, Yang BH, Guo GZ, Shi H, Tian Y, He MH, et al. The study of noble metal nanoparticles applied on third-order nonlinear optical nanocomposite materials. The European Physical Journal Applied Physics. 2011;**56**(1):10402-p1-10402-p5

[68] Jagannath G, Eraiah B, Naga Krishnakanth K, Rao SV. Linear and nonlinear optical properties of gold nanoparticles doped borate glasses. Journal of Non-Crystalline Solids. 2018;**482**:160-169

[69] Nagatoshi K, Robert WJZ, Richard AVD, Nobuyuki H, Ben LF. Light-driven monodirectional molecular rotor. Nature. 1999;**401**:152-155

[70] Qiao L, Tang W, Chu T. 32× 32 silicon electro-optic switch with built-in monitors and balanced-status units. Scientific Reports. 2017;**7**:42306

[71] Tu X, Song C, Huang T, Chen Z, Fu H. State of the art and perspectives on silicon photonic switches. Micromachines. 2019;**10**(1):51

[72] Cheng Q, Bahadori M, Rumley S, Bergman K. Highly-scalable, low-crosstalk architecture for ring-based optical space switch fabrics. In: 2017 IEEE Optical Interconnects Conference (OI). IEEE; 2017. pp. 41-42

[73] Lu L, Zhou L, Li Z, Li D, Zhao S, Li X, et al. 16×16 silicon optical switches based on double-ring-assisted Mach–Zehnder interferometers. IEEE Photonics Technology Letters. 2015;**27**(23):2457-2460

[74] Xing J, Li Z, Zhou P, Xiao X, Yu J, Yu Y. Nonblocking 4× 4 silicon electro-optic switch matrix with push–pull drive. Optics Letters. 2013;**38**(19):3926-3929

[75] Chua SJ, Li BJ. Introduction to optical switches. In: Optical Switches: Materials and Design. Cambridge, UK: Woodhead Publishing Limited; 2010

[76] Soref R. Design of low-energy on-chip electro-optical 1× M wavelength-selective switches. Photonics Research. 2017;**5**(4):340-345

[77] Sun J, Wang X, Yue Y, Meng J, Sun X, Chen C, et al. Nanosecond response polymer electro-optic switch with high poling efficiency and good long-term stability. Modern Physics Letters B. 2013;**27**(22):1350154

[78] Shoman H, Jayatilleka H, Park AH, Mistry A, Jaeger NA, Shekhar S, et al. Compact wavelength-and bandwidth-tunable microring modulator. Optics Express. 2019;**27**(19):26661-26675

[79] Abel S, Stöferle T, Marchiori C, Caimi D, Czornomaz L, Stuckelberger M, et al. A hybrid barium titanate–silicon photonics platform for ultraefficient electro-optic tuning. Journal of Lightwave Technology. 2016;**34**(8):1688-1693

[80] Huang B, Zhang X, Wang W, Dong Z, Guan N, Zhang Z, et al. CMOS monolithic optoelectronic integrated circuit for on-chip optical interconnection. Optics Communication. 2011;**284**(16-17):3924-3927

[81] Lu Z, Murray K, Jayatilleka H, Chrostowski L. Michelson interferometer thermo-optic switch on SOI with a 50-μW power consumption. IEEE Photonics Technology Letters. 2015;**27**(22):2319-2322

[82] Judy JW. Microelectromechanical systems (MEMS): Fabrication, design and applications. Smart Materials and Structures. 2001;**10**:1115-1134

[83] Wang N, Zhang H, Zhou L, Lu L, Chen J, Rahman BM. Design

of ultra-compact optical memristive switches with GST as the active material. Micromachines. 2019;**10**(7):453

[84] Miller DA. Silicon photonics: Meshing optics with applications. Nature Photonics. 2017;**11**(7):403-404

[85] Sattari H, Graziosi T, Kiss M, Seok TJ, Han S, Wu MC, et al. Silicon photonic MEMS phase-shifter. Optics Express. 2019;**27**(13):18959-18969

[86] Goldstein EL, Lin LY, Walker JA. Lightwave micromachines for optical networks. Optics & Photonics News. 2001;**12**(3):60-65

[87] Seok TJ, Kwon K, Henriksson J, Luo J, Wu MC. Wafer-scale silicon photonic switches beyond die size limit. Optica. 2019;**6**(4):490-494

[88] Ashcroft CM, Jacqueline MC. Molecular engineering of organic and organometallic second-order nonlinear optical materials. In: Handbook of Organic Materials for Electronic and Photonic Devices. Cambridge, UK: Woodhead Publishing; 2019. pp. 139-176

[89] Scott JC, Bozano LD. Nonvolatile memory elements based on organic materials. Advanced Materials. 2007;**19**(11):1452-1463

Chapter 10

Bright, Dark, and Kink Solitary Waves in a Cubic-Quintic-Septic-Nonical Medium

Mati Youssoufa, Ousmanou Dafounansou and Alidou Mohamadou

Abstract

In this chapter, evolution of light beams in a cubic-quintic-septic-nonical medium is investigated. As the model equation, an extended form of the well-known nonlinear Schrödinger (NLS) equation is taken into account. By the use of a special ansatz, exact analytical solutions describing bright/dark and kink solitons are constructed. The existence of the wave solutions is discussed in a parameter regime. Moreover, the stability properties of the obtained solutions are investigated, and by employing Stuart and DiPrima's stability analysis method, an analytical expression for the modulational stability is found.

Keywords: higher-order nonlinear Schrödinger equation, spatial solitons, stability analysis method, modulational instability, optical fibers

1. Introduction

The study of spatial solitons in the field of fiber-optical communication has attracted considerable interest in recent years. In a uniform nonlinear fiber, soliton can propagate over relatively long distance without any considerable attenuation. The formation of optical solitons in optical fibers results from an exact balancing between the diffraction and/or group velocity dispersion (GVD) and the self-phase modulation (SPM). The theorical prediction of a train of soliton pulses from a continuous-wave (CW) light in optical fibers was first suggested by Hasegawa and Tappert [1, 2] and first experimentally demonstrated by Mollenauer et al. [3] in single-mode fibers in the case of negative GVD, in liquid CS_2 by Barthelemy et al. in 1985 [4]. In nonlinear optic, optical solitons are localized electromagnetic waves that transmit in nonlinear Kerr or non-Kerr media with dispersion or (and) diffraction without any change in shapes. In nonlinear media, the dynamics of spatial optical solitons is governed by the well-known nonlinear Schrödinger (NLS) equation. Depending on the signs of GVD, the NLS equation admits two distinct types of soliton, namely, bright and dark solitons. The bright soliton exists in the regime of anomalous GVD, and the dark soliton arises in the regime of normal GVD. The physics governing the soliton differs depending on whether one considers a bright or a dark soliton and accordingly features distinct applications [5–8]. The unique property of optical solitons, either bright or dark, is their particle-like behavior in interaction [9].

In addition to fundamental bright and dark solitons, various other forms and shapes of solitary waves can appear in nonlinear media. Kink solitons, for example, are an important class of solitons which may propagate in nonlinear media exhibiting higher-order effects such as third-order dispersion, self-steepening, higher-order nonlinearity, and intrapulse stimulated Raman scattering. In the setting of nonlinear optics, a kink soliton represents a shock front that propagates undistorted inside the dispersive nonlinear medium [10]. This type of solitons has been studied extensively, both analytically and numerically [11–13]. These spatial soliton solutions can maintain their overall shapes but allow their widths and amplitudes and the pulse center to change according to the management of the system's parameters, such as the dispersion, nonlinearity, gain, and so on [14].

The cubic nonlinear Schrödinger equation (CNLSE) has been widely used to model the propagation of light pulse in material's systems involving third-order susceptibility $\chi^{(3)}$, though, for moderate pulse intensity, the higher-order nonlinearities are related to higher-order nonlinear susceptibilities (nonlinear responses) of a material. For example, the cubic-quintic-nonlinear Schrödinger equation (CQNLSE) models materials with fifth-order susceptibility $\chi^{(5)}$. This kind of nonlinearity (cubic-quintic CQ) is named as *parabolic law nonlinearity* and existing in nonlinear media such as the p-toluene sulfonate (PTS) crystals. The parabolic law can closely describe the nonlinear interaction between the high-frequency Langmuir waves and the ion acoustic waves by ponderomotive forces [15, 16], in a region of reduced plasma density, and the nonlinear interaction between Langmuir waves and electrons. In addition, CQ was experimentally proposed as an empirical description of special semiconductor (e.g., AlGaAs, CdS, etc.) waveguides and semiconductor-doped glasses, particularly for the CdS_xSe_{1-x}-doped glass, which exhibit a significant fifth-order susceptibilities $\chi^{(5)}$ as experimentally reported earlier [17, 18]. Moreover, using high laser intensity, the saturation of nonlinearity has been established experimentally in many materials such as nonlinear organic polymers, semiconductor-doped glasses, and so on, which have the property that their absorption coefficient decreases [19]. More generally, a self-defocusing $\chi^{(5)}$ usually accounts for the saturation of $\chi^{(3)}$.

In recent years, many influential works have devoted to construct exact analytical solutions of CQNLSE, such as the pioneering work of Serkin et al. [20]. In particular, Dai et al. [21–25] obtained exact self-similar solutions (similaritons), their nonlinear tunneling effects of the generalized CQNLSE, and their higher-dimensional forms with spatially inhomogeneous group velocity dispersion, cubic-quintic nonlinearity, and amplification or attenuation.

Since the measurement of third-, fifth-, and seventh-order nonlinearities of silver nanoplatelet colloids using a femtosecond laser [26], an extension of nonlinear Schrödinger equation including the cubic-quintic-septic nonlinearity was used to model the propagation of spatial solitons. In [27], for example, the authors performed numerical calculations based on higher-order nonlinearity parameters including seventh-order susceptibility $\chi^{(7)}$ (a chalcogenide glass is an example). This seeds several motivations to discover new features of solitons with combined effects of higher-order nonlinear parameters. In this regard, Houria et al. [28] constructed dark spatial solitary waves in a cubic-quintic-septic-nonlinear medium, with a profile in a functional form given in terms of "$sech^{2/3}$". They have also investigated chirped solitary pulses for a derivative nonical-NLS equation on a CW background [29]. It is obvious to notice that the contributions of the higher-order nonlinearities can give way to generate stable solitons in homogeneous isotropic media and influence many aspects of filamentation in gases and condensed matters [30–33].

Recently, the study of modulational instability (MI) in non-Kerr media has receiving particular attention. MI is a fundamental and ubiquitous process that appears in most nonlinear systems [6, 9, 34–37]. This instability is referred to as modulation instability because it leads to a spontaneous temporal modulation of the CW beam and transforms it into a pulse train. During this process, small perturbations upon a uniform intensity beam grow exponentially due to the interplay between nonlinearity and dispersion or diffraction. As a result, under specific conditions, a CW light often breaks up into trains of ultrashort solitons like pulses [9]. To date, there has not been any report of MI in the cubic-quintic-septic-nonical-nonlinear Schrödinger equation (CQSNNLSE).

Our study will be focused on the analysis of solitary wave's solutions of systems described by the higher-order NLSE named CQSNNLSE. We will discuss the model with higher-order nonlinearities and explore the dynamics of bright, dark, and kink soliton solutions. Finally, the linear stability analysis of the MI is formulated, and the analytical expression of the gain of MI is obtained. Moreover, the typical outcomes of the nonlinear development of the MI are reported.

2. Model equation

The dynamics of (1 + 1)-dimensional (one spatial and one temporal variables) spatial optical solitons is the well-known nonlinear Schrödinger equation. If we consider the higher-order effects, an extended model is required, and the propagation of optical pulses through the highly nonlinear waveguides can be described by the CQSNNLSE:

$$E_z = i\alpha_1 E_{tt} + i\alpha_2 |E|^2 E + i\alpha_3 |E|^4 E + i\alpha_4 |E|^6 E + i\alpha_5 |E|^8 E, \tag{1}$$

where $E(z,t)$ is the slowly varying envelope of the electric field, the subscripts z and t are the spatial and temporal partial derivatives in the frame moving with the pulsed solutions, α_1 is the parameter of diffraction or dispersion, and α_2, α_3, α_4, and α_5 are the cubic, quintic, septic, and nonical nonlinear terms, respectively. This model is relevant to some applications in which higher-order nonlinearities are important.

For example, Eq. (1) with $\alpha_1 = \frac{1}{2}$, $\alpha_4 = 1$, and $\alpha_5 = 0$ was used to study numerically the stability conditions of one-dimensional spatial solitons [38]. Recently, Eq. (1) with $\alpha_5 = 0$ was analyzed for systems that are valid for several types of septic nonlinear materials [28]. Here, we consider arbitrary parameters $\alpha_j (\, j = 1, 2, 3, 4, 5)$ for the sake of a general analysis that is valid for several types of nonical media.

To obtain the exact analytic optical solitary-wave solutions of Eq. (1), we can employ the following transformation:

$$E(\, z, t\,) = \theta(t + \beta z) e^{i(kz - \omega t)} = \theta(\zeta) e^{i(kz - \omega t)}. \tag{2}$$

Here, $\theta(\zeta)$ is a real function and β is a real constant to be determined.

Upon substituting Eq. (2) into Eq. (1) and separating the real and imaginary parts, one obtains

$$\beta = 2\alpha_1 \omega, \tag{3}$$

$$\theta_{\zeta\zeta} = \frac{k + \alpha_1 \omega^2}{\alpha_1} \theta - \frac{\alpha_2}{\alpha_1} \theta^3 - \frac{\alpha_3}{\alpha_1} \theta^5 - \frac{\alpha_4}{\alpha_1} \theta^7 - \frac{\alpha_5}{\alpha_1} \theta^9, \tag{4}$$

Eq. (4) represents the evolution of an anharmonic oscillator with an effective potential energy V [28] defined by

$$V(\theta) = -\frac{k+\alpha_1\omega^2}{2\alpha_1}\theta^2 + \frac{\alpha_2}{4\alpha_1}\theta^4 + \frac{\alpha_3}{6\alpha_1}\theta^6 + \frac{\alpha_4}{8\alpha_1}\theta^8 + \frac{\alpha_5}{10\alpha_1}\theta^{10}. \tag{5}$$

Integrating Eq. (4) yields

$$(\theta_\zeta)^2 = a_1\theta^2 - a_2\theta^4 - a_3\theta^6 - a_4\theta^8 - a_5\theta^{10} + 2\xi, \tag{6}$$

where

$$a_1 = \frac{k+\alpha_1\omega^2}{2\alpha_1}, a_n = \frac{\alpha_n}{n\alpha_1}(n = 2, 3, 4, 5), \tag{7}$$

and ξ is the constant of integration, which can represent the energy of the anharmonic oscillator [39].

In order to get the exact soliton solutions, we first rewrite Eq. (6) in a simplified form by using transformation:

$$\theta(\zeta) = u^{1/2}(\zeta). \tag{8}$$

By substituting Eq. (8) into Eq. (6), we obtain a new auxiliary equation possessing a sixth-degree nonlinear term:

$$\frac{1}{4}(u_\zeta)^2 = a_1u^2 - a_2u^3 - a_3u^4 - a_4u^5 - a_5u^6 + 2\xi. \tag{9}$$

To solve Eq. (9), we will employ three types of localized solutions named bright, dark, and kink solitons. In the following, we solve Eq. (9) by using appropriate ansatz and obtain alternative types of solitary-wave solutions on a CW background and investigate parameter domains in which these optical spatial solitary waves exist.

3. Exact solitary-wave solutions

In this section, we find bright-, dark-, and kink-solitary-wave localized solutions of Eq. (9), by using a special ansatz:

3.1 Bright solitary-wave solutions

The bright solitary solutions of Eq. (9) have the form:

$$u_b(\zeta) = \frac{A_b}{\sqrt{1+N_b\,cosh\,(\alpha_b\zeta)}}, \tag{10}$$

where A_b, N_b, and α_b are real constants which represent wave parameters (A_b and α_b related to the amplitude and pulse width of the bright wave profiles, respectively) to be determined by the physical coefficients of the model.

Substituting the ansatz Eq. (10) into Eq. (9), we obtain the unknown parameters A_b, N_b, α_b, and energy ξ:

$$\begin{cases} A_b = \sqrt{\dfrac{2a_1}{a_3}} \\ \alpha_b = \sqrt[4]{a_1} \\ N_b = \sqrt{\dfrac{a_3{}^2 + 4a_1a_5}{a_3{}^2}}, \\ \xi = \dfrac{a_1a_2}{2a_3} \end{cases} \tag{11}$$

with parametric conditions

$$a_2a_3 + 4a_1a_4 = 0, a_1 > 0, a_3 > 0, a_3{}^2 + 4a_1a_5 > 0. \tag{12}$$

Thus, the exact bright solitary-wave solutions on a CW background of Eq. (1) are of the form:

$$E_b(z,t) = \left\{ \frac{\frac{2a_1}{a_3}}{1 + \sqrt{\frac{a_3{}^2 + 4a_1a_5}{a_3{}^2}} cosh\left[\sqrt[4]{a_1}(t + 2\alpha_1\omega z)\right]} \right\}^{\frac{1}{4}} \times e^{i(kz-\omega t)}. \tag{13}$$

3.2 Dark solitary-wave solutions

The dark solitary solutions of Eq. (9) take the form [40]:

$$u_d(\zeta) = \frac{A_d\, sinh\,(\alpha_d\zeta)}{\sqrt{1 + N_d\, sinh^2(\alpha_d\zeta)}}. \tag{14}$$

Here N_d is a real constant supposed to be positive. Real parameters A_d and α_d are related to the amplitude and pulse width of the dark wave profiles, respectively.

By substituting the ansatz Eq. (14) into Eq. (9), we get the unknown parameters A_d, α_d, and energy ξ:

$$\begin{cases} A_d = \sqrt{\dfrac{2a_1N_d}{a_3}} \\ \alpha_d = \sqrt[2]{a_1} \\ \xi = \dfrac{a_1}{2a_3} \end{cases}, \tag{15}$$

subject to the parametric conditions

$$2(a_2a_3 + 2a_1a_4) - a_3 = 0, a_1 > 0, a_3 > 0, a_3{}^2 + 4a_1a_5 > 0. \tag{16}$$

The exact dark solitary-wave solutions on a CW background of Eq. (1) are of the form:

$$E_d(z,t) = \left\{ \frac{2a_1N_d}{a_3} \frac{sinh^2\left[\sqrt[2]{a_1}(t + \beta z)\right]}{1 + N_d\, sinh^2\left[\sqrt[2]{a_1}(t + 2\alpha_1\omega z)\right]} \right\}^{\frac{1}{4}} \times e^{i(kz-\omega t)}. \tag{17}$$

3.3 Kink solitary-wave solutions

The kink solitary solutions of Eq. (9) are in the following form:

$$u_k(\zeta) = A_k\sqrt{1 + tanh\,(\alpha_k\zeta)}, \tag{18}$$

where A_k and α_k are real parameters related to the amplitude and pulse width of the kink wave profiles, respectively.

Substituting Eq. (18) into Eq. (9), we get

$$\begin{cases} A_k = \sqrt{-\dfrac{a_3}{4a_5}} \\ \alpha_k = \sqrt{-\dfrac{{a_3}^2}{4a_5}}, \\ \xi = -\dfrac{{a_2}^2}{4a_4} \end{cases} \tag{19}$$

under the parametric conditions

$$2a_2a_5 - a_3a_4 = 0, {a_3}^2 + 4a_1a_5 = 0, a_3 > 0, a_5 < 0. \tag{20}$$

Thus, the exact bright solitary-wave solutions on a CW background of Eq. (1) are of the form:

$$E_k(z,t) = \left\{-\frac{a_3}{4a_5}\left[1 + tanh\left(\left(\sqrt{-\frac{{a_3}^2}{4a_5}}(t + 2\alpha_1\omega z)\right)\right)\right]\right\}^{\frac{1}{4}} \times e^{i(kz-\omega t)}. \tag{21}$$

The previous three exact solitary-wave solutions (13), (17), and (21) exist for the governing nonical-NLS model due to a balance among diffraction (or dispersion) and competing cubic-quintic-septic-nonical nonlinearities. For better insight, we plot in **Figure 1** the intensity profile on top of the related first two exact solution solitons named bright and dark, corresponding to the CQNLS models

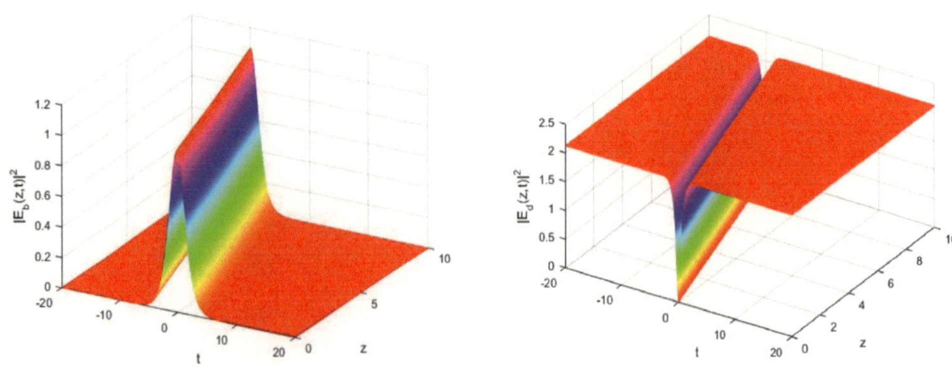

(a) (b)

Figure 1.
Intensity $|E_j(z,t)|^2$ distribution of the (a) bright and (b) dark solitons given by Eqs. (14) and (17), respectively, with the parameter values corresponding to CQNLS models as $\alpha_1 = 0.5$, $\alpha_2 = 0$, $\alpha_3 = 1$, $\alpha_4 = 0$, $\alpha_5 = 0$, $k = 1$, and $\omega = 1$.

(with $\alpha_4 = 0, \alpha_5 = 0$) that is available in the current literature. As we can see from Eq. (21), the kink solitons exist only if $a_5 \neq 0$, consequently $\alpha_5 \neq 0$; thus, we cannot plot the corresponding CQNLS kink solution.

4. Modulational instability of the CW background

One of the essential aspects of solitary waves is their stability on propagation, in particular their ability to propagate in a perturbed environment over an appreciable distance [41]. Unlike the conventional pulses of different forms, the solitons are relatively stable, even in an environment subjected to external perturbations.

The previous three exact solitary-wave solutions given by the expressions (13), (17), and (21) are sitting on a CW background, which may be subject to MI. If this phenomenon occurs, then the CW background will be quickly destroyed, which will inevitably cause the destruction of the soliton. It is therefore of paramount importance to verify whether the condition of the existence of the soliton can be compatible with the condition of the stability of the CW background. Since MI properties can be used to understand the different excitation patterns on a CW in nonlinear systems, in this section, we perform the standard linear stability analysis [9, 34] on a generic CW:

$$E_0(z,t) = \sqrt{P_0} e^{i\phi_{nl}}, \tag{22}$$

in the system modeled by Eq. (1), where $\phi_{nl} = P_0\left(\alpha_2 + \alpha_3 P_0 + \alpha_4 {P_0}^2 + \alpha_5 {P_0}^3\right) z$ is the nonlinear phase shift induced by self-phase modulation and non-Kerr quintic-septic-nonical nonlinear terms, P_0 being the initial power inside a medium exhibiting optical nonlinearities up to the ninth order. A perturbed nonlinear background plane-wave field for the CQSNNLSE (Eq. (1)) can be written as

$$E(z,t) = \left[\sqrt{P_0} + a(z,t)\right] e^{i\phi_{nl}}, \tag{23}$$

where $a(z,t)$ is a small perturbation field which is given by collecting the Fourier modes as

$$a(z,t) = a_+ e^{i(Kz - \Omega t)} + a_- e^{-i(Kz - \Omega t)}, \tag{24}$$

a_+ and a_- are much less than the background amplitude P_0, and Ω represents the perturbed frequency. Here, the complex field $|a(z,t)| \ll P_0$. Thus, if the perturbed field grows exponentially, the steady state (CW) becomes unstable. Inserting the expression of a perturbed nonlinear background Eq. (23) into Eq. (1), with respect to Eq. (24), we obtain after linearization the following dispersion relation:

$$K = |\alpha_1||\Omega|\sqrt{\Omega^2 - sgn\,(\alpha_1){\Omega_c}^2}, \tag{25}$$

where $\Omega_c = \frac{1}{\sqrt{\alpha_1}}\sqrt{2\alpha_2 P_0 + 4\alpha_3 {P_0}^2 + 6\alpha_4 {P_0}^3 + 8\alpha_5 {P_0}^4}$ and $sgn\,(\alpha_1) = \pm 1$ depending on the sign of α_1 [$sgn\,(\alpha_1) = +1$, for $\alpha_1 > 0$, and $sgn\,(\alpha_1) = -1$, for $\alpha_1 < 0$]. The dispersion relation (25) shows that the steady-state stability depends critically on whether the light experiences normal or anomalous GVD inside the fiber. In the case of normal GVD ($\alpha_1 < 0$), the wave number K is real for all Ω, and the steady state is stable against small perturbations. By contrast, in the case of anomalous

GVD ($\alpha_1 > 0$), K becomes imaginary for $\Omega < \Omega_c$, and the perturbation $a(z, t)$ grows exponentially with z. As a result, the CW solution E_0 is inherently unstable for $\alpha_1 > 0$. This instability is referred to as modulation instability because it leads to a spontaneous temporal modulation of the CW beam and transforms it into a pulse train. Similar instabilities occur in many other nonlinear systems and are often called self-pulsing instabilities [9, 42–45]. Then, by setting $sgn\ (\alpha_1) = 1$, one can obtain the MI gain $G = 2\,Im(K)$, where the factor 2 convert G to power gain. The gain exists only if for $|\Omega| < \Omega_c$ and is given by

$$G(\Omega) = 2|\alpha_1\Omega|\sqrt{\Omega_c{}^2 - \Omega^2}. \tag{26}$$

The gain attains its peak values when the modulated frequency reaches its optimum value, i.e., its optimum modulation frequency (OMF). The OMF corresponding to the gain spectrum (26) is given by

$$\Omega_{op} = \pm\frac{\Omega_c}{\sqrt{2}}, \tag{27}$$

and the peak value given by

$$G_{op} = G(\Omega_{op}) = |\alpha_1\Omega_c{}^2|. \tag{28}$$

In **Figure 2**, we have shown the variation of OMF, computed from Eq. (27) as a function of the GVD parameter (α_1). The parameter values we have used are given as [34]

$$\begin{aligned} &P_0 = 15W, \alpha_2 = 2736W^{-1}/km, \alpha_3 = 2.63W^{-2}/km, \\ &\alpha_4 = -9.12\times 10^{-4}W^{-3}/km, \alpha_5 = 0.5W^{-4}/km. \end{aligned} \tag{29}$$

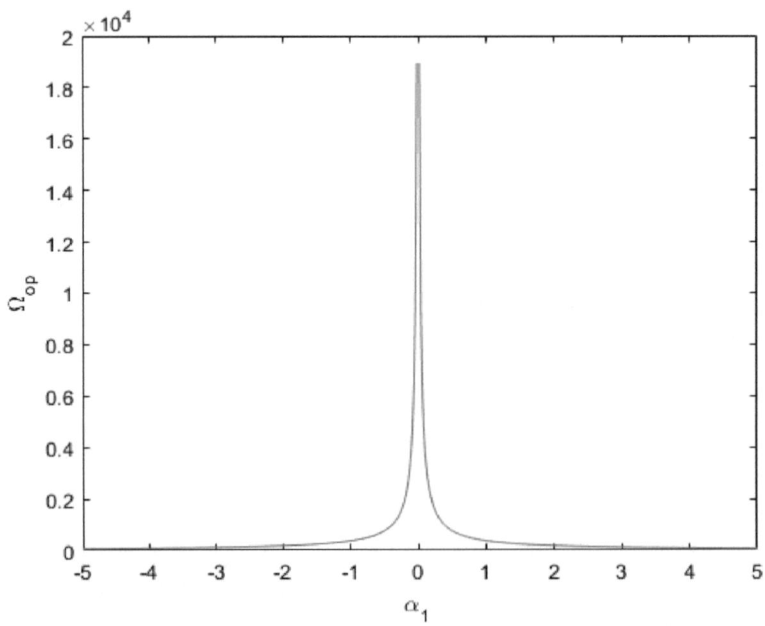

Figure 2.
Variation of optimum modulation frequency Ω_{op} as a function of second-order dispersion α_1.

We can observe that the OMF increases (respectively decreases) with the increasing $\alpha_1 < 0$ (respectively with the increasing $\alpha_1 > 0$).

Figure 3 shows the variation of MI gain as a function of the nonic nonlinearity α_5. The MI gain increases with the decreasing nonic nonlinearity. In **Figure 4**, as the input power increases, the maximum gain also increases.

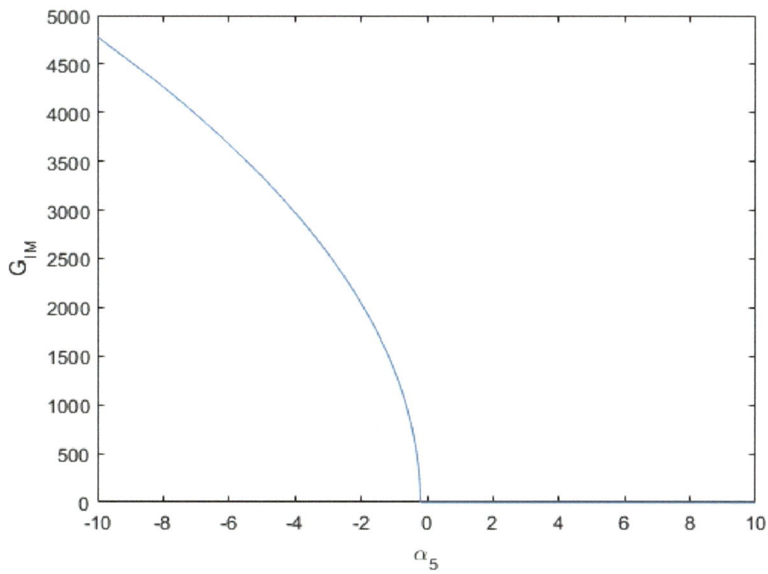

Figure 3.
Variation of the MI gain G as a function of the nonic nonlinearity α_5, with the same parameter values as in ***Figure 2****.*

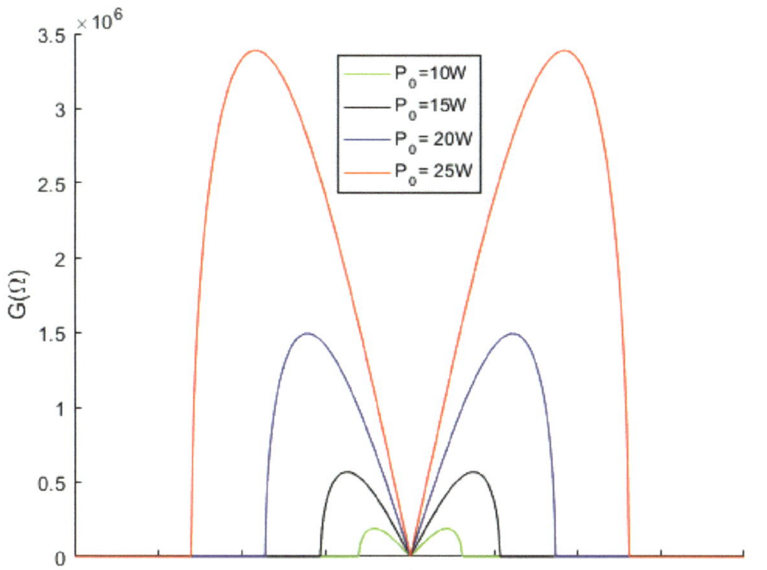

Figure 4.
Variation of the MI gain G (km^{-1}) as a function of frequency Ω (Hz), at a four-power level P_o for an optical fiber. The other parameters are $\alpha_5 = 0.5ps^2/km$, $\alpha_2 = 2736W^{-1}/km$, $\alpha_3 = 2.63W^{-2}/km$, $\alpha_4 = -9.12 \times 10^{-4}W^{-3}/km$, $\alpha_5 = 0.5W^{-4}/km$.

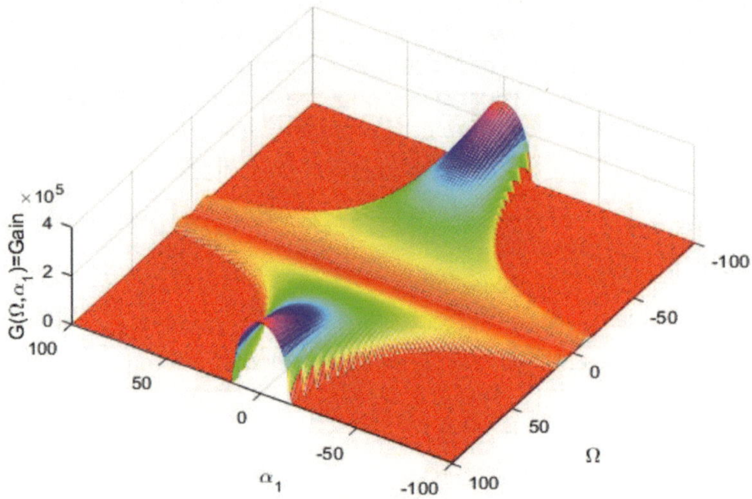

Figure 5.
Variation of the MI gain $G(km^{-1})$ *as a function of frequency* Ω *and the GVD* α_1*. The other parameters are* $P_o = 15W, \alpha_5 = -0.5ps^2/km, \alpha_2 = 2736W^{-1}/km, \alpha_3 = 2.63W^{-2}/km, \alpha_4 = -9.12 \times 10^{-4}W^{-3}/km,$ $\alpha_5 = 5W^{-4}/km.$

The MI gain spectrum in **Figure 5** is a constitutive of two symmetrical sidebands which stand symmetrically along the line $\Omega = 0$. The maximum gain is nil at the zero perturbation frequency $\Omega = 0$; thus, there is no instability at the zero perturbation frequency.

5. Conclusion

In this chapter, we have investigated the higher-order nonlinear Schrödinger equation involving nonlinearity up to the ninth order. We have constructed exact solutions of this equation by means of a special ansatz. We showed the existence of a family of solitonic solutions: bright, dark, and kink solitons. The conditions on the physical parameters for the existence of this propagating envelope have also been reported. These conditions show a subtle balance among the diffraction or dispersion, Kerr nonlinearity, and quintic-septic-nonical non-Kerr nonlinearities, which has a profound implication to control the wave dynamics. Moreover, by employing Stuart and DiPrima's stability analysis method, an analytical expression for the MI gain has been obtained. The outcomes of the instability development depend on the nonlinearity and dispersion (or diffraction) parameters. Results may find straightforward applications in nonlinear optics, particularly in fiber-optical communication.

Conflict of interest

The authors declare no conflict of interest.

Author details

Mati Youssoufa[1*], Ousmanou Dafounansou[1,2*] and Alidou Mohamadou[1,3,4,5*]

1 Faculty of Science, Department of Physics, University of Douala, Douala, Cameroon

2 Faculty of Science, Department of Physics, University of Maroua, Maroua, Cameroon

3 National Advanced School of Engineering, University of Maroua, Maroua, Cameroon

4 Max Planck Institute for the Physics of Complex Systems, Dreseden, Germany

5 The Abdus Salam International Centre for Theoretical Physics, Trieste, Italy

*Address all correspondence to: mati.youss@yahoo.fr, dagaf10@yahoo.fr and mohdoufr@yahoo.fr

References

[1] Hasegawa A, Tappert FD. Transmission of stationary nonlinear optical pulses in dispersive dielectric fibers. I. Anomalous dispersion. Applied Physics Letters. 1973;**23**:142

[2] Hasegawa A, Tappert FD. Transmission of stationary nonlinear optical pulses in dispersive dielectric fibers. II. Normal dispersion. Applied Physics Letters. 1973;**23**:171

[3] Mollenauer LF, Stolen RH, Gordon JP. Experimental observation of picosecond pulse narrowing and solitons in optical fibers. Physical Review Letters. 1980;**45**:1095

[4] Barthelemy A, Maneuf S, Froehly C. Soliton propagation and self-confinement of laser-beams by Kerr optical nonlinearity. Optics Communication. 1985;**55**:201

[5] Hasegawa A, Kodama Y. Solitons in Optical Communications. Oxford: Oxford University Press; 1995

[6] Agrawal GP. Nonlinear Fiber Optics. New York: Academic Press; 2013

[7] Abdullaev F, Darmanyan S, Khabibullaev P. Optical Solitons. Berlin: Springer-Verlag; 1991

[8] Kivshar YS, Luther-Davies B. Dark optical solitons: Physics and applications. Physics Reports. 1998;**298**:81

[9] Kivshar YS, Agrawal GP. Optical Solitons: "From Fibers to Photonic Crystal". San Diego: Academic Press; 2003

[10] Agrawal GP, Headley C III. Kink solitons and optical shocks in dispersive nonlinear media. Physical Review A. 1992;**46**:1573

[11] Raju TS, Panigrahi PK. Self-similar propagation in a graded-index nonlinear-fiber amplifier with an external source. Physical Review A. 2010;**81**:043820

[12] Goyal A, Gupta R, Kumar CN, Raju TS. Chirped femtosecond solitons and double-kink solitons in the cubic-quintic nonlinear Schrödinger equation with self-steepening and self-frequency shift. Physical Review A. 2011;**84**: 063830

[13] Porubov AV, Andrievsky BR. Kink and solitary waves may propagate together. Physical Review E. 2012;**85**: 046604

[14] Agrawal GP. Optical Solitons, Autosolitons, and Similaritons. NY: Institute of Optics, University of Rochester; 2008

[15] Zhou Q, Liu L, Zhang H, Wei C, Lu J, Yu H, et al. Analytical study of Thirring optical solitons with parabolic law nonlinearity and spatio-temporal dispersion. The European Physical Journal - Plus. 2015;**130**:138

[16] Jiang Q, Su Y, Nie H, Ma Z, Li Y. New type gray spatial solitons in two-photon photorefractive media with both the linear and quadratic electro-optic effects. Journal of Nonlinear Optical Physics & Materials. 2017;**26**(1): 1750006 (9 pp)

[17] Topkara E, Milovic D, Sarma AK, Zerrad E, Biswas A. Optical solitons with non-Kerr law nonlinearity and inter-modal dispersion with time-dependent coefficients. Communications in Nonlinear Science and Numerical Simulation. 2010;**15**: 2320-2330

[18] Jovanoski Z, Roland DR. Variational analysis of solitary waves in a homogeneous cubic-quintic nonlinear medium. Journal of Modern Optics. 2001;**48**:1179

[19] Stegeman GI, Stolen RH. Waveguides and fibers for nonlinear optics. Journal of the Optical Society of America B: Optical Physics. 1989;**6**:652

[20] Serkin VN, Belyaeva TL, Alexandrov IV, Melchor GM. Optical pulse and beam propagation III. In: Band YB, editor. SPIE Proceedings. Vol. 4271. Bellingham: SPIE; 2001. p. 292

[21] Dai CQ, Wang YY, Zhang JF. Analytical spatiotemporal localizations for the generalized (3+1)-dimensional nonlinear Schrödinger equation. Optics Letters. 2010;**35**:1437

[22] Dai CQ, Zhang JF. Exact spatial similaritons and rogons in 2D graded-index waveguides. Optics Letters. 2010; **35**:2651

[23] Dai CQ, Zhu SQ, Wang LL, Zhang JF. Exact spatial similaritons for the generalized (2+1)-dimensional nonlinear Schrödinger equation with distributed coefficients. Europhysics Letters. 2010;**92**:24005

[24] Dai CQ, Wang XG, Zhang JF. Nonautonomous spatiotemporal localized structures in the inhomogeneous optical fibers: interaction and control. Annals of Physics (NY). 2011;**326**:645

[25] Dai CQ, Yang Q, He JD, Wang YY. Nonlinear tunneling effect in the (2+1)-dimensional cubic-quintic nonlinear Schrödinger equation with variable coefficients. The European Physical Journal. 2011;**D63**:141

[26] Jayabalan J, Singh A, Chari R, Khan S, Srivastava H, Oak SM. Transient absorption and higher-order nonlinearities in silver nanoplatelets. Applied Physics Letters. 2009;**94**: 181902

[27] Reyna AS, Jorge KC, de Araújo CB. Two-dimensional solitons in a quintic-septimal medium. Physical Review A. 2014;**90**:063835

[28] Triki H, Porsezian K, Dinda PT, Grelu P. Dark spatial solitary waves in a cubic-quintic-septimal nonlinear medium. Physical Review A. 2017;**95**: 023837

[29] Triki H, Porsezian K, Choudhuri A, Tchofo Dinda P. Chirped solitary pulses for a nonic nonlinear Schrödinger equation on a continuous-wave background. Physical Review A. 2016; **93**:063810

[30] Boyd RW, Lukishova SG, Shen YR, editors. Self-focusing: "Past and present (fundamentals and prospects)". In: Topics in Applied Physics. Vol. 114. Berlin: Springer; 2009

[31] Zeng J, Malomed BA. Bright solitons in defocusing media with spatial modulation of the quintic nonlinearity. Physics Review. 2012;**E86**:036607

[32] Couairon A, Mysyrowicz A. Femtosecond filamentation in transparent media. Physics Reports. 2007;**441**:47

[33] Liu W, Petit S, Becker A, Akцozbek N, Bowden CM, Chin SL. Intensity clamping of a femtosecond laser pulse in condensed matter. Optics Communication. 2002;**202**:189

[34] Mohamadou A, Latchio Tiofack CG, Kofané TC. Wave train generation of solitons in systems with higher-order nonlinearities. Physical Review E. 2010; **82**:016601

[35] Shukla PK, Rasmussen JJ. Modulational instability of short pulses in long optical fibers. Optics Letters. 1986;**11**:171

[36] Potasek MJ. Modulation instability in an extended nonlinear Schrödinger equation. Optics Letters. 1987;**12**:921

[37] Porsezian K, Nithyanandan K, Vasantha Jayakantha Raja R, Shukla PK. Modulational instability at the proximity of zero dispersion wavelength in the relaxing saturable nonlinear system. Journal of the Optical Society of America. 2012;**B29**:2803

[38] Reyna AS, Malomed BA, de Araújo CB. Stability conditions for one-dimensional optical solitons in cubic-quintic septimal media. Physical Review A. 2015;**92**:033810

[39] Palacios SL, Guinea A, Fernández-Díaz JM, Crespo RD. Dark solitary waves in the nonlinear Schrodinger equation with third order dispersion, self-steepening, and self-frequency shift. Physics Review. 1999;**E60**:R45

[40] Youssoufa M, Dafounansou O, Mohamadou A. W-shaped, dark and grey solitary waves in the nonlinear Schrödinger equation competing dual power-law nonlinear terms and potentials modulated in time and space. Journal of Modern Optics. 2019;**66**(5): 530-540

[41] Tang XY, Shukla PK. Solution of the one-dimensional spatially inhomogeneous cubic-quintic nonlinear Schrodinger equation with an external potential. Physical Review A. 2007;**76**: 013612

[42] Boyd RW, Raymer MG, Narducci LM, editors. Optical Instabilities. London: Cambridge University Press; 1986

[43] Arecchi FT, Harrison RG, editors. Instabilities and Chaos in Quantum Optics. Berlin: Springer-Verlag; 1987

[44] Weiss CO, Vilaseca R. Dynamics of Lasers. New York: Weinheim; 1991

[45] van Tartwijk GHM, Agrawal GP. Progress in Quantum Electronics. 1998; **22**:43

Chapter 11

Emergence of Raman Peaks Due to Septic Nonlinearity in Noninstantaneous Kerr Media

Michel-Rostand Soumo Tchio, Saïdou Abdoulkary and Alidou Mohamadou

Abstract

We analyze the modulation instability induced by cross-phase modulation of two co-propagating optical beams in nonlinear fiber with the effect of higher-order dispersion and septic nonlinearity. We investigate in detail the effect of relaxation nonlinear response to the gain spectrum both in normal group velocity dispersion (GVD) and anomalous dispersion regime. We show that the walk-off, the relaxation nonlinear response time as well as the higher-order process particularly influence the generation of the modulation instability gain. Our results shows that the emerging Raman peaks is observable both in the case of weak dispersion and in a higher-order dispersion for mixed GVD regime with slow response time. These Raman peaks are shifted toward higher frequencies with the decrease of their magnitude, when the walk-off increases.

Keywords: septic nonlinearity, higher-order dispersions, walk-off effects, delay response time, cross-phase modulation

1. Introduction

The generation of a wave train is a preoccupying subject in the realm of nonlinear science. This is mainly due to two effects: nonlinearity and dispersion. These two notions are essential in the propagation of the wave over long distances and the optical pulse resulting from this interaction gives rise to an optical soliton. The dynamic evolution of nonlinear pulses in nonlinear optical systems can be modeled by the well known nonlinear Schrödinger (NLS) equation which represents the lowest-order nontrivial condition describing the propagation process [1]. The co-propagation of two nonlinear waves in nonlinear optical Kerr media under a slowly varying amplitude approximation is made by using extensions of the NLS equation, whose analytical results provide the dispersion relation, the unstable conditions, as well as the gain spectra. This extension of NLS equation can take into account a large variety of physical properties such as higher-orders dispersion (like third-order dispersion (TOD) and fourth-order dispersion (FOD)) [2–9]; multiple optical beams [10]; negative index material [11]; saturable nonlinearity [12]; and non-instantaneous nonlinear response [13]. Third-order dispersion is used to describe the proprieties of ultrashort pulses in the subpicosecond to femtosecond domain. Usually in nonlinear optic, Kerr nonlinearity is used to compensate the dispersion effects leading to the formation of soliton. Due to this property, focusing Kerr-type media promote the stable propagation of bright solitons [14]. Despite this, the diffraction effect is not enough to balance the self-focusing in focusing quintic nonlinear media and therefore the pulse undergoes critical collapse [15]. Thus, higher-order nonlinearities (HON) play an important role in the description and the propagation of the pulses in Kerr media. Consequently, the extension of NLS equation can also take into account the effect of HON. Reyna et al. [16–18] have experimentally and numerically investigated the nonlinearity management and spatial modulation instability for cubic, quintic, and septic nonlinearity for optical beams propagation in metal-dielectric nanocomposites.

The study of the propagation of an intense optical beams through a nonlinear and dispersive media may lead to fascinating effects such as exponential growth of amplitude due to modulation in amplitude and frequency. This effect is called modulation instability (MI). Modulation instability is a fundamental phenomenon firstly detected in hydrodynamic systems [19] and appears in most nonlinear wave systems. In nonlinear Kerr media such as optical fiber, MI results from the interaction between the nonlinear and dispersive effects [20] characterized by the instability undergone by a continuous wave (cw) or quasi-cw when it propagates inside a nonlinear dispersive system with low noise [21]. Modulation instability has been studied for waves in fluids dynamics [22], plasmas physic [23], dielectric media [24], electrodynamics [25], and atomic Bose-Einstein condensates [26, 27] and was first analyzed theoretically in glass fiber by Hasegawa and Brinkman [28] in 1984 to study the generation of ultrashort optical beams. This phenomenon is generally studied in the anomalous dispersion regime, but it can also be observed in normal dispersion regime when the pumping is carried out close to the zero dispersion subject to certain conditions on the higher order dispersion (HOD) coefficients [29, 30]. Cavalcanti et al. [31] predicted the possibility of MI to occur even in normal group velocity dispersion regime due to the negative values of fourth-order dispersion (FOD). The effects of FOD was also been investigated by many authors who show their inclusion leads to generation of new spectral window. Tchofo et al. have analytically and numerically investigated the behavior of MI under the combined effects of HOD and delayed Raman response [5, 6]. In Ref. [7], the authors shown that FOD shifts the MI peak gain to the higher frequency side and also increases the instability region. Nithyanandan et al. [8] analyzed that the cumulative effect of HOD and walk-off brings new characteristic spectral bands at a definite frequency window.

Wang et al. have shown in their work that the distribution of speed of system flow and the speed of sound determining the occurrence of the sonic horizon is in agreement with the corresponding quantities obtained from a pure numerical evaluation for quantum system incorporating septic nonlinearity modeled by NLS equation [32]. An essential manifestation of the intensity dependence of the refractive index in nonlinear optical media rises through self-phase modulation (SPM) [33], which leads to spectral spreading of optical pulses. It is well known that the co-propagation of two optical waves in nonlinear Kerr media is coupled due to refractive index of the media through the nonlinear phenomenon called cross-phase modulation (XPM) [29, 34].

In this paper, we study the MI in non-instantaneous Kerr media with cubic-quintic-septic nonlinearity, described by a system of two-coupled NLS equation. We mainly focus on revealing the contribution of group velocity mismatch δ, relaxing Kerr nonlinearity, delay response time, cross-phase modulation, and higher-order terms. Analyzing the interplay between instantaneous and non-instantaneous Kerr response for the case of the beams is experiencing normal GVD, and the order beams undergo anomalous GVD. In Section 2, we present the model equation and the linear stability analysis approach; Section 3 features the investigation of MI in the case of weak dispersion regime with the effects of HON. Section 4 is devoted to the analysis of MI gain spectrum by considering the relation between higher-order effects and delay response time. And finally, the conclusion is given in Section 5.

2. Model equations and linear stability analysis

The optical electromagnetic field propagations are described, under a slowly varying amplitude approximation by a coupled nonlinear Schrödinger equation (CNLSE) with higher order dispersion [8] in a single-mode optical Kerr media where higher-order nonlinearities [16] are taken into account. These equations result from two optical beams at different frequencies and the same polarizations where time-dependent nonlinear response is incorporated in the system. The governing systems are read as:

$$i\left(\frac{\partial A_1}{\partial z}+\frac{\delta}{2}\frac{\partial A_1}{\partial t}\right)=\frac{\beta_{21}}{2}\frac{\partial^2 A_1}{\partial t^2}+i\frac{\beta_{31}}{6}\frac{\partial^3 A_1}{\partial t^3}-\frac{\beta_{41}}{24}\frac{\partial^4 A_1}{\partial t^4}-\gamma_1 N_1 A_1, \tag{1}$$

$$i\left(\frac{\partial A_2}{\partial z}-\frac{\delta}{2}\frac{\partial A_2}{\partial t}\right)=\frac{\beta_{22}}{2}\frac{\partial^2 A_2}{\partial t^2}+i\frac{\beta_{32}}{6}\frac{\partial^3 A_2}{\partial t^3}-\frac{\beta_{42}}{24}\frac{\partial^4 A_2}{\partial t^4}-\gamma_2 N_2 A_2. \tag{2}$$

$$\begin{aligned}\frac{\partial N_1}{\partial t}=\frac{1}{\tau}[&-N_1+\kappa_1\left(|A_1|^2+2|A_2|^2\right)+\kappa_2\left(|A_1|^4+6|A_1|^2|A_2|^2+3|A_2|^4\right)\\&+\kappa_3\left(|A_1|^6+18|A_1|^2|A_2|^4+12|A_1|^4|A_2|^2+4|A_2|^6\right)],\end{aligned} \tag{3}$$

$$\begin{aligned}\frac{\partial N_2}{\partial t}=\frac{1}{\tau}[&-N_2+\kappa_4\left(2|A_1|^2+|A_2|^2\right)+\kappa_5\left(3|A_1|^4+6|A_1|^2|A_2|^2+|A_2|^4\right)\\&+\kappa_6\left(4|A_1|^6+12|A_1|^2|A_2|^4+18|A_1|^4|A_2|^2+|A_2|^6\right)]\end{aligned} \tag{4}$$

(κ_1, κ_4), (κ_2, κ_5), and (κ_3, κ_6) are respectively cubic, quintic, and septic nonlinearities parameters.

The nonlinear Kerr parameters γ_i can be expressed as $\gamma_1 = \gamma_2 = n_2/n_0$, the cubic parameters κ_i defined by $\kappa_1 = \kappa_4 = 1$ are control parameters and the quintic-septic parameters can be given, respectively, by $\kappa_2 = \kappa_5 = n_4/n_2$ and $\kappa_3 = \kappa_6 = n_6/n_2$. Where $k = 2\pi n_0/\lambda$ with λ being the wavelength of the laser pump, n_0 is the linear refractive index of the host medium, n_2, n_4, and n_6 are, respectively, the characteristic nonlinear coefficients of the material related to third, fifth and seventh-order susceptibility.

Reyna et al. [16] have proposed this class of HON to study the spatial phase modulation induced by quintic and septic nonlinearities for optical beams propagation in metal colloids, where septic nonlinearity arises from the development up to seventh-order susceptibility of theoretical treatment of Maxwell-Garnett model. Quintic nonlinearity has been considered by Kumar [35] to show the influence of spatial delay on the modulational instability in a composite system with a controllable nonlinearity.

We extend the previous works on MI taking into account the CNLS with higher-order effects and using the linear stability approach. Therefore, we study the stability of the steady-state solution of the above dynamical equations against the small harmonic perturbations; the steady-state solution of the continuous wave (cw) is obtained by setting the time derivative in Eqs. (1)–(4) to zero. This leads the exacts solution in the form: $A_1 = A_1^0 e^{[i\gamma_1 N_1^{cw} z]}$, $A_2 = A_2^0 e^{[i\gamma_2 N_2^{cw} z]}$, where N_1^{cw} and N_2^{cw} reads as: $N_1^{cw} = \kappa_1\left(|A_1^0|^2 + 2|A_2^0|^2\right) + \kappa_2\left(|A_1^0|^4 + 6|A_1^0|^2|A_2^0|^2 + 3|A_2^0|^4\right) + \kappa_3\left(|A_1^0|^6 + 18|A_1^0|^2|A_2^0|^4 + 12|A_2^0|^4|A_2^0|^2 + 4|A_2^0|^6\right)$, $N_1^{cw} = \kappa_4\left(2|A_1^0|^2 + |A_2^0|^2\right) + \kappa_5\left(3|A_1^0|^4 + 6|A_1^0|^2|A_2^0|^2 + |A_2^0|^4\right) + \kappa_6\left(4|A_1^0|^6 + 12|A_1^0|^2|A_2^0|^4 + 18|A_1^0|^4\,|A_2^0|^2 + |A_2^0|^6\right)$ and corresponding to nonlinear phase shift. The dynamic of the system is studied by taking into account the small perturbation using the linear stability analysis theory. Hence, we impose a slight modulation on plane wave as:

$$A_1 = \left[A_1^0 + a_1(z,t)\right]e^{[i\gamma_1 N_1^{cw} z]}, \tag{5}$$

$$A_2 = \left[A_2^0 + a_2(z,t)\right]e^{[i\gamma_2 N_2^{cw} z]}. \tag{6}$$

and

$$N_1 = n_1(z,t) + N_1^{cw}, \tag{7}$$

$$N_2 = n_2(z,t) + N_2^{cw}. \tag{8}$$

where $a_j(z,t)$ are the complex functions and assumed to be small in comparison with the amplitude of the carrier wave ($|a_j(z,t)|^2 \ll \left|A_j^0\right|^2$), and $n_j(z,t)$ are a small deviation from the stationary solution of the nonlinear index. Then inserting Eqs. (5)–(8) in Eqs. (1)–(4), we obtain a set of coupled complex linearized nonlinear Schrödinger equations satisfying the perturbation $a_j(z,t)$ and $n_j(z,t)$ as follows:

$$i\frac{\partial a_1}{\partial z} + \frac{i}{v_{g1}}\frac{\partial a_1}{\partial t} = \frac{1}{2}\beta_{21}\frac{\partial^2 a_1}{\partial t^2} + i\frac{1}{6}\beta_{31}\frac{\partial^3 a_1}{\partial t^3} - \frac{1}{24}\beta_{41}\frac{\partial^4 a_1}{\partial t^4} - \gamma_1 n_1 A_1^0, \tag{9}$$

$$i\frac{\partial a_2}{\partial z} + \frac{i}{v_{g2}}\frac{\partial a_2}{\partial t} = \frac{1}{2}\beta_{22}\frac{\partial^2 a_2}{\partial t^2} + i\frac{1}{6}\beta_{32}\frac{\partial^3 a_2}{\partial t^3} - \frac{1}{24}\beta_{42}\frac{\partial^4 a_2}{\partial t^4} - \gamma_2 n_2 A_2^0. \tag{10}$$

$$\begin{aligned}\frac{\partial n_1}{\partial t} = \frac{1}{\tau}\Big[&-n_1 + \kappa_1\left[A_1^0(a_1+a_1^*) + 2A_2^0(a_2+a_2^*)\right] + \kappa_2[2|A_1^0|^3(a_1+a_1^*)\\ &+6|A_1^0|^2A_2^0(a_2+a_2^*) + 6|A_2^0|^2A_1^0(a_1+a_1^*) + 6|A_2^0|^3(a_2+a_2^*)] + X_1\Big],\end{aligned} \tag{11}$$

$$\begin{aligned}\frac{\partial n_2}{\partial t} = \frac{1}{\tau}\Big[&-n_2 + \kappa_4\left[2A_1^0(a_1+a_1^*) + A_2^0(a_2+a_2^*)\right] + \kappa_5[6|A_1^0|^3(a_1+a_1^*)\\ &+6|A_1^0|^2A_2^0(a_2+a_2^*) + 6|A_2^0|^2A_1^0(a_1+a_1^*) + 2|A_2^0|^3(a_2+a_2^*)] + X_2\Big]\end{aligned} \tag{12}$$

with X_1 and X_2 given by:

$$\begin{aligned}X_1 = \kappa_3[&3|A_1^0|^5(a_1+a_1^*) + 36|A_1^0|^3|A_2^0|^2(a_2+a_2^*) + 18|A_2^0|^4A_1^0(a_1+a_1^*)\\ &+12|A_1^0|^4A_2^0(a_2+a_2^*) + 24|A_1^0|^3|A_2^0|^2(a_1+a_1^*) + 12|A_2^0|^5(a_2+a_2^*)],\\ X_2 = \kappa_6[&12|A_1^0|^5(a_1+a_1^*) + 24|A_2^0|^3|A_1^0|^2(a_2+a_2^*) + +12|A_2^0|^4A_1^0(a_1+a_1^*)\\ &+18|A_1^0|^4A_2^0(a_2+a_2^*) + 36|A_1^0|^3|A_2^0|^2(a_1+a_1^*) + 3|A_2^0|^5(a_2+a_2^*)]\end{aligned}$$

where a_j^* is the complex conjugate of a_j. At this step, to study the stability of the above set of linear equations, we used the Fourier and Laplace transform of the perturbation as follows:

$$a_j(z,t) = \frac{1}{\sqrt{2\pi}} \int a^{-ikz} a^{i\Omega t} \hat{a}_j(\Omega, k) dk d\Omega, \tag{13}$$

$$n_j(z,t) = \frac{1}{\sqrt{2\pi}} \int a^{-ikz} a^{i\Omega t} \hat{n}_j(\Omega, k) dk d\Omega, \tag{14}$$

where $\hat{a}_j(\Omega, k)$ and $\hat{n}_j(\Omega, k)$ stands as Fourier transform in time and Laplace transform in space, Ω and k are, respectively, the frequency pump and wave number of perturbation. Putting Eq. (13) in the previous set of equations (Eqs. (9)–(12)) and after eliminating $(\hat{n}_j(\Omega, k))$, we obtain a set of two homogeneous equations for a_1, a_1^*, a_2, and a_2^*. Then by considering their conjugates, we obtain a set of four homogeneous equations. The resulting matrix presents a nontrivial solution only when the dispersion relation satisfies the following relation:

$$\left[\left(k - \frac{\Omega}{v_{g1}}\right)^2 - f_1\right]\left[\left(k - \frac{\Omega}{v_{g2}}\right)^2 - f_2\right] = C_{XPM}, \tag{15}$$

where the parameters of this equation are defined by the following relation:

$$\begin{aligned} f_1 = B_1(\Omega)[&B_1(\Omega) + 2\tilde{\gamma}_1\kappa_1|A_1^0|^2 + 4\tilde{\gamma}_1\kappa_2\left(|A_1^0|^4 + 3|A_1^0|^2|A_2^0|^2\right) \\ &+6\tilde{\gamma}_1\kappa_3\left(|A_1^0|^6 + 6|A_1^0|^2|A_2^0|^4 + 8|A_1^0|^4|A_2^0|^2\right)], \end{aligned} \tag{16}$$

$$\begin{aligned} f_2 = B_2(\Omega)[&B_2(\Omega) + 2\tilde{\gamma}_2\kappa_4|A_2^0|^2 + 4\tilde{\gamma}_2\kappa_5\left(|A_2^0|^4 + 3|A_1^0|^2|A_2^0|^2\right) \\ &+6\tilde{\gamma}_1\kappa_6\left(|A_2^0|^6 + 6|A_1^0|^4|A_2^0|^2 + 8|A_1^0|^2|A_2^0|^4\right)] \end{aligned} \tag{17}$$

$$\begin{aligned} C_{XPM} = &16B_1(\Omega)B_2(\Omega)\tilde{\gamma}_1\kappa_1\tilde{\gamma}_2\kappa_4|A_1^0|^2|A_2^0|^2 \\ &+16B_1(\Omega)B_2(\Omega)[9\tilde{\gamma}_1\kappa_2\tilde{\gamma}_2\kappa_5\left(|A_1^0|^2|A_2^0|^6 + |A_1^0|^6|A_2^0|^2 + 2|A_1^0|^4|A_2^0|^4\right) \\ &+3\tilde{\gamma}_1\kappa_1\tilde{\gamma}_2\kappa_5\left(|A_1^0|^2|A_2^0|^4 + |A_1^0|^4|A_2^0|^2\right) + 3\tilde{\gamma}_1\kappa_2\tilde{\gamma}_2\kappa_4\left(|A_1^0|^4|A_2^0|^2 + |A_1^0|^2|A_2^0|^4\right) \\ &+6\tilde{\gamma}_1\kappa_1\tilde{\gamma}_2\kappa_6\left(|A_1^0|^6|A_2^0|^2 + |A_1^0|^2|A_2^0|^6 + 3|A_1^0|^4|A_2^0|^4\right) + 6\tilde{\gamma}_1\kappa_3\tilde{\gamma}_2\kappa_4(|A_1^0|^6|A_2^0|^2 + |A_1^0|^2|A_2^0|^6 \\ &+3|A_1^0|^4|A_2^0|^4) + 18\tilde{\gamma}_2\kappa_2\tilde{\gamma}_2\kappa_6\left(|A_1^0|^8|A_2^0|^2 + |A_1^0|^2|A_2^0|^8 + 4|A_1^0|^4|A_2^0|^6 + 4|A_1^0|^6|A_2^0|^4\right) \\ &+18\tilde{\gamma}_1\kappa_3\tilde{\gamma}_2\kappa_5\left(|A_1^0|^8|A_2^0|^2 + |A_1^0|^2|A_2^0|^8 + 4|A_1^0|^6|A_2^0|^4 + 4|A_1^0|^4|A_2^0|^6\right) \\ &+36\tilde{\gamma}_1\kappa_3\tilde{\gamma}_2\kappa_6\left(6|A_1^0|^8|A_2^0|^4 + 6|A_1^0|^4|A_2^0|^8 + |A_1^0|^{10}|A_2^0|^2 + |A_1^0|^2|A_2^0|^{10} + 11|A_1^0|^6|A_2^0|^6\right)] \end{aligned} \tag{18}$$

With the parameters $B_1(\Omega) = \beta_{21}\frac{\Omega^2}{2} - \beta_{31}\frac{\Omega^3}{6} + \beta_{41}\frac{\Omega^4}{24}$ and $B_2(\Omega) = \beta_{22}\frac{\Omega^2}{2} - \beta_{32}\frac{\Omega^3}{6} + \beta_{42}\frac{\Omega^4}{24}$. Eq. (15) looks similar to the case of dispersion relation obtained from nonlinear Schrödinger equation with second order dispersion. However, the difference here arising from the definition of the parameters f_1, f_2, and C_{XPM}. Hence, the dispersion relation obtained above [Eqs. (15)–(18)] regulates the stability condition for the steady-state solution against harmonic perturbations. This stability condition depends on XPM, HOD, and HON. The perturbations grow exponentially if the wave vector k acquires an imaginary part, along the medium with the MI gain given by the relation $g(\Omega) = 2Im(k)$. Consequently, we can examine qualitatively and quantitatively the role played by the delayed nonlinear response (τ) and the group velocity mismatch (δ).

Hence, for usual Kerr approach ($\tau = 0\,ps$) and for non null XPM ($C_{XPM} \neq 0$), it is straightforward to notice that the dispersion relation is fourth order polynomial with real coefficients in k which yields four solutions. From these four solutions, two are always real and thus, irrelevant to investigations of MI. However, the two other can probably be a complex conjugate pair, thereby could affect the MI dynamics and leading to only one unstable gain sideband in the case of cubic nonlinearity as extensively studied in diverse previous works [8, 20, 29, 33, 34, 36–38]. On the other hand, all the four solutions are complex conjugate pairs as far as higher-order effects are concerned and therefore can participate in the MI dynamics; this feature leads to the possibility of two unstable gain sidebands [9]. As the GVM is defined by $\delta = |v_{g1}^{-1} - v_{g2}^{-1}|$, then while considering that the two optical beams are so close to each other by assuming $v_{g1} \approx v_{g2}$, that is, the GVM is negligible [29], these four solutions are given by:

$$k = \frac{\Omega}{v_{g1}} \pm \frac{1}{2}\sqrt{f_1 + f_2 \pm 2\sqrt{(f_1 + f_2)^2 + 4(C_{XPM} - f_1 f_2)}}, \tag{19}$$

The condition so that MI can occur read as: $C_{XPM} > f_1 f_2$. Thus using Eqs. (16), (17), and (18), this condition reads as:

$$\begin{aligned}
&\left[\frac{B_1(\Omega)}{2\tilde{\gamma}_1\kappa_1|A_1^0|^2} + \sum sgn\left(\beta_{ji}\right) + \frac{2\kappa_2\left(|A_1^0|^2 + 3|A_2^0|^2\right)}{\kappa_1} + \frac{6\kappa_3\left(|A_1^0|^4 + 6|A_2^0|^4 + 8|A_1^0|^2|A_2^0|^2\right)}{\kappa_1}\right] \\
&\times\left[\frac{B_2(\Omega)}{2\tilde{\gamma}_2\kappa_4|A_2^0|^2} + \sum sgn\left(\beta_{ji}\right) + \frac{2\kappa_5\left(3|A_1^0|^2 + |A_2^0|^2\right)}{\kappa_4} + \frac{6\kappa_6\left(6|A_1^0|^4 + |A_2^0|^4 + 8|A_1^0|^2|A_2^0|^2\right)}{\kappa_4}\right] < 4 \\
&+\frac{4}{\kappa_1\kappa_4}\left\{\left(|A_1^0|^2 + |A_2^0|^2\right)\left[9\kappa_2\kappa_5\left(|E_1^0|^2 + |A_2^0|^2\right) + 3(\kappa_1\kappa_5 + \kappa_2\kappa_4)\right] + [6(\kappa_1\kappa_6 + \kappa_3\kappa_4)(|A_1^0|^6|A_2^0|^2\right. \\
&+|A_1^0|^2|A_2^0|^6 + 3|A_1^0|^4|A_2^0|^4) + 18(\kappa_2\kappa_6 + \kappa_3\kappa_5)\left(|A_1^0|^8|A_2^0|^2 + |A_1^0|^2|A_2^0|^8 + 4|A_1^0|^4|A_2^0|^6 + 4|A_1^0|^6|A_2^0|^4\right) \\
&\left.+36\kappa_3\kappa_6\left(6|A_1^0|^8|A_2^0|^4 + 6|A_1^0|^4|A_2^0|^8 + |A_1^0|^{10}|A_2^0|^2 + |A_1^0|^2|A_2^0|^{10} + 11|A_1^0|^6|A_2^0|^6\right)\right]\right\}
\end{aligned} \tag{20}$$

This condition shows that MI depends on the sign of β_{ji} and there is a range of frequency (Ω) over which the MI gain can exist. Thus at these frequencies, the steady-state solution becomes unstable to perturbations. Then, irrespective of the sign of dispersion parameters, this condition can lead to instability because of the presence of XPM that enhanced the MI.

By considering the effects of relaxation time ($\tau \neq 0$), from dispersion relation [Eqs. (15)–(18)], one can observe that the terms C_{XPM} and f_j are complex and consequently produce an imaginary part to the wave number k at any frequency, despite the sign of dispersion parameters. The dispersion relation [Eq. (15)] gives rise to fourth-order polynomial equation with complex coefficients for any finite value of delay response time (τ). Due to the fact that the complex roots do not really appear in conjugate pairs, one has the possibility to obtain until four unstable modes for a given frequency Ω [38]. Therefore, the combine effect of relaxation time and XPM gives rise to four unstable modes contrarily to the case of an instantaneous system. Hence, the role of septic nonlinearity inducing MI is illustrated taking into consideration the effects of walk-off and delay response time for the case in which one of the beams is experiencing normal GVD ($\beta_{ij} > 0$) and the other beam undergoes anomalous GVD ($\beta_{21} < 0, \beta_{22} < 0$) as well as mixed GVD ($\beta_{21} < 0, \beta_{22} > 0$) and "total mixed" GVD ($\beta_{i1} < 0, \beta_{i2} > 0$), ($i = 2, 3, 4; j = 1, 2$). We will separately discuss these cases. Firstly, we illustrate directly the effect of septic nonlinearity on MI gain spectrum for low dispersion and discuss the role of walk-off, the delay response time, and the combine effect of these two parameters. Secondly, we investigate the impact of HOD and HON as well as the nonlinear response time on MI gain spectrum. We end our analysis with the investigation of the septic parameters κ_3. In numerical calculation, we use the following value parameters: the nonlinear terms are given by $\gamma_1 = \gamma_2 = 0.015\ W^{-1}m^{-1}$, the dispersion parameters are $\beta_{21} = \beta_{22} = \pm 0.060\, ps^2 m^{-1}$, $\beta_{41} = \beta_{42} = \pm 0.00010\, ps^4 m^{-1}$, the TOD is giving in the range $\delta_1 = (0 - 8) \times 10^{-4}\, ps^3 m^{-1}$, and the input power is given by $\left|E_j^0\right|^2 = 100\ V.m^{-1}$.

3. Effects of weak dispersion on XPM-induced modulation instability

We investigate the effect of HON on MI gain in the case of low dispersion regime ($\beta_{31} = \beta_{32} = \beta_{41} = \beta_{42} = 0$), considering the effects of walk-off, nonlinear response time, and their combination. In this regime, several studies have been conducted and authors have shown the interplay between walk-off and delay response time [8, 9, 13, 34, 38, 39].

3.1 Roles of walk-off and relaxation of nonlinear response on modulation instability gain spectra

First, let us briefly consider the case of instantaneous response time ($\tau = 0$). In this case, Eq. (15) gives rise to four roots as defined by Eq. (19) which turns to two for null GVM ($\delta = 0$) in normal dispersion regime and to one for mixed dispersion regime, while these equations also turn to two for non null GVM and as GVM reaches certain values, these two roots turned to one. This feature is given in **Figure 1(a)** where one can note different MI gain spectra. Obviously, we noted from this figure that, for null GVM in normal and mixed GVD, the behavior of MI

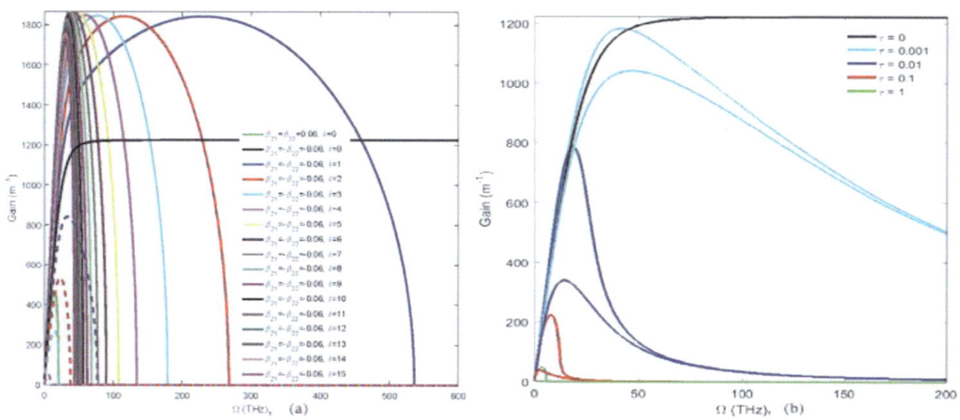

Figure 1.
MI gain spectra g (m^{-1}) as a function of the frequency Ω (THz): (a) for different value of GVM with $\tau = 0$ and (b) in non-instantaneous response time ($\tau \neq 0$) system with null GVM.

gain spectrum is as described in Ref. [39]. Its mean that, taking into account the effect of GVM, the description remain; however, we showed that, as GVM increases the MI gain increase till reach certain values ($\delta = 7$) and started decrease. We also got two gain maxima for more value of delta than in [39] (see blue to magenta curves). This aspect is due to septic nonlinearity.

Now, considering the effect of delay response time by setting GVM null, Eq. (15) leads to four unstable modes which turns to two as given in **Figure 1(b)**. We note from this figure that, for ultra-fast response time, the behavior is as for fast response time for cubic nonlinearity, and the rest of the curves is as described in quintic nonlinearity [39]. As already mentioned, the non-instantaneous response time brings new bands attributed to Raman effect involves in the fibers.

3.2 Combined effects of walk-off and delay response time on modulation instability gain spectra

Taking into account the combined effects of delay response time and GVM, Eq. (15) leads to four unstable modes that turns to two for response time reaching certain value ($\tau = 1$) and coalesce into one at high frequencies. This is given in **Figure 2** where we have fixed the value of δ and varied the nonlinear response time. For mixed GVD, our results shows that, for fast response time ($\tau = 0.01ps$, see blue curve), the conventional band increases when GVM increases, while the Raman band decreases and their width shift toward lower frequencies. The spectra seem the same as in the case of quintic nonlinearity [39].

Now considering that GVM varies with fixed nonlinear response time, we have observed that when the GVM increases, the behavior of the MI gain spectra changes drastically as given in **Figure 3** for mixed GVD. In this figure, the four unstable modes coalesce into one at high frequencies.

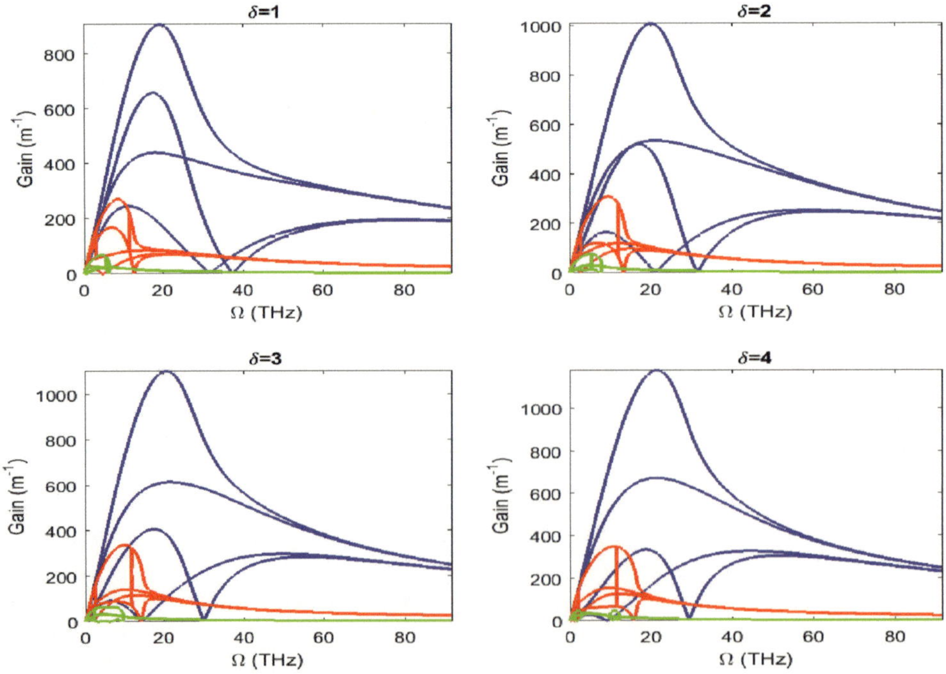

Figure 2.
MI gain spectra g (m^{-1}) as a function of the frequency Ω (THz) for non null GVM and nonlinear response time.

However, for fast response time gives in **Figure 3(a)**, we observe that for conventional and Raman band, the magnitude increases when δ increases; it is the inverse for the sporadic band while their width slightly enlarge and shift toward higher frequencies. The situation is different in the case of slow response time ($\tau = 0.1ps$) as given in **Figure 3(b)**. We observe that as δ increases, the magnitude of conventional MI gain decreases with the appearance of the Raman peaks that shift toward higher frequencies as GVM increases. The behavior of the Raman band remains as in panel (a). However for the sporadic band, their magnitude increases as δ increases and their width enlarges with the appearance of the sporadic peaks that shift toward higher frequencies. These Raman and sporadic peaks are probably brought by the combined effect of delay response time and opposite sign of SOD.

In the case of anomalous dispersion regime, the four unstable mode turns to two modes as given in **Figure 4**. One can observe from this figure that, for conventional spectral band, as GVM increases, the MI gain decreases, while for Raman band (second spectral band), it is inverted. However, the behavior of the MI gain spectra is different from these panels and we can see that as nonlinear response time increases, the magnitude of the gain decreases. In panel (a), we note that before extended in all frequencies, the curves are neutralized at a given frequency and

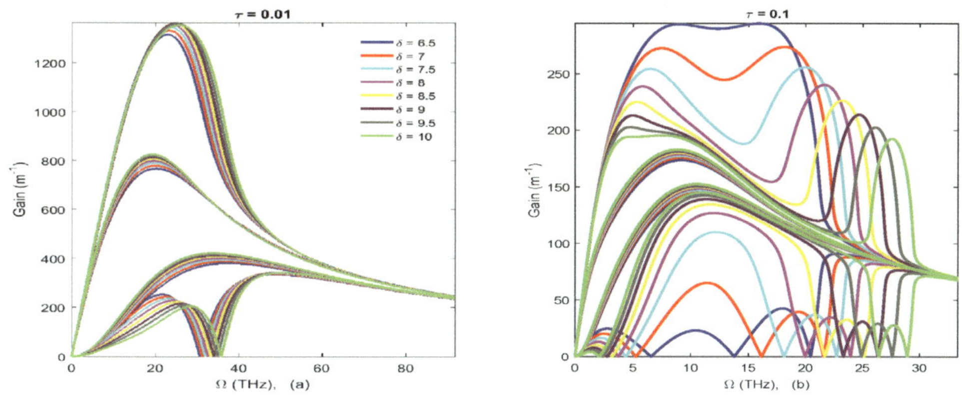

Figure 3.
MI gain spectra g (m^{-1}) as a function of the frequency Ω (THz) for different values of δ in mixed GVD. (a) for τ = 0.01ps, (b) for τ = 0.1ps.

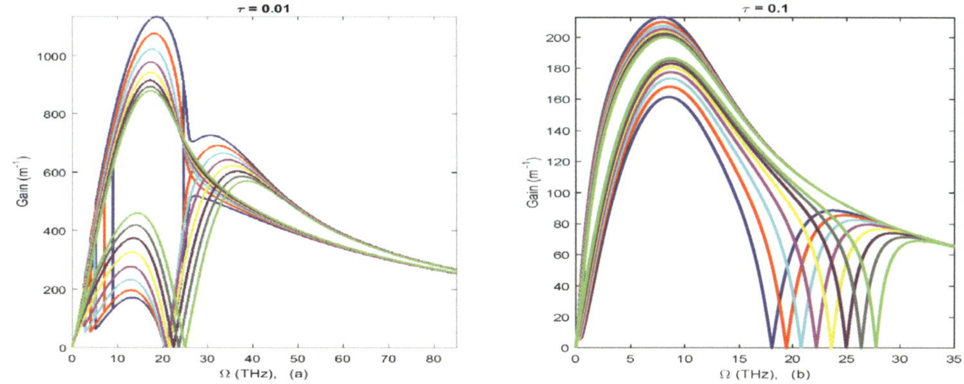

Figure 4.
*MI gain spectra g (m^{-1}) as a function of the frequency Ω (THz) for different values of δ in anomalous GVD. (a) for τ = 0.01ps, (b) for τ = 0.1ps. The colors and the δ values are the same in **Figure 3**.*

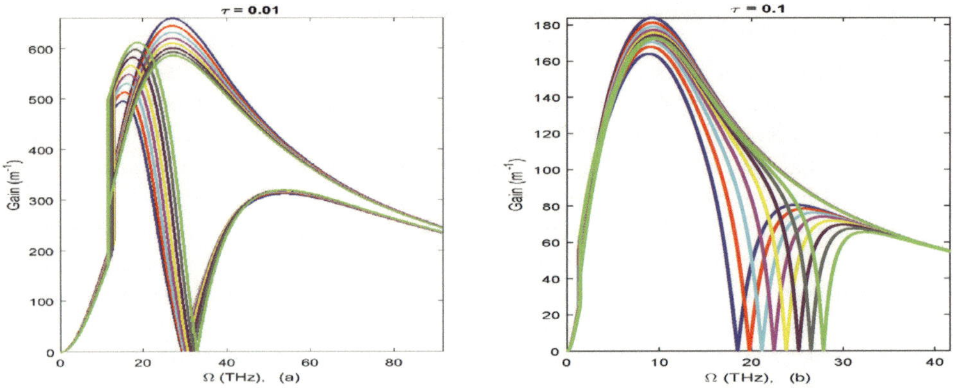

Figure 5.
*MI gain spectra g (m^{-1}) as a function of the frequency Ω (THz) for different values of δ in normal GVD. (a) for τ = 0.01ps, (b) for τ = 0.1ps. The colors and the δ values are the same in **Figure 3**.*

move slowly toward high frequencies while in panel (b), they move rapidly as GVM increases.

In the case of normal dispersion regime, for slow response time [panel (a)], the two unstable modes turn to one for lower frequencies and to two at characteristic frequency until colliding to one at high frequencies. The previous description in anomalous case remains the same. Meanwhile for fast response time [panel (b)], the situation is slightly different where one can observe two unstable modes that collid to one as GVM increases. **Figure 5** portrays this feature. The Raman and sporadic peaks observed previously in **Figure 3** disappeared in these figures due probably to the sign of SOD (they have the same sign).

4. Impact of higher-order dispersion and septic nonlinearity on XPM-induced modulation instability

It has been shown that HOD effects such as FOD and TOD are most essential and give rise to a wide diversity of information concerning the MI dynamics.

4.1 Effect of fourth-order dispersion on modulation instability

The inclusion of FOD brings a great change to the MI spectrum and we obtain for each GVM value four gain spectra. In the case of mixed dispersion regime, from **Figure 6** we note that, the spectrum presents two regions of instability which are connected to each other. The first one at the pump frequency is separated to the second one due to the effect of FOD. Thus in the case of fast delay response time gives in panel (a), one can observe that for the first instability region, the primary band remains unchanged for all GVM values; while for the Raman band as GVM increases, the MI gain spectra increase. For the second instability region, the situation remains the same as in the mixed case for weak dispersion regime (**Figure 3(a)**). In panel (b), our work unveiled that for first instability region, the conventional spectral MI gain decreases as GVM increases, while it is inverted for Raman bands. According to the second instability region, we revealed the presence of Raman and sporadic peaks that magnitude and width decrease as GVM increases and they move toward high frequencies. The fourth unstable mode turns to one at high frequencies. In addition at around $\Omega = 15Thz$, we observe a singularity brought by FOD. In panel (a), this effect neutralized the spectra at this point.

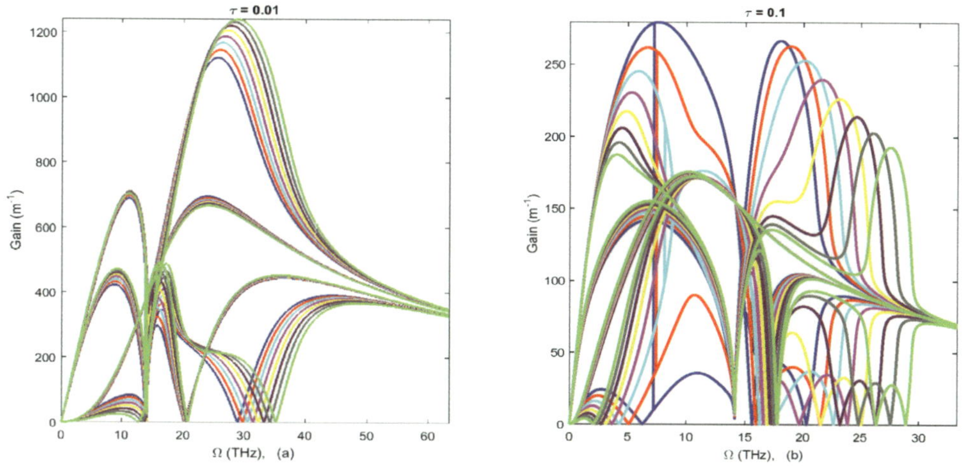

Figure 6.
MI gain spectra g (m^{-1}) as a function of the frequency Ω (THz) for different values of δ in mixed GVD with the inclusion of FOD. (a) for τ = 0.01ps, (b) for τ = 0.1ps. The colors and the δ values are the same in ***Figure 3****.*

Taking into account the propagation with anomalous GVM qualitatively changes the MI spectrum, like shape as in **Figure 7** that for the primary spectral band as GVM increases, the MI gain spectra decrease, contrarily to the Raman bands which increase as GVM increases. In addition, when increasing the delay response time, the magnitude and the width of MI gain spectra decrease considerably with a singularity around $\Omega = 14Thz$ brought by the effect of FOD. Meanwhile in the case of normal GVD as in **Figure 8**, contrarily to what are observed in previous figures, the inclusion of FOD hardly brings any changes to the MI spectrum due mainly to the dominance of normal GVD over FOD. However, the magnitude of the MI gain has considerably increases precisely for the second spectral band.

4.2 Effect of third-order dispersion on modulation instability

We now shift our analysis to the effect of walk-off and delay response time in virtue of TOD. In the case of mixed GVD as seen in **Figure 9**, the spectra hardly

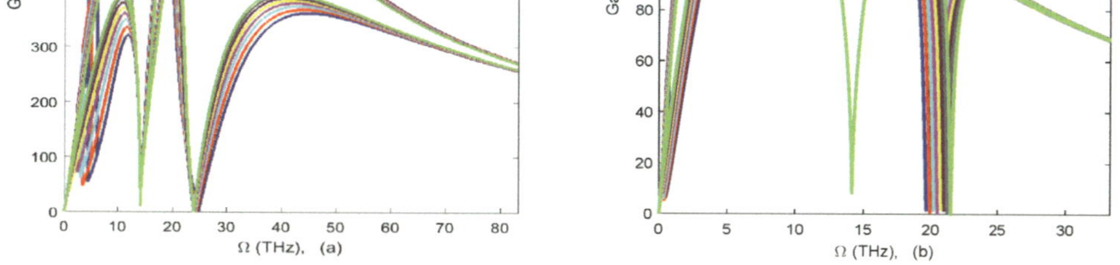

Figure 7.
MI gain spectra g (m^{-1}) as a function of the frequency Ω (THz) for different values of δ in anomalous GVD with the inclusion of FOD. (a) for τ = 0.01ps, (b) for τ = 0.1ps. The colors and the δ values are the same in ***Figure 3****.*

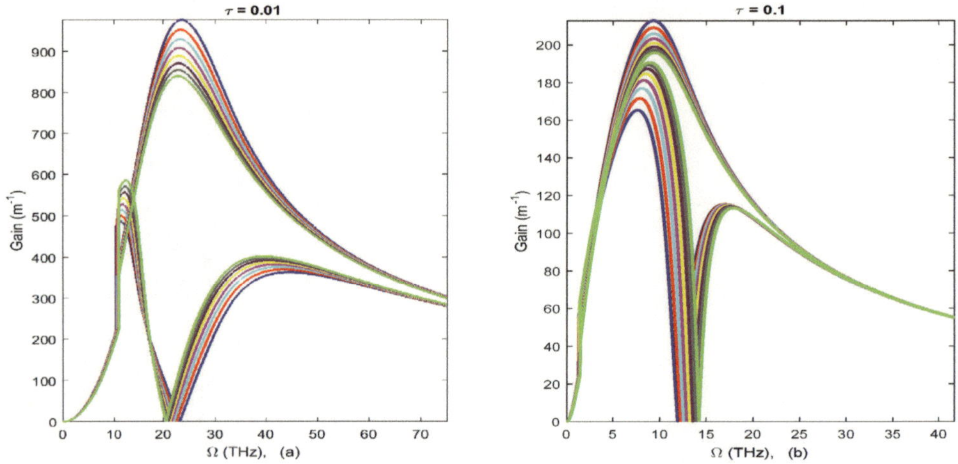

Figure 8.
*MI gain spectra g (m^{-1}) as a function of the frequency Ω (THz) for different values of δ in normal GVD with the inclusion of FOD. (a) for τ = 0.01ps, (b) for τ = 0.1ps. The colors and the δ values are the same in **Figure 3**.*

change for $\tau = 0.01$ [panel (a)] in comparison to **Figure 6** where we considered FOD effect. However, for first instability region, concerning the primary spectral band the magnitude of MI gain slowly increases when GVM increases and their width slightly shifts toward higher frequencies, whereas it is the reverse for second spectral band in this region. In addition, in the case of quintic nonlinearity as done in Ref. [39], the spectrum extended over all frequencies while in this case, we find that the gain spectrum is neutralized three times before extend over all frequencies. The second instability region remains as described in **Figure 6**. Then at high frequencies, the four gain spectra coalesce into one. For slow response time ($\tau = 0.1$) in panel (b), the first instability region remains unchanged, whereas for the second instability region, we note that as GVM increases, the MI gain spectra increase until it reaches a value $\delta = 8$ before it starts decreasing.

In the case of anomalous dispersion regime, the four obtained solution leads to four unstable modes that coalesce to one at higher frequencies. **Figure 10** depicts

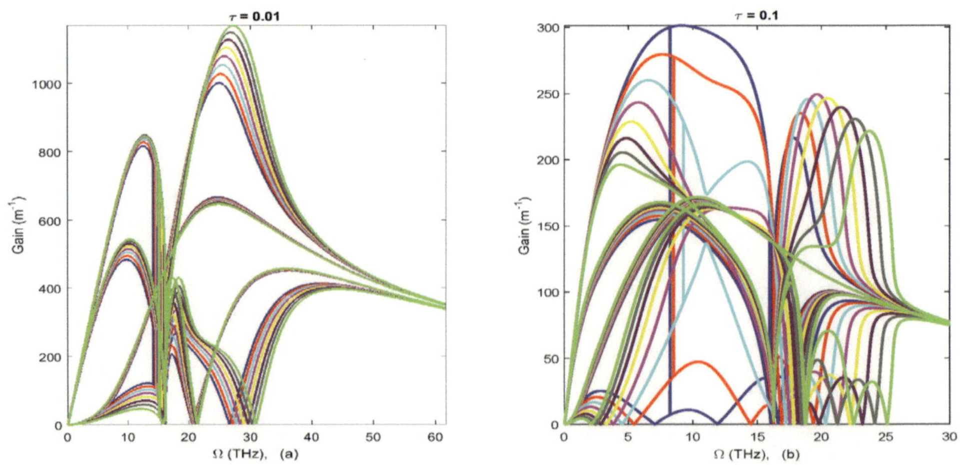

Figure 9.
*MI gain spectra g (m^{-1}) as a function of the frequency Ω (THz) for different values of δ in mixed GVD with the inclusion of TOD. (a) for τ = 0.01ps, (b) for τ = 0.1ps. The colors and the δ values are the same in **Figure 3**.*

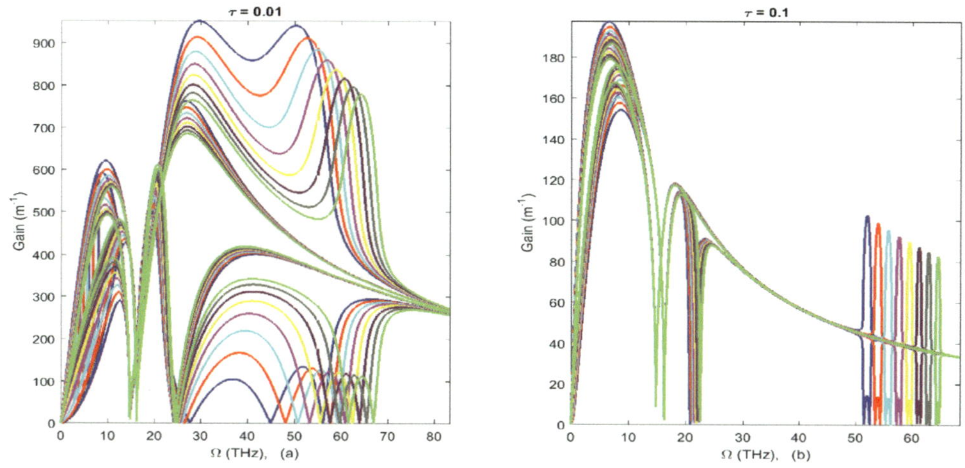

Figure 10.
MI gain spectra g (m^{-1}) as a function of the frequency Ω (THz) for different values of δ in anomalous GVD with the inclusion of TOD. (a) for τ = 0.01ps, (b) for τ = 0.1ps. The colors and the δ values are the same in ***Figure 3****.*

the MI gain spectrum for some GVM value. One observed two singularities due certainly by HOD effects. In the case of fast response time, for first spectral region in panel (a), the primary and Raman bands decrease as GVM increases while for spectra obtained at low MI gain in this region the situation is inverted. In the second region, the inclusion of TOD brings Raman and sporadic peaks that shifted toward higher frequencies due to GVM effect. For slow response time given in panel (b), the situation is almost different as earlier. The sporadic and Raman peaks get shrink and still move toward higher frequencies. It is to explain the behavior of the obtained modes at lower frequencies coalesce to two at middle frequencies and to one at high frequencies. Contrarily to what observed in the case of quintic nonlinearity, in these figures (**Figure 10**), HOD acts in the case of septic nonlinearity.

We now turn our attention to the case of normal dispersion regime as given in **Figure 11**. We note that, the second spectral region remains almost the same as described in previous figure. However, in the first spectral region, the four solution

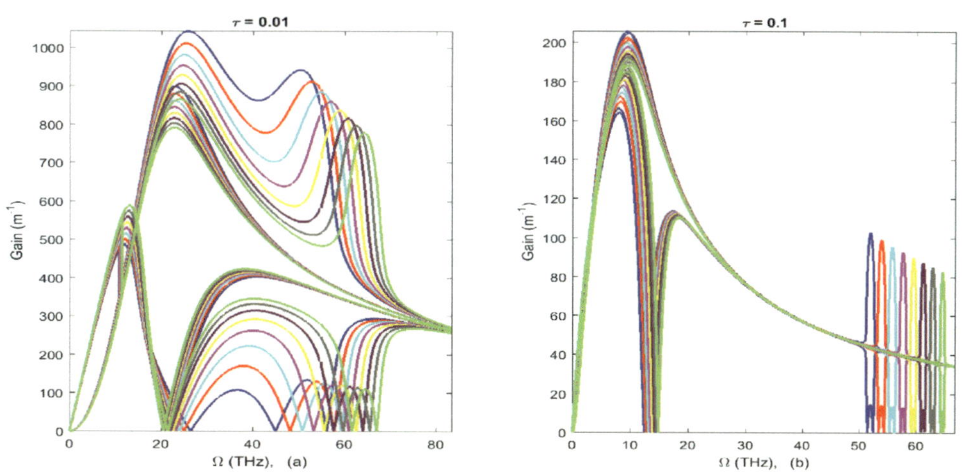

Figure 11.
MI gain spectra g (m^{-1}) as a function of the frequency Ω (THz) for different values of δ in normal GVD with the inclusion of TOD. (a) for τ = 0.01ps, (b) for τ = 0.1ps. The colors and the δ values are the same in ***Figure 3****.*

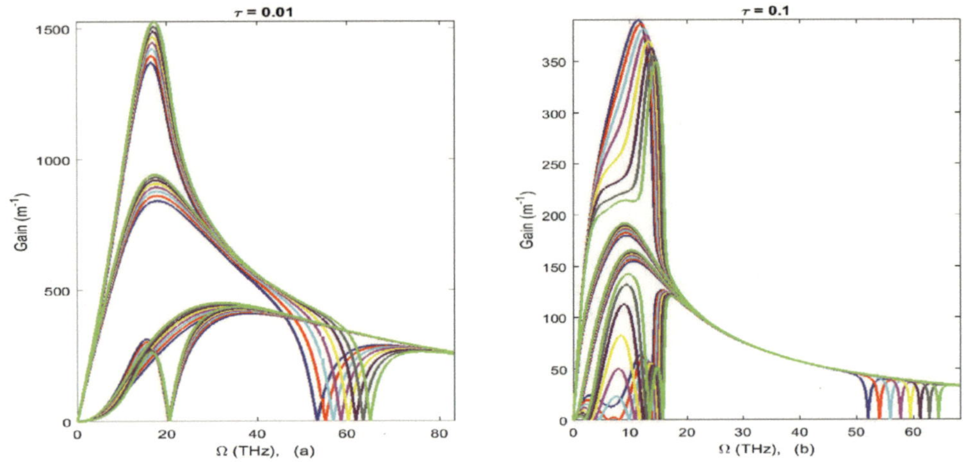

Figure 12.
MI gain spectra g (m^{-1}) as a function of the frequency Ω (THz) for different values of δ in "total mixed" GVD with the inclusion of TOD. (a) for τ = 0.01ps, (b) for τ = 0.1ps. The colors and the δ values are the same in ***Figure 3***.

leads to two unstable modes in panel (a) and to one in panel (b) for lower frequencies. The effect of FOD vanishes due probably to the dominance of SOD and HON. Obviously, we note that as response time increases, the Raman and sporadic peaks get shrink.

Figure 12 depicts the evolution of MI gain spectra in the case of "total mixed" GVD. It is obvious to see from panel (a) (fast response time) that as GVM increases, the conventional bands decrease while it is inverted for Raman bands. The effects of TOD is visible around $\Omega = 60Thz$ where the spectra are neutralized before extended over all frequencies. According to panel (b) for slow response time, we observe the appearance of Raman peaks in conventional bands brought by the effect of SOD that magnitude decreases as GVM increases and they shifted toward high frequencies however their width get shrink over what observe in **Figure 12(a)**. Also we note the existence of non-conventional bands between $\Omega = (11 - 14)Thz$ due to the effect of FOD that decreases and shift toward high frequencies when GVM increases while is it inverted for the other bands between $\Omega = (2 - 13)Thz$ as the magnitude increases, the width widens toward both the low and high frequencies. The effect of TOD is visible around $\Omega = 60Thz$ with the appearance of singularity. It is important to note from this figure that the Raman peaks observe at higher frequencies in previous figures vanish due probably to the sign of used value.

We generally note from these different figures that, the inclusion of TOD and FOD increases considerably the MI gain spectrum and their width. In addition, the formation of Raman and sporadic peaks observed is due on the one hand to the combined effect of SOD and relaxation time and on the other hand due to the combined effect of TOD. They move toward high frequencies due to the effect of GVM. From fast to slow response time, we note that the magnitude of MI gain spectra decreases considerably. The MI occurs in the case of HOD and septic nonlinearity only when the interactions between beams are less (gives here by walk-off parameter δ).

4.3 Effect of septic nonlinearity parameter κ_3 on modulation instability gain

To better appreciate the effect of septic nonlinearity on MI gain, we portrayed the latter versus κ_6 in different cases above. In other to do this, the solution of Eq. (15) gives rise to four roots that turn to two pairs of roots due to HON effect.

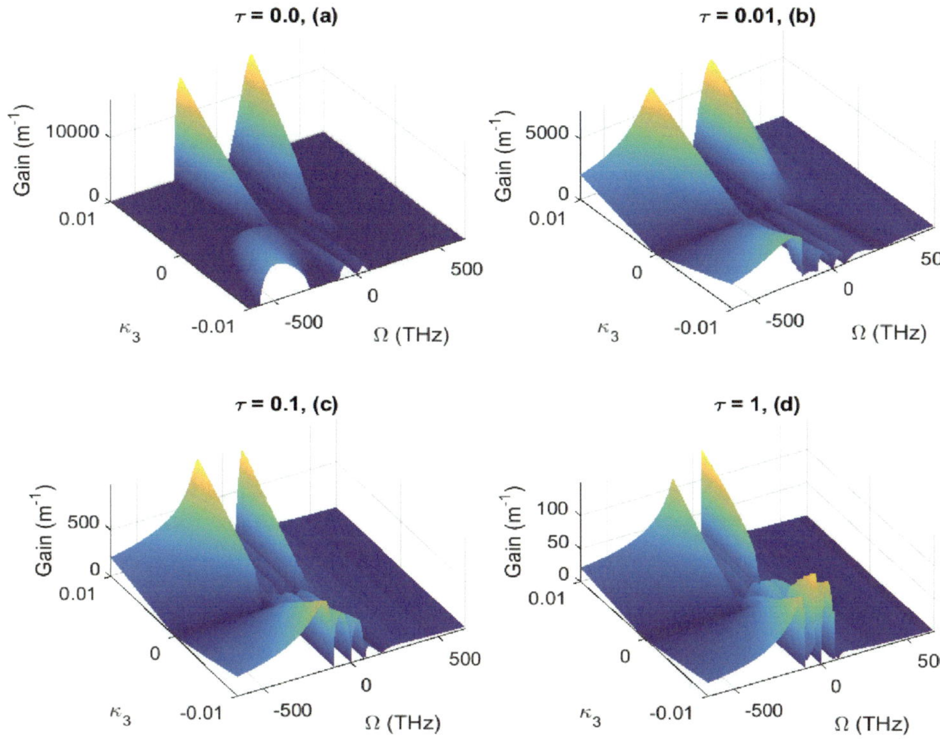

Figure 13.
MI gain spectra g (m^{-1}) *as a function of the frequency* Ω *(THz) and the nonlinear parameter* κ_3 *in the case of normal GVD for null GVM* $(\delta = 0)$*. (a)* $\tau = 0.0ps$*, (b)* $\tau = 0.01ps$*, (c)* $\tau = 0.1ps$*, (d)* $\tau = 1ps$*.*

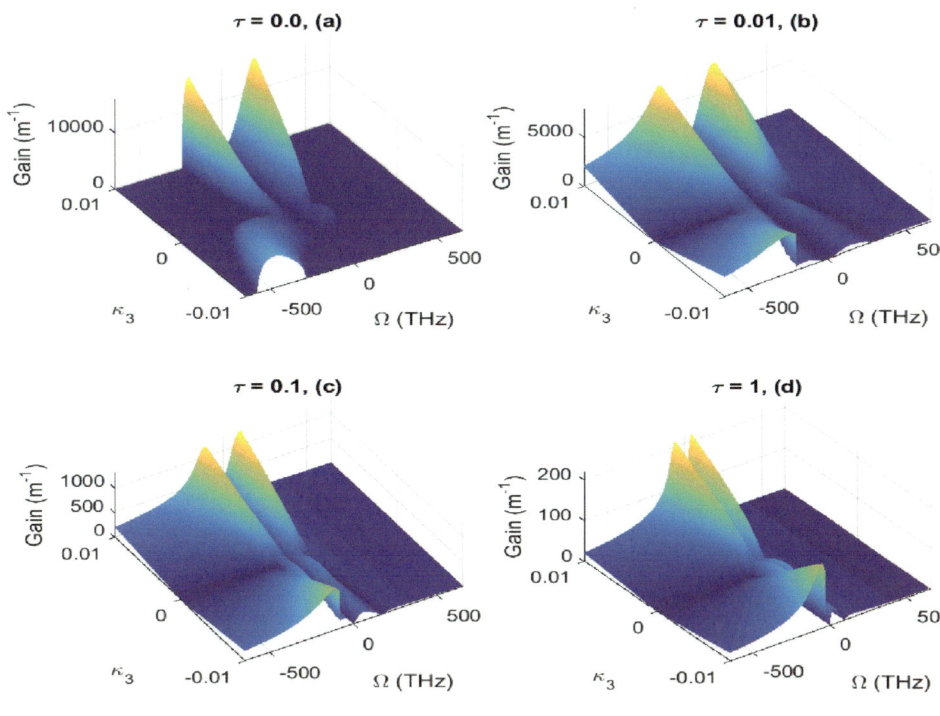

Figure 14.
MI gain spectra g (m^{-1}) *as a function of the frequency* Ω *(THz) and the nonlinear parameter* κ_3 *in the case of anomalous GVD for null GVM* $(\delta = 0)$*. (a)* $\tau = 0.0ps$*, (b)* $\tau = 0.01ps$*, (c)* $\tau = 0.1ps$*, (d)* $\tau = 1ps$*.*

We have plotted the behavior of the MI gain in **Figure 13** for normal GVD. We observe from this figure that MI appears on both side of κ_3. Each panel shows that the distinct side bands are antisymmetric. As κ_3 increases, the MI gain increases and the bandwidth enlarges. The high value of MI gain is obtained for $\kappa_3 > 0$ and the bandwidth increases when delay response time increases for $\Omega < 0$, while it is inverted for $\Omega > 0$. For instantaneous response time [see panel (a)], we note the appearance of two high side bands for $\kappa_3 > 0$ due to the effect of SOD, while there are three side bands for $\kappa_3 < 0$ brought by TOD around $\Omega = -500$ *Thz* and by FOD around $\Omega = 0$ *Thz* that slowly increases for negative value of κ_3. Considering non-instantaneous response time [see panels (b), (c), and (d)], we note that the side bands are unstable for $\Omega < 0$, while it is slightly unstable for $\Omega > 0$. In addition, as delay response time increases, we note more side bands. The latter information remains in the case of "total mixed" dispersion regime; however, the bandwidth is reversed due probably to opposite signs of the dispersion parameter.

In the case of anomalous dispersion regime plotted in **Figure 14**, we notice that the side bands brought by FOD are suppressed due to their opposite sign with the sign of SOD, the latter dominates and leads to the observed spectrum. Apart from this difference, the information obtained in the normal case remains almost the same. We now turns our attention in the case of mixed dispersion regime gives in **Figure 15**. Our results show that contrarily to what are observed in previous figures, the bandwidth is clearly visible for $\kappa_3 > 0$ where the band rapidly increases as κ increases by enlarging their bandwidth meanwhile for $\kappa_3 < 0$, we notice that the bandwidth progressively appears as delay response time increases. In each panels, the effect of TOD is suppressed due probably to the opposite sign of SOD; this leads to symmetric bandwidth observed. For instantaneous response time in panel (a), the MI gain spectrum appears only for $\kappa_3 < 0$. For non-instantaneous response time,

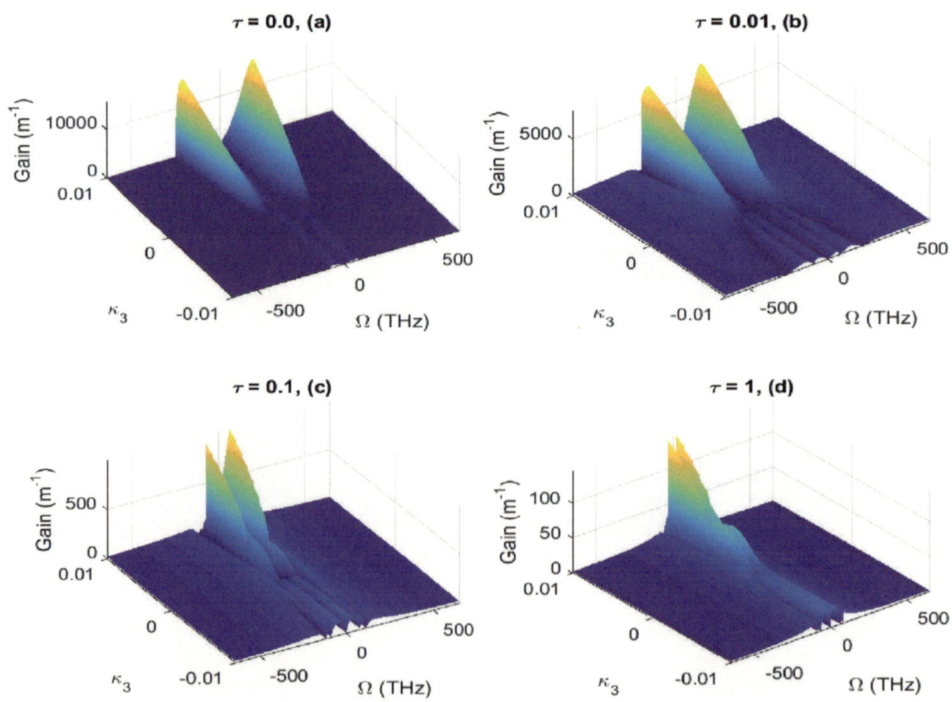

Figure 15.
MI gain spectra g (m^{-1}) as a function of the frequency Ω (THz) and the nonlinear parameter κ_3 in the case of mixed GVD for null GVM ($\delta = 0$). (a) τ = 0.0ps, (b) τ = 0.01ps, (c) τ = 0.1ps, (d) τ = 1ps.

we observe that for $\kappa_3 < 0$, the band slightly appears due to the effect of FOD that bandwidth decreases as delay response time increases.

5. Conclusion

In this work, we have investigated analytically and numerically the MI process in nonlinear Kerr media with delayed Kerr nonlinearity and subjected to XPM, described by a system of two coupled NLS equation with cubic-quintic-septic nonlinearities, experiencing diverse GVD regime. The system is considered by incorporating the Debye relaxational time-dependent nonlinearities in NLS equation with combined high-order effects modeling the propagation of ultrashort optical pulses in highly nonlinear Kerr media. By considering a small harmonic perturbations to the stationary solutions, we derive the exact dispersion relation for the components of the perturbation fields that include both the XPM and relaxation effects. Our study revealed the effects of septic nonlinearity in the case of the pulses co-propagating in diverse GVD regime. Obviously from our work, the incorporation of septic nonlinearity increases considerably the MI gain spectrum with the emergence of new instabilities frequencies contrarily to that was observed in Ref. [8, 9, 38, 39], while the increase of delay response time decreases the magnitude and the width of the MI gain.

Our study unveiled in the case of weak dispersion regime that, in instantaneous Kerr media, the typical frequency converges to finite value for mixed GVD regime with two gain maxima for certain GVM, meanwhile the MI gain continuously grows with the GVM until it reaches certain values before decays. In non-instantaneous Kerr media, we note the appearance of Raman band with the emergence of Raman peaks for slow response time ($\tau = 0.1ps$) in the case of mixed dispersion regime contrarily to what was done in [39]. We note that, these emerged Raman peaks decrease in width and magnitude when the GVM increases and shifts toward higher frequencies.

Now considering the effect of HOD, we observed that their incorporation hardly change the spectrum. The inclusion of FOD does't affect the MI gain in normal dispersion regime contrarily to the works done in the case of cubic and quintic nonlinearities [8, 9, 39] where the effect of FOD vanishes in the case of anomalous GVD. The effects of TOD vanish in the case of mixed dispersion regime. However, in the case of anomalous and normal dispersion regime, we note the appearance of Raman peaks that width and magnitude decrease as GVM increases and get shrink as response time increases. In addition, the four unstable modes observed in fast response time leads to two unstable modes for slow response time. For "total mixed" dispersion regime, we also note the appearance of Raman peaks that slightly shifted toward higher frequencies due to the effects of GVM. The effects of septic nonlinearity was also portrayed and our results unveil the appearance of instability in both side of septic nonlinearity parameter. This aspect shows that MI can be investigated as well as in focusing nonlinearity as in defocusing nonlinearity.

Due to the fact that the MI gain increases when taking into account the effect of HON, we can conclude that the system is more stable against harmonic perturbations. In addition, the incorporation of higher order effects lead to the emergence of Raman peaks on the Raman bands. This means that the incident particles after collision with the molecules of the medium, diffuse with a new energy thus creating the Raman peaks observed resulting to the generation of new frequencies range. The study of MI gain has aroused great enthusiasm within the scientific community, motivated in large part by their potential in innovative applications in information technology and telecommunications.

Author details

Michel-Rostand Soumo Tchio[1*], Saïdou Abdoulkary[2] and Alidou Mohamadou[1,3]

1 Faculty of Sciences, University of Maroua, Maroua, Cameroon

2 Département des SF, Faculté des Mines et des Industries Pétrolieres, Maroua, Cameroon

3 The Abdus Salam International Centre for Theoretical Physics, Trieste, Italy

*Address all correspondence to: rostandsoumotchio@yahoo.fr

References

[1] Bourkoff E, Zhao W, Joseph RI. Evolution of femtosecond pulses in single-mode fibers having higher-order nonlinearity and dispersion. Optics Letters. 1987;**12**:272–274

[2] Abdoulkary S, Aboubakar AD, Aboubakar M, Mohamadou A, Kavitha L. Solitary wave solutions and modulational instability analysis of the nonlinear Schröndinger equation with higher-order nonlinear terms in the left-handed nonlinear transmission lines. Communications in Nonlinear Science and Numerical Simulation. 2015;**22**:1288

[3] Saha M, Sarma AK. Modulation instability in nonlinear metamaterials induced by cubic-quintic nonlinearities and higher order dispersive effects. Optical Communication. 2013;**291**:321–325

[4] Benabid F, Biancalana F, Light PS, Couny F, Luiten A, Roberts PJ, et al. Fourth-order dispersion mediated solitonic radiations in HC-PCF cladding. Optics Letters. 2008;**33**:2680–2682

[5] Tchofo DP, Ngabireng CM, Porsezian K, Kalithasan B. Modulational instability in optical fibers with arbitrary higher-order dispersion and delayed Raman response. Optic Communication. 2006;**266**:142–150

[6] Tchofo DP, Porsezian K. Impact of fourth-order dispersion in the modulational instability spectra of wave propagation in glass fibers with saturable nonlinearity. Journal of the Optical Society of America B. 2010;**27**: 1143–1152

[7] Nithyanandan K, Vasantha Jayakantha Raja R, Uthayakumar T, Porsezian K. Impact of higher-order dispersion in the modulational instability spectrum of a relaxing coupled saturable media. Pramana Journal of Physics. 2014;**82**:339-345

[8] Nithyanandan K, Vasantha Jayakantha Raja R, Porsezian K, Kalithasan B. Modulational instability with higher-order dispersion and walk-off in Kerr media with cross-phase modulation. Physical Review A. 2012; **86**:023827

[9] Soumo Tchio MR, Abdoulkary S, Mohamadou A. Modulation instability induced by high-order dispersion to a coupled nonlinear Schrödinger equation in a single-mode optical fiber with Kerr nonlinearity. Physica Scripta. 2019;**94**: 035207

[10] Chow KW, Wong KKY, Lam K. Modulation instabilities in a system of four coupled nonlinear Schrödinger equations. Physics Letters A. 2008;**372**: 4596–4600

[11] Zhang L, Xiang Y, Dai X, Wen S. Modulation instability in an array of positive- and negative-index waveguides. Journal of the Optical Society of America B. 2014;**31**:3029–3037

[12] Zhong X, Cheng K, Chiang S. Modulation instability with arbitrarily high perturbation frequencies in metamaterials with nonlinear dispersion and saturable nonlinearity. Journal of the Optical Society of America B. 2014; **31**:1484–1493

[13] da Silva GL, Canabarro AA, de Lima Bernardo B. Stochastic noise amplification of partially coherent CW beam in the region of minimum GVD in lossless Kerr media with delayed response. Annals of Physics. 2015;**363**: 476–484

[14] Aitchison JS, Weiner AM, Silberberg Y, Oliver MK, Jackel JL, Leaird DE, et al. Observation of optical solitons in a nonlinear glass waveguide. Optics Letters. 1990;**15**:471–474

[15] Chung Y, Lushnikov PM. Strong collapse turbulence in a quintic nonlinear Schrödinger equation. Physical Review E. 2011;**84**:036602

[16] Reyna AS, de Araújo CB. Spatial phase modulation due to quintic and septic nonlinearities in metal colloids. Optics Express. 2014;**22**:22456–224569

[17] Reyna AS, de Araújo CB. Nonlinearity management of photonic composites and observation of spatial-modulation instability due to quintic nonlinearity. Physical Review A. 2014; **89**:063803

[18] Reyna AS, de Araújo CB. High-order optical nonlinearities in plasmonic nanocomposites-a review. Advances in Optics and Photonics. 2017;**9**:720–761

[19] Benjamin TB, Feir JE. The desintegration of wave trains on deep water. Journal of Fluid Mechanics. 1967; **27**:417–430

[20] Agrawal GP. Nonlinear Fiber Optic. 3rd ed. University of Rochester: John Wiley and Sons, Academic Press; 2001

[21] Xiang Y, Wen S, Dai X, Tang Z, Su W, Fan D. Modulation instability induced by nonlinear dispersion in nonlinear metamaterials. Journal of the Optical Society of America B. 2007;**24**: 3058–3063

[22] Whitham GB. A general approach to linear and non-linear dispersive waves using a Lagrangian. Journal of Fluid Mechanics. 1965;**22**:273–283

[23] Hasegawa A. Theory and computer experiment on self-trapping instability of plasma cyclotron waves. Physics of Fluids. 1972;**15**:870–881

[24] Ostrovsky LA. Propagation of wave packets and space-time self-focusing in a nonlinear. Soviet Physics—JETP. 1968; **24**:797–800

[25] Zakharov VE, Ostrovsky LA. Modulation instability: The beginning. Physica D. 2009;**238**:540–548

[26] Wu B, Niu Q. Landau and dynamical instabilities of the superflow of Bose-Einstein condensates in optical lattices. Physical Review A. 2001;**64**: 061603(R)

[27] Salasnich L, Parola A, Reatto L. Modulational instability and complex dynamics of confined matter-wave solitons. Physical Review Letters. 2003; **91**:080405

[28] Hasegawa A, Brinkman WF. Tunable coherent IR and FIR sources utilizing modulational instability. IEEE Journal of Quantum Electronics. 1980; **16**:694–697

[29] Agrawal GP. Modulation instability induced by cross-phase modulation. Physical Review Letters. 1987;**59**:880–883

[30] Boucon A. Instabilité modulationnelle et génération de supercotinuum en regime d'exitation quasi-continue dans les fibres optiques hautement nonlinéaires et microstructurées [Thesis]. Besançon France: Université de Franche-Comté; 2008

[31] Cavalcanti SB, Cressoni JC, da Cruz HR, Gouveia-Neto AS. Modulation instability in the region of minimum group-velocity dispersion of single-mode optical fibers via an extended nonlinear Schrödinger equation. Physical Review A. 1991;**43**:6162–6165

[32] Wang Y, Cheng Q, Guo J, Wang W. Sonic horizon dynamics for quantum systems with cubic-quintic-septic nonlinearity. AIP Advances. 2019;**9**: 075206

[33] Agrawal GP. Nonlinear Fiber Optics. SanDiego: Academic; 2001

[34] Canabarro AA, Santos B, Gleria I, Lyra ML, Sombra ASB. Interplay of XPM and nonlinear response time in the modulational instability of copropagating optical pulses. Journal of the Optical Society of America B. 2010; **27**:1878–1885

[35] Kumar M, Nithyanandan K, Porsezian K. Influence of spatial delay on the modulational instability in a composite system with a controllable nonlinearity. Physical Review E. 2018; **97**:062208

[36] Zhong X, Xiang A. Cross-phase modulation induced modulation instability in single-mode optical fiber swith saturable nonlinearity. Optical Fiber Technology. 2007;**13**:271–279

[37] Liu X, Haus JW, Shahriar SM. Modulation instability for a relaxational Kerr medium. Optic Communication. 2008;**281**:2907–2912

[38] Canabarro AA, Santos B, de Lima Bernardo B, Moura AL, Soares WC, de Lima E, et al. Modulation instability in noninstantaneous Kerr media with walk-off and cross-phase modulation for mixed group-velocity-dispersion regimes. Physical Review A. 2016;**93**: 023834

[39] Soumo Tchio MR, Abdoulkary S, Mohamadou A. Raman peaks shift due to walk-off in non-instantaneous Kerr media with higher-order effects. Journal of the Optical Society of America B. 2019;**36**:3479–3491